MODERN ALGEBRA
ELEMENTS OF A

JIMMIE GILBERT and LINDA GILBERT
Louisiana Tech University

PRINDLE, WEBER & SCHMIDT, Boston

PWS PUBLISHERS

Prindle. Weber & Schmidt · ☙ · Willard Grant Press · **wG** · Duxbury Press · ♠

Statler Office Building · 20 Providence Street · Boston. Massachusetts 02116 · 617-482-2344 · telex 940-491

PWS Publishers is a division of Wadsworth, Inc.

84 85 86 87 88   —   10 9 8 7 6 5 4 3 2 1

Library of Congress Cataloging in Publication Data

Gilbert, Jimmie D.,
   Elements of modern algebra.

   Includes index.
   1. Algebra, Abstract. I. Gilbert, Linda. II. Title.
QA162.G527 1984   512'.02   83-21598
ISBN 0-87150-458-8

ISBN 0-87150-458-8

Cover image: "Everything Set"—watercolor by Nelson W. Aldrich.
Reproduced by permission of the artist.
Cover and text design: Michael Michaud
Typesetter: Polyglot Pte. Ltd.
Printing and binding: Haddon Craftsmen
Cover printing: New England Book Components

PREFACE

This book is intended as a text for a one-semester or one-quarter introductory abstract algebra course at the undergraduate level. To provide some flexibility, it contains more material than would normally be taught in one course. The book's main objectives are to provide a knowledge of basic algebraic structures (groups, rings, fields, etc.), and to prepare students for higher-level mathematics or computer science courses.

The basic algebraic structures are introduced in a gradual and natural fashion. An introductory chapter presents the fundamental concepts of set, mapping, binary operation, and relation. The second chapter provides a fairly brief and rigorous development of the fundamental properties of the integers. This development includes the topics of mathematical induction, the well-ordering theorem, divisibility, and congruence of integers.

In the sequel, groups are introduced before rings. It is the authors' opinion that a student can develop the ability to construct and present proofs of theorems more easily when working with a system in which only one binary operation is under consideration. The standard topics in elementary group theory are treated in Chapter 3. These include subgroups, cyclic groups, permutation groups, and isomorphisms. The first two sections in Chapter 4 are devoted to normal subgroups, quotient groups, and homomorphisms. The last two sections of the chapter present some results on finite abelian groups, and they give a sample of the flavor of more advanced work.

Rings, integral domains, and fields are introduced in Chapter 5. The field of quotients of an integral domain is constructed, and ordered integral domains are considered. The basic theorems on quotient rings and homomorphisms are presented in Chapter 6 along with a section on the characteristic of a ring.

v

The material on real and complex numbers in Chapter 7 is included for the benefit of those students who would not see it in some other course, and it could be skipped in many classes.

Chapter 8 presents the elementary theory of polynomials over a field. Topics included are the division algorithm, greatest common divisor, factorization theorems, and simple algebraic extensions.

The treatment of the set Z_n of congruence classes modulo n is somewhat unorthodox in that it threads throughout most of the book. The first contact with Z_n is early in Chapter 2, where it appears as a set of equivalence classes. Binary operations of addition and multiplication are defined in Z_n at a later point of that chapter. Both the additive and multiplicative structures are drawn upon for examples in Chapters 3 and 4, but it is not until Chapter 5 that Z_n appears in its proper context as a ring. The final description of Z_n as a quotient ring of the integers by the principal ideal (n) does not appear until Chapter 6.

A minimal amount of mathematical maturity is assumed in the text; a major goal is to develop mathematical maturity. The material is presented in a theorem-proof format, with definitions and major results easily located. The treatment is rigorous and self-contained, in keeping with the objective of training the student in the techniques of algebra.

The text has been developed from materials used in undergraduate courses in abstract algebra at Louisiana Tech University over a period of several years. Most of it has been class tested in xerographic form to ensure clarity and teachability. Many detailed examples are included, and enough exercises are provided to allow different assignments of the same approximate difficulty to be made. Many of these problems are elementary or of a computational nature.

The important features of the book include:

1. A gradual introduction and development of concepts, proceeding from the simplest structures to the more complex;
2. A careful and rigorous treatment of definitions and proofs;
3. An abundance of examples, designed to develop the student's intuition;
4. A large number and variety of exercises, designed to develop the student's maturity and ability to construct proofs;
5. A format that presents material in an easily referenced form;
6. A clear and readable presentation of the material;
7. A summary of key words and phrases at the end of each chapter;
8. A list of references at the end of each chapter, and a bibliography at the end of the book. The references are intended to point out the abundance of literature on the subject and to encourage further reading.

ACKNOWLEDGMENTS

We are grateful to the following persons for their helpful reviews:
Gordon Brown, University of Colorado, Boulder;
Marshall Cates, California State University, Los Angeles;

Howard Frisinger, Colorado State University;
Nickolas Heerema, Florida State University;
Robert P. Webber, Longwood College.

Our sincere thanks go to Margaret Dunn for her excellent typing of the entire manuscript; to Connie Caldwell for suggesting the project in the beginning, to Theron Shreve for initiating the project, and to Dave Pallai for his guidance and encouragement in developing the book. In addition, we express our appreciation to Debbie Schneider and Michael Michaud for their cooperation during the production phase.

Thanks are also due to the administration of Louisiana Tech University for its help. Finally, we would like to thank our colleague, Dallas Lankford, for his interest and involvement.

Jimmie Gilbert
Linda Gilbert

CONTENTS

CONTENTS

1 FUNDAMENTALS

1.1 SETS

Abstract algebra had its beginnings in attempts to solve mathematical problems such as the solution of polynomial equations by radicals and geometric constructions with straightedge and compass. From the solutions of specific problems, general techniques evolved that could be used to solve problems of the same type, and treatments were generalized to deal with whole classes of problems rather than particular ones.

In our study of abstract algebra, we shall make use of our knowledge of the various number systems. At the same time, in many cases we wish to examine how certain properties are consequences of other known properties. This sort of examination deepens our understanding of the system. As we proceed, we shall be careful to distinguish between the properties we have assumed and made available for use, and those that must be deduced from these properties. We must accept without definition some terms that are basic objects in our mathematical systems. Initial assumptions about each system are formulated using these undefined terms.

One such undefined term is **set**. We think of a set as a collection of objects about which it is possible to determine whether or not a particular object is a member of the set. Sets are usually denoted by capital letters and are described by a list of their elements, as illustrated in the following examples.

EXAMPLE 1   We write

$$A = \{0, 1, 2, 3\}$$

to indicate that the set A contains the elements 0, 1, 2, 3, and no other elements.
□

EXAMPLE 2   The set B consisting of all the nonnegative integers is written

$$B = \{0, 1, 2, 3, \ldots\}.$$

The ... means that the pattern established before the dots continues indefinitely.
□

A standard way of describing sets is called *set-builder notation*. Set-builder notation uses braces to enclose a property that is the qualification for membership in the set.

EXAMPLE 3   The set B in Example 2 can be described using set-builder notation as

$$B = \{x \mid x \text{ is a nonnegative integer}\}.$$

The vertical slash is shorthand for "such that," and we read "B is the set of all x such that x is a nonnegative integer."
□

There is also a shorthand notation for "is an element of." We write "$x \in A$" to mean "x is an element of the set A." We write "$x \notin A$" to mean "x is not an element of the set A." For the set A in Example 1 we can write

$$2 \in A \quad \text{and} \quad 7 \notin A.$$

DEFINITION 1.1   If every element of a set A is an element of a set B, then A is called a **subset** of B and we write $A \subseteq B$ or $B \supseteq A$.

The notation $A \subseteq B$ is read "A is a subset of B" or "A is contained in B." Also, $B \supseteq A$ is read as "B contains A." The symbol $\in$ is reserved for elements whereas the symbol $\subseteq$ is reserved for subsets.

EXAMPLE 4   We write

$$a \in \{a, b, c, d\} \quad \text{or} \quad \{a\} \subseteq \{a, b, c, d\}.$$

However,

$$a \subseteq \{a,b,c,d\} \quad \text{and} \quad \{a\} \in \{a,b,c,d\}$$

are both **incorrect** uses of set notation.   □

DEFINITION 1.2   Two sets are **equal** if they contain exactly the same elements.

The sets A and B are equal, and we write $A = B$, if each member of A is also a member of B, and if each member of B is also a member of A. Typically, a proof that two sets are equal is presented in two parts: the first shows that $A \subseteq B$ and the second that $B \subseteq A$. We then conclude that $A = B$. We shall have an example of this type of proof shortly.

DEFINITION 1.3   If A and B are sets such that $A \subseteq B$ and $A \neq B$, then A is a **proper subset** of B.

We sometimes write $A \subset B$ to denote that A is a proper subset of B.

EXAMPLE 5   The following statements illustrate the notation for proper subsets and equality of sets.

$$\{1,2,4\} \subset \{1,2,3,4,5\}$$
$$\{a,c\} = \{c,a,c\}$$ □

There are two basic operations, *union* and *intersection*, that are used to combine sets. These operations are defined as follows.

DEFINITION 1.4   If A and B are sets, the **union** of A and B is the set $A \cup B$ given by

$$A \cup B = \{x \mid x \in A \text{ or } x \in B\}.$$

The **intersection** of A and B is the set $A \cap B$ given by

$$A \cap B = \{x \mid x \in A \text{ and } x \in B\}.$$

The union of two sets A and B is the set whose elements are either in A or in B or in both A and B. The intersection of sets A and B is the set of those elements common to both A and B.

EXAMPLE 6   Suppose $A = \{2,4,6\}$ and $B = \{4,5,6,7\}$. Then

$$A \cup B = \{2,4,5,6,7\}$$

and

$$A \cap B = \{4, 6\}. \qquad \square$$

It is easy to find sets that have no elements at all in common. For example, the sets

$$A = \{1, -1\} \quad \text{and} \quad B = \{0, 2, 3\}$$

have no element in common. Hence, there is no element in their intersection, $A \cap B$, and we say that the intersection is *empty*. Thus, it is logical to introduce the **empty set**, the set that has no members. We denote the empty set by $\varnothing$ or $\{\ \}$. Two sets A and B are said to be **disjoint** if $A \cap B = \varnothing$. The sets $\{1, -1\}$ and $\{0, 2, 3\}$ are disjoint since

$$\{1, -1\} \cap \{0, 2, 3\} = \varnothing.$$

The empty set, $\varnothing$, is considered to be a subset of every set. For a set A with n elements (n a nonnegative integer), we can write out all the subsets of A. For example, if

$$A = \{a, b, c\},$$

then the subsets of A are

$$\varnothing, \{a\}, \{b\}, \{c\}, \{a, b\}, \{a, c\}, \{b, c\}, A.$$

DEFINITION 1.5 For any set A, the **power set** of A, denoted by $\mathscr{P}(A)$, is the set of all subsets of A and is written

$$\mathscr{P}(A) = \{X \mid X \subseteq A\}.$$

EXAMPLE 7 For $A = \{a, b, c\}$, the power set of A is

$$\mathscr{P}(A) = \{\varnothing, \{a\}, \{b\}, \{c\}, \{a, b\}, \{a, c\}, \{b, c\}, A\}. \qquad \square$$

It is often helpful to draw a picture or diagram of the sets under discussion. When we do this, we must assume that all the sets we are dealing with, along with all possible unions and intersections of those sets, are subsets of some **universal set**, denoted by U. In Figure 1.1, we let two overlapping circles represent the two sets A and B. The sets A and B are subsets of the universal set U, represented by the rectangle. Hence the circles are contained in the rectangle. The intersection of A and B, $A \cap B$, is the crosshatched region where the two circles overlap. This type of pictorial representation is called a **Venn diagram**. Another special subset is defined next.

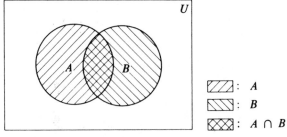

FIGURE 1.1

DEFINITION 1.6 For arbitrary subsets A and B of U, the **complement** of B in A is

$A - B = \{x \in U \mid x \in A \text{ and } x \notin B\}.$

The special notation A' is reserved for a particular complement, $U - A$:

$A' = U - A = \{x \in U \mid x \notin A\}.$

We read A' as simply "the complement of A" rather than as "the complement of A in U."

EXAMPLE 8 Let

$U = \{x \mid x \text{ is an integer}\},$

$A = \{x \mid x \text{ is an even integer}\},$

$B = \{x \mid x \text{ is a positive integer}\}.$

Then

$\begin{aligned} B - A &= \{x \mid x \text{ is a positive odd integer}\} \\ &= \{1, 3, 5, 7, \ldots\}; \end{aligned}$

$\begin{aligned} A - B &= \{x \mid x \text{ is a nonpositive even integer}\} \\ &= \{0, -2, -4, -6, \ldots\}; \end{aligned}$

$\begin{aligned} A' &= \{x \mid x \text{ is an odd integer}\} \\ &= \{\ldots, -3, -1, 1, 3, \ldots\}; \end{aligned}$

$\begin{aligned} B' &= \{x \mid x \text{ is a nonpositive integer}\} \\ &= \{0, -1, -2, -3, \ldots\}. \end{aligned}$ □

EXAMPLE 9 The overlapping circles representing the sets A and B separate the interior of the rectangle representing U into four regions, labeled 1, 2, 3, and 4, in the Venn diagram in Figure 1.2. Each region represents a particular subset of U.

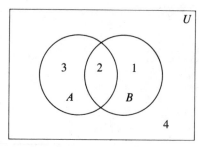

FIGURE 1.2

Region 1: $B - A$
Region 2: $A \cap B$
Region 3: $A - B$
Region 4: $(A \cup B)'$ □

The operations of union and intersection can be applied repeatedly. For instance, we might form the intersection of A and B, obtaining $A \cap B$, and then form the intersection of this set with a third set C: $(A \cap B) \cap C$.

EXAMPLE 10   The sets $(A \cap B) \cap C$ and $A \cap (B \cap C)$ are equal since

$$(A \cap B) \cap C = \{x \mid x \in A \text{ and } x \in B\} \cap C$$
$$= \{x \mid x \in A \text{ and } x \in B \text{ and } x \in C\}$$
$$= A \cap \{x \mid x \in B \text{ and } x \in C\}$$
$$= A \cap (B \cap C).$$

In analogy with the associative property

$$x + (y + z) = (x + y) + z$$

for addition of numbers, we say that the operation of intersection is **associative**. When we work with numbers, for convenience we drop the parentheses and write

$$x + y + z = x + (y + z) = (x + y) + z.$$

Similarly, for sets A, B, and C we write

$$A \cap B \cap C = A \cap (B \cap C) = (A \cap B) \cap C.$$ □

Just as simply, we can show (see Problem 14) that the union of sets is an associative operation. We write

$$A \cup B \cup C = A \cup (B \cup C) = (A \cup B) \cup C.$$

A separation of a nonempty set A into mutually disjoint nonempty subsets is called a **partition** of the set A. For example, if

$$A = \{a, b, c, d, e, f\},$$

then one partition of A is

$$X_1 = \{a, d\}, \qquad X_2 = \{b, c, f\}, \qquad X_3 = \{e\},$$

since

$$A = X_1 \cup X_2 \cup X_3$$

with

$$X_1 \cap X_2 = \varnothing, \qquad X_1 \cap X_3 = \varnothing, \qquad X_2 \cap X_3 = \varnothing.$$

The operations of intersection, union, and forming complements can be combined in all sorts of ways, and several nice equalities can be obtained relating some of these results. As illustrations, it can be shown that

$$A \cap (B \cup C) = (A \cap B) \cup (A \cap C)$$

and that

$$A \cup (B \cap C) = (A \cup B) \cap (A \cup C).$$

Because of the resemblance between these equations and the familiar distributive property $x(y + z) = xy + xz$ for numbers, we call these equations **distributive laws**.

We shall prove the first of these distributive laws in the next example and leave the last one as an exercise. To prove the first law, we shall show that $A \cap (B \cup C) \subseteq (A \cap B) \cup (A \cap C)$ and that $(A \cap B) \cup (A \cap C) \subseteq A \cap (B \cup C)$. This illustrates the point made earlier in the discussion of equality of sets, immediately after Definition 1.2.

The symbol $\Rightarrow$ is shorthand for "implies," and $\Leftarrow$ is shorthand for "is implied by." We use them in the next example.

EXAMPLE 11   To prove

$$A \cap (B \cup C) = (A \cap B) \cup (A \cap C),$$

we first let $x \in A \cap (B \cup C)$. Now

$$
\begin{aligned}
x \in A \cap (B \cup C) &\Rightarrow x \in A \quad \text{and} \quad x \in (B \cup C) \\
&\Rightarrow x \in A, \quad \text{and} \quad x \in B \quad \text{or} \quad x \in C \\
&\Rightarrow x \in A \quad \text{and} \quad x \in B, \quad \text{or} \quad x \in A \quad \text{and} \quad x \in C \\
&\Rightarrow x \in A \cap B, \quad \text{or} \quad x \in A \cap C \\
&\Rightarrow x \in (A \cap B) \cup (A \cap C).
\end{aligned}
$$

Thus $A \cap (B \cup C) \subseteq (A \cap B) \cup (A \cap C)$.

Conversely, suppose $x \in (A \cap B) \cup (A \cap C)$. Then

$$x \in (A \cap B) \cup (A \cap C) \Rightarrow x \in A \cap B, \quad \text{or} \quad x \in A \cap C$$
$$\Rightarrow x \in A \quad \text{and} \quad x \in B,$$
$$\text{or} \quad x \in A \quad \text{and} \quad x \in C$$
$$\Rightarrow x \in A, \quad \text{and} \quad x \in B \quad \text{or} \quad x \in C$$
$$\Rightarrow x \in A \quad \text{and} \quad x \in (B \cup C)$$
$$\Rightarrow x \in A \cap (B \cup C).$$

Therefore $(A \cap B) \cup (A \cap C) \subseteq A \cap (B \cup C)$, and we have shown that $A \cap (B \cup C) = (A \cap B) \cup (A \cap C)$.

It should be evident that the second part of the proof can be obtained from the first simply by reversing the steps. That is, when each $\Rightarrow$ is replaced by $\Leftarrow$, a valid implication results. In fact, then, we could obtain a proof of both parts by replacing $\Rightarrow$ by $\Leftrightarrow$, where $\Leftrightarrow$ is short for "if and only if." Thus,

$$x \in A \cap (B \cup C) \Leftrightarrow x \in A \quad \text{and} \quad x \in (B \cup C)$$
$$\Leftrightarrow x \in A, \quad \text{and} \quad x \in B \quad \text{or} \quad x \in C$$
$$\Leftrightarrow x \in A \quad \text{and} \quad x \in B, \quad \text{or} \quad x \in A \quad \text{and} \quad x \in C$$
$$\Leftrightarrow x \in A \cap B, \quad \text{or} \quad x \in A \cap C$$
$$\Leftrightarrow x \in (A \cap B) \cup (A \cap C). \qquad \square$$

In proving an equality of sets S and T, we can frequently use the technique of showing that $S \subseteq T$ and then check to see if the steps are reversible. In many cases, the steps are indeed reversible, and we obtain the other part of the proof easily. However, this method should not obscure the fact that there are still two parts to the argument: $S \subseteq T$ and $T \subseteq S$.

There are some interesting relations between complements and unions or intersections. For example, it is true that

$$(A \cup B)' = A' \cap B'.$$

This statement is known as **De Morgan's Theorem**. The proof of this fact and other similar results are requested in the exercises.

EXERCISES 1.1

1 For each set A, describe A by indicating a property that is a qualification for membership in A.
   (a) $A = \{0, 2, 4, 6, 8, 10\}$ (b) $A = \{1, -1\}$
   (c) $A = \{-1, -2, -3, \ldots\}$ (d) $A = \{1, 4, 9, 16, 25, \ldots\}$

2 Decide whether or not each statement is true for $A = \{2, 7, 11\}$ and $B = \{1, 2, 9, 10, 11\}$.
   (a) $2 \subseteq A$ (b) $\{11, 2, 7\} \subseteq A$
   (c) $2 = A \cap B$ (d) $\{7, 11\} \in A$
   (e) $A \subseteq B$ (f) $\{2, 7, 2, 11, 2\} = A$

3   Decide whether or not each statement is true, where A and B are arbitrary sets.

(a) $B \cup A \subseteq A$ (b) $B \cap A \subseteq A \cup B$

(c) $\varnothing \subseteq A$ (d) $0 \in \varnothing$

(e) $\varnothing \in \{\varnothing\}$ (f) $\varnothing \subseteq \{\varnothing\}$

(g) $\{\varnothing\} \subseteq \varnothing$ (h) $\{\varnothing\} = \varnothing$

(i) $\varnothing \in \varnothing$ (j) $\varnothing \subseteq \varnothing$

4   Evaluate each of the following sets, where

$U = \{0, 1, 2, 3, \ldots, 10\},$

$A = \{0, 1, 2, 3, 4, 5\},$

$B = \{0, 2, 4, 6, 8, 10\},$

$C = \{2, 3, 5, 7\}.$

(a) $A \cup B$ (b) $A \cap C$

(c) $A' \cup B$ (d) $A \cap B \cap C$

(e) $A' \cap B \cap C$ (f) $A \cup (B \cap C)$

(g) $A \cap (B \cup C)$ (h) $(A \cup B')'$

(i) $A - B$ (j) $B - A$

(k) $A - (B - C)$ (l) $C - (B - A)$

(m) $(A - B) \cap (C - B)$ (n) $(A - B) \cap (A - C)$

5   Write each of the following as either A, A', U, or $\varnothing$, where A is an arbitrary subset of the universal set U.

(a) $A \cap A$ (b) $A \cup A$ (c) $A \cap A'$

(d) $A \cup A'$ (e) $A \cup \varnothing$ (f) $A \cap \varnothing$

(g) $A \cap U$ (h) $A \cup U$ (i) $U \cup A'$

(j) $A - \varnothing$ (k) $\varnothing'$ (l) U'

(m) $(A')'$

6   Write out the power set, $\mathscr{P}(A)$, for each set A.

(a) $A = \{a\}$ (b) $A = \{0, 1\}$

(c) $A = \{a, b, c\}$ (d) $A = \{1, 2, 3, 4\}$

7   Describe two partitions of each of the following sets.

(a) $\{x \mid x \text{ is an integer}\}$ (b) $\{a, b, c, d\}$

(c) $\{1, 5, 9, 11, 15\}$ (d) $\{x \mid x \text{ is a complex number}\}$

8   State the most general conditions on the subsets A and B of U under which the given equality holds.

(a) $A \cap B = A$ (b) $A \cup B' = A$

(c) $A \cup B = A$ (d) $A \cap B' = A$

(e) $A \cap B = U$ (f) $A' \cap B' = \varnothing$

(g) $A \cup \varnothing = U$ (h) $A' \cap U = \varnothing$

In Problems 9–24, prove each statement.

9 $A \subseteq A \cup B$

10 $A \cap B \subseteq A \cup B$

11 $(A')' = A$

12   If $A \subseteq B$ and $B \subseteq C$, then $A \subseteq C$.

13 $A \subseteq B$ if and only if $B' \subseteq A'$

14 $A \cup (B \cup C) = (A \cup B) \cup C$

15   $(A \cup B)' = A' \cap B'$

16   $(A \cap B)' = A' \cup B'$

17   $A \cup (B \cap C) = (A \cup B) \cap (A \cup C)$

18   $A \cap (A' \cup B) = A \cap B$

19   $A \cup (A \cap B) = A \cap (A \cup B)$

20   If $A \subseteq B$, then $A \cup C \subseteq B \cup C$.

21   If $A \subseteq B$, then $A \cap C \subseteq B \cap C$.

22   $B - A = B \cap A'$

23   $A \cap (B - A) = \emptyset$

24   $A \cup (B - A) = A \cup B$

25   Express $(A \cup B) - (A \cap B)$ in terms of A, A', B, and B'.

26   Let the operation of addition be defined on subsets A and B of U by

$A + B = (A \cup B) - (A \cap B)$.

Use a Venn diagram with labeled regions to illustrate each of the following statements.

(a) $A + B = (A - B) \cup (B - A)$

(b) $A + (B + C) = (A + B) + C$

(c) $A \cap (B + C) = (A \cap B) + (A \cap C)$.

1.2   MAPPINGS

The concept of a function is fundamental to nearly all areas of mathematics. The term *function* is the one most widely used for the concept that we have in mind, but it has become traditional to use the terms *mapping* and *transformation* in algebra. It is likely that these words are used because they express an intuitive feel for the association between the elements involved. The basic idea is that we wish to speak of correspondences of a certain type between the elements of two sets. There is to be a rule of association between the elements of a first set and those of a second set. The association is to be such that for each element in the first set, there is one and only one associated element in the second set. This rule of association leads to a natural pairing of the elements that are to correspond, and thence to the formal statement in Definition 1.8.

By "an ordered pair of elements" we mean a pairing (a, b) where there is to be a distinction between the pair (a, b) and the pair (b, a), if a and b are different. That is, there is to be a first position and a second position such that $(a, b) = (c, d)$ if and only if both $a = c$ and $b = d$. This ordering is altogether different from listing the elements of a set, for there the order of listing is of no consequence at all. The sets $\{1, 2\}$ and $\{2, 1\}$ have exactly the same elements, and $\{1, 2\} = \{2, 1\}$. When we speak of ordered pairs, however $(1, 2)$ and $(2, 1)$ are not considered to be equal. With these ideas in mind, we make the following definition.

DEFINITION 1.7 For two nonempty sets A and B, the **Cartesian product** $A \times B$ is the set of all ordered pairs (a, b) of elements $a \in A$ and $b \in B$. That is,

$$A \times B = \{(a, b) \mid a \in A \text{ and } b \in B\}.$$

EXAMPLE 1 If $A = \{1, 2\}$ and $B = \{3, 4, 5\}$, then

$$A \times B = \{(1, 3), (1, 4), (1, 5), (2, 3), (2, 4), (2, 5)\}.$$

We observe that the order in which the sets appear is important. In this example,

$$B \times A = \{(3, 1), (3, 2), (4, 1), (4, 2), (5, 1), (5, 2)\},$$

so $A \times B$ and $B \times A$ are quite distinct from each other. ☐

We can now make our formal definition of a mapping.

DEFINITION 1.8 Let A and B be nonempty sets. A **mapping** f from A to B is a subset of $A \times B$ such that for each $a \in A$, there is a unique (one and only one) element $b \in B$ such that $(a, b) \in f$. If f is a mapping from A to B, and the pair (a, b) belongs to f, we write $b = f(a)$ and call b the **image** of a under f.

A mapping f from A to B, pictured in Figure 1.3, is the same as a function from A to B, and the image of $a \in A$ under f is the same as the value of the function f at a. Two mappings f from A to B and g from A to B are **equal** if $f(x) = g(x)$ for all $x \in A$.

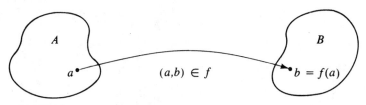

FIGURE 1.3

EXAMPLE 2 Let $A = \{-2, 1, 2\}$, and let $B = \{1, 4, 9\}$. The set f given by

$$f = \{(-2, 4), (1, 1), (2, 4)\}$$

is a mapping from A to B, since for each $a \in A$ there is a unique element $b \in B$ such that $(a, b) \in f$. As is frequently the case, this mapping can be efficiently described by giving the rule for the image under f. In this case, $f(a) = a^2$, $a \in A$. This mapping is pictured in Figure 1.4.

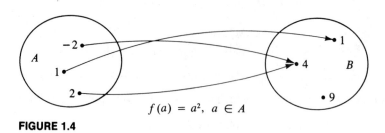

$$f(a) = a^2, \ a \in A$$

FIGURE 1.4   □

When it is possible to describe a mapping by giving a simple rule for the image of an element, it is certainly desirable to do so. We must keep in mind, however, that the set A, the set B, and the rule must all be known before the mapping is determined.

If f is a mapping from A to B, we write $f: A \rightarrow B$ or $A \xrightarrow{f} B$ to indicate this. The set A is called the **domain** of f, and B is called the **codomain** of f. The **range** of f is the set

$$C = \{y \mid y \in B \text{ and } y = f(x) \text{ for some } x \in A\}.$$

The range of f is denoted by $f(A)$.

EXAMPLE 3   Let $A = \{-2, 1, 2\}$ and $B = \{1, 4, 9\}$, and let f be the mapping described in the previous example:

$$f = \{(a, b) \mid f(a) = a^2, a \in A\}.$$

The domain of f is A, the codomain of f is B, and the range of f is $\{1, 4\} \subset B$.   □

More generally, if $f: A \rightarrow B$ and $S \subseteq A$, then

$$f(S) = \{y \mid y \in B \text{ and } y = f(x) \text{ for some } x \in S\}.$$

Note that $f(S)$ denotes a set of elements (a subset of B), and does *not* indicate a single value of f. This set, $f(S)$, is called the **image** of S under f. For any subset T of B, the **inverse image** of T is denoted by $f^{-1}(T)$ and defined by

$$f^{-1}(T) = \{x \mid x \in A \text{ and } f(x) \in T\}.$$

Once again, this is a set of elements, not the value of a mapping. We illustrate these notations in the next example.

EXAMPLE 4   Let $f: A \to B$ as in Example 3. If $S = \{1, 2\}$, then $f(S) = \{1, 4\}$, as shown in Figure 1.5.

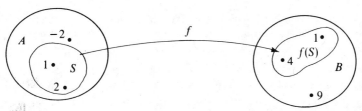

FIGURE 1.5

With $T = \{4, 9\}$, $f^{-1}(T)$ is given by $f^{-1}(T) = \{-2, 2\}$, as shown in Figure 1.6.

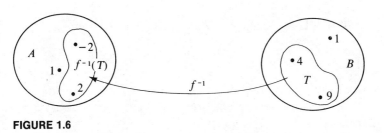

FIGURE 1.6

Among the various mappings from a nonempty set A to a nonempty set B, there are some that have properties worthy of special designation. We make the following definition.

DEFINITION 1.9   Let $f: A \to B$. If $B = f(A)$, then f is called **surjective**, or **onto**. If different elements of A always have different images, then f is called **injective**, or **one-to-one**. If f is both surjective and injective, it is called **bijective**. A bijective mapping from A to B is called a **bijection** from A to B, or a **one-to-one correspondence** from A to B.

EXAMPLE 5   Suppose $f: A \to B$ where $A = \{-1, 0, 1\}$, $B = \{1, -1\}$, and $f = \{(-1, 1), (0, 1), (1, 1)\}$. The mapping f is not surjective since there is no $a \in A$ such that $f(a) = -1 \in B$. We see that f is not injective since

$$f(-1) = f(0) = 1 \quad \text{but} \quad -1 \neq 0.$$

We also note that f is not a bijection.

According to our definition, a mapping f from A to B is surjective if and only if every element of B is the image of at least one element in A. A standard way to demonstrate that $f: A \to B$ is surjective is to take an arbitrary element b in B and show (usually by some kind of formula) that there exists an element $a \in A$ such that $b = f(a)$.

A mapping $f: A \to B$ is injective if and only if it has the property that $a_1 \neq a_2$ in A always implies that $f(a_1) \neq f(a_2)$ in B. This is just a precise statement of the fact that different elements always have different images. The trouble with this statement is that it is formulated in terms of unequal quantities, whereas most of the manipulations in mathematics deal with equalities. For this reason, we take the logically equivalent statement "$f(a_1) = f(a_2)$ always implies $a_1 = a_2$" as our working form of the definition. That is, we usually show that f is injective by assuming that $f(a_1) = f(a_2)$ and proving that this implies that $a_1 = a_2$.

EXAMPLE 6 Let $g: \mathbf{Z} \to \mathbf{Z}$, where $\mathbf{Z}$ is the set of integers. If g is defined by

$$g = \{(a, -a) \mid a \in \mathbf{Z}\},$$

then we write $g(a) = -a$, $a \in \mathbf{Z}$.

To show that g is surjective (onto), we choose an arbitrary element $a \in \mathbf{Z}$. Then there exists $-a \in \mathbf{Z}$ such that

$$(-a, a) \in g$$

since $g(-a) = -(-a) = a$,

and hence g is surjective.

To show that g is injective, we assume for $a_1 \in \mathbf{Z}$ and $a_2 \in \mathbf{Z}$,

$$g(a_1) = g(a_2)$$

or $-a_1 = -a_2$.

Hence, we have $a_1 = a_2$ and g is injective (one-to-one). Since g is both surjective and injective, then g is said to be bijective.   ☐

Examples of mappings that are injective but not surjective and vice versa can be found in the exercises.

The next definition describes how two appropriately chosen mappings can be combined.

DEFINITION 1.10 Let $g: A \to B$ and $f: B \to C$. The **composite mapping** $f \circ g$ is the mapping from A to C defined by

$$(f \circ g)(x) = f(g(x))$$

for all $x \in A$.

The composite mapping $f \circ g$ is diagrammed in Figure 1.7. Note that the domain of f must contain the range of g before the composition $f \circ g$ is defined.

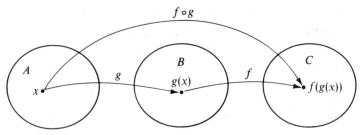

FIGURE 1.7

EXAMPLE 7   Let $\mathbf{Z}$ be the set of integers, Z_1 the set of nonnegative integers, and Z_2 the set of nonpositive integers. Suppose the mappings g and f are defined as

$$g: \mathbf{Z} \to Z_1, \qquad g(x) = x^2;$$
$$f: Z_1 \to Z_2, \qquad f(x) = -x.$$

Then the composition $f \circ g$ is a mapping from $\mathbf{Z}$ to Z_2 with

$$(f \circ g)(x) = f\big(g(x)\big) = f(x^2) = -x^2.$$

Note that $f \circ g$ is not surjective since $-3 \in Z_2$, but there is no integer x such that

$$(f \circ g)(x) = -x^2 = -3.$$

Note that $f \circ g$ is not injective since

$$(f \circ g)(-2) = -(-2)^2 = -2^2 = (f \circ g)(2)$$

and

$$-2 \neq 2. \qquad\qquad \square$$

When the composition mapping can be formed, we have an operation defined that is associative. If $h: A \to B$, $g: B \to C$, and $f: C \to D$, then

$$\begin{aligned}
((f \circ g) \circ h)(x) &= (f \circ g)\big(h(x)\big) \\
&= f\big[g(h(x))\big] \\
&= f\big((g \circ h)(x)\big) \\
&= (f \circ (g \circ h))(x)
\end{aligned}$$

for all $x \in A$. Thus the compositions $(f \circ g) \circ h$ and $f \circ (g \circ h)$ are the same mapping from A to D.

In many cases, we will be dealing with mappings of a set into itself; that is, the domain and codomain of the mappings are the same. This is the case in our next example.

EXAMPLE 8   Let $\mathbf{Z}$ be the set of all integers, and let the mappings $f: \mathbf{Z} \to \mathbf{Z}$ and $g: \mathbf{Z} \to \mathbf{Z}$ be defined for each $n \in \mathbf{Z}$ by

$$f(n) = 2n,$$

$$g(n) = \begin{cases} \dfrac{n}{2} & \text{if } n \text{ is even} \\ 4 & \text{if } n \text{ is odd.} \end{cases}$$

In this case, the composition mappings $f \circ g$ and $g \circ f$ are both defined. We have, on the one hand,

$$\begin{aligned} (g \circ f)(n) &= g(f(n)) \\ &= g(2n) \\ &= n, \end{aligned}$$

so $(g \circ f)(n) = n$ for all $n \in \mathbf{Z}$. On the other hand,

$$\begin{aligned} (f \circ g)(n) &= f(g(n)) \\ &= \begin{cases} f\left(\dfrac{n}{2}\right) = n & \text{if } n \text{ is even} \\ f(4) \;\; = 8 & \text{if } n \text{ is odd,} \end{cases} \end{aligned}$$

so $f \circ g \neq g \circ f$. This example shows that mapping composition is not commutative.   □

As in Examples 7 and 8, $\mathbf{Z}$ will be used as the standard notation for the set of all integers. We will use $\mathbf{Z}^+$ to denote the set of all positive integers, and $\mathbf{E}$ to denote the set of all even integers.

There are several interesting properties of the composition mapping. We state one as a theorem, and others can be found in the exercises.

THEOREM 1.11   Let $g: A \to B$ and $f: B \to C$. If f and g are both surjective, then $f \circ g$ is surjective.

Proof   The composition $f \circ g$ maps A to C. Suppose $c \in C$. Since f is surjective, there exists $b \in B$ such that

$$f(b) = c.$$

Since g is surjective, for every $b \in B$ there exists $a \in A$ such that

$g(a) = b$.

Hence, for $c \in C$ there exists $a \in A$ such that

$(f \circ g)(a) = f(g(a)) = f(b) = c$,

and $f \circ g$ is surjective. ∎

EXERCISES 1.2

1 For the given sets, form the indicated Cartesian product.
   (a) $A \times B$; $A = \{a, b\}$, $B = \{0, 1\}$
   (b) $B \times A$; $A = \{a, b\}$, $B = \{0, 1\}$
   (c) $A \times B$; $A = \{2, 4, 6, 8\}$, $B = \{2\}$
   (d) $B \times A$; $A = \{1, 5, 9\}$, $B = \{-1, 1\}$
   (e) $B \times A$; $A = B = \{1, 2, 3\}$

2 For each of the following mappings, state the domain, the codomain, and the range, where $f : \mathbf{E} \to \mathbf{Z}$.
   (a) $f(x) = x/2$, $x \in \mathbf{E}$
   (b) $f(x) = x$, $x \in \mathbf{E}$
   (c) $f(x) = |x|$, $x \in \mathbf{E}$
   (d) $f(x) = x + 1$, $x \in \mathbf{E}$

3 For each of the following mappings, write out $f(S)$ and $f^{-1}(T)$ for the given S and T, where $f : \mathbf{Z} \to \mathbf{Z}$.
   (a) $f(x) = |x|$; $S = \{\ldots, -5, -3, -1, 1, 3, 5, \ldots\}$, $T = \{1, 3, 4\}$

   (b) $f(x) = \begin{cases} x + 1 & \text{if } x \text{ is even} \\ x & \text{if } x \text{ is odd} \end{cases}$; $S = \{0, 1, 5, 9\}$, $T = \{\ldots, -3, -1, 1, 3, \ldots\}$

   (c) $f(x) = x^2$; $S = \{-2, -1, 0, 1, 2\}$, $T = \{2, 7, 11\}$

4 For each of the mappings $f : \mathbf{Z} \to \mathbf{Z}$ given below, determine if the mapping is surjective, injective, or bijective. Justify your answer.

   (a) $f(x) = \begin{cases} x & \text{if } x \text{ is even} \\ 2x - 1 & \text{if } x \text{ is odd} \end{cases}$

   (b) $f(x) = 2x$
   (c) $f(x) = x^2$
   (d) $f(x) = x^3$

   (e) $f(x) = \begin{cases} x & \text{if } x \text{ is even} \\ x - 1 & \text{if } x \text{ is odd} \end{cases}$

   (f) $f(x) = |x|$
   (g) $f(x) = x - |x|$

   (h) $f(x) = \operatorname{sgn} x = \begin{cases} 1 & \text{if } x > 0 \\ 0 & \text{if } x = 0 \\ -1 & \text{if } x < 0 \end{cases}$

   (This function is known as the *signum function*, or the *sign function*.)

5   For the given subsets A and B of $\mathbf{Z}$, let $f(x) = 2x$ and determine whether $f: A \to B$ is surjective, injective, or bijective. Justify your answer.
  (a) $A = \mathbf{Z}, B = \mathbf{E}$
  (b) $A = \mathbf{E}, B = \mathbf{E}$

6   For the given subsets A and B of $\mathbf{Z}$, let $f(x) = |x|$ and determine whether $f: A \to B$ is surjective, injective, or bijective. Justify your answer.
  (a) $A = \mathbf{Z}, B = \mathbf{Z}^+$
  (b) $A = \mathbf{Z}^+, B = \mathbf{Z}$
  (c) $A = \mathbf{Z}^+, B = \mathbf{Z}^+$

7   Let $f: \mathbf{Z} \to \mathbf{Z}$ and $g: \mathbf{Z} \to \mathbf{Z}$ be as defined below. In each case, compute $(f \circ g)(x)$ and $(g \circ f)(x)$ for arbitrary $x \in \mathbf{Z}$.

  (a) $f(x) = 2x, \quad g(x) = \begin{cases} x & \text{if } x \text{ is even} \\ 2x - 1 & \text{if } x \text{ is odd} \end{cases}$

  (b) $f(x) = 2x, \quad g(x) = x^3$

  (c) $f(x) = x + |x|, \quad g(x) = \begin{cases} \dfrac{x}{2} & \text{if } x \text{ is even} \\ -x & \text{if } x \text{ is odd} \end{cases}$

  (d) $f(x) = \begin{cases} \dfrac{x}{2} & \text{if } x \text{ is even} \\ x + 1 & \text{if } x \text{ is odd} \end{cases}, \quad g(x) = \begin{cases} x - 1 & \text{if } x \text{ is even} \\ 2x & \text{if } x \text{ is odd} \end{cases}$

  (e) $f(x) = x^2, \quad g(x) = x - |x|$

8   For each pair f, g given in Problem 7, is $f \circ g$ surjective or injective? Is $g \circ f$ surjective or injective? Justify your answers.

9   For each part below, give an example of a mapping from $\mathbf{E}$ to $\mathbf{E}$ that satisfies the given conditions.
  (a) Injective and surjective
  (b) Injective and not surjective
  (c) Surjective and not injective
  (d) Not injective and not surjective

10 (a) Give an example of mappings f and g such that one of f or g is not surjective but $f \circ g$ is surjective.
  (b) Give an example of mappings f and g such that one of f or g is not injective but $f \circ g$ is injective.

11 For the given $f: A \to B$, determine whether f is surjective, injective, or bijective, and justify your answer.
  (a) $A = \mathbf{Z} \times \mathbf{Z}, B = \mathbf{Z} \times \mathbf{Z}, f(x, y) = (y, x)$
  (b) $A = \mathbf{Z} \times \mathbf{Z}, B = \mathbf{Z}, f(x, y) = x + y$
  (c) $A = \mathbf{Z} \times \mathbf{Z}, B = \mathbf{Z}, f(x, y) = x$
  (d) $A = \mathbf{Z}, B = \mathbf{Z} \times \mathbf{Z}, f(x) = (x, 1)$

12 (a) Show that the mapping f given in Example 2 is neither surjective nor injective.
  (b) For this mapping f, show that if $S = \{1, 2\}$, then $f^{-1}(f(S)) \neq S$.
  (c) For this same f and $T = \{4, 9\}$, show that $f(f^{-1}(T)) \neq T$.

13 (a) Let $f: A \to B$, where A and B are nonempty. Prove that f has the property that $f^{-1}(f(S)) = S$ for every subset S of A if and only if f is injective. (Compare Problem 12(b).)

     (b) Let $f: A \to B$, where A and B are nonempty. Prove that f has the property that $f(f^{-1}(T)) = T$ for every subset T of B if and only if f is surjective. (Compare Problem 12(c).)

14 Let $g: A \to B$ and $f: B \to C$. Prove that $f \circ g$ is injective if each of f and g is injective.

15 Let $f: A \to B$ and $g: B \to A$. Prove that f is bijective if $f \circ g$ is injective and $g \circ f$ is surjective.

1.3 BINARY OPERATIONS

We are familiar with the operations of addition, subtraction, and multiplication on real numbers. These are examples of *binary operations*. When we speak of a binary operation on a set, we have in mind a process that combines two elements of the set to produce a third element of the set. This third element, the result of the operation on the first two, must be unique. That is, there must be one and only one result from the combination. Also, it must always be possible to combine the two elements, no matter which two are chosen. This discussion is admittedly somewhat vague, in that the terms "process" and "combine" are somewhat indefinite. To eliminate this vagueness we make the following formal definition.

> **DEFINITION 1.12** A **binary operation** on a nonempty set A is a mapping f from $A \times A$ into A.

We now have a precise definition of the term "binary operation," but some of the feel for the concept may have been lost. The definition will, however, give us what we want. Suppose f is a mapping from $A \times A$ into A. Then $f(x, y)$ is defined for every ordered pair (x, y) of elements of A, and the image $f(x, y)$ is unique. In other words, any two elements x and y of A can be combined to obtain a unique third element of A by finding the value $f(x, y)$. The result of performing the binary operation on x and y is $f(x, y)$, and the only thing unfamiliar about this is the notation for the result. We are accustomed to indicating results of binary operations by symbols such as $x + y$ and $x - y$. We can use a similar notation and write $x * y$ in place of $f(x, y)$. Thus $x * y$ represents the result of an arbitrary binary operation on A, just as $f(x, y)$ represents the value of an arbitrary mapping from $A \times A$ to A.

EXAMPLE 1   Two examples of binary operations from $\mathbf{Z} \times \mathbf{Z}$ to $\mathbf{Z}$ are:

(i) $x * y = x + y - 1$,   for $(x, y) \in \mathbf{Z} \times \mathbf{Z}$.
(ii) $x * y = 1 + xy$,   for $(x, y) \in \mathbf{Z} \times \mathbf{Z}$.   □

EXAMPLE 2   The operation of forming the intersection $A \cap B$ of subsets A and B of a universal set U is a binary operation on the collection of all subsets of U. This is also true of the operation of forming the union.   □

Since we are dealing with ordered pairs in connection with a binary operation, the results $x * y$ and $y * x$ may well be different.

DEFINITION 1.13   If $*$ is a binary operation on the nonempty set A, then $*$ is called **commutative** if $x * y = y * x$ for all x and y in A. If $x * (y * z) = (x * y) * z$ for all x, y, z in A, we say the binary operation is **associative**.

EXAMPLE 3   The usual binary operations of addition and multiplication on the integers are both commutative and associative. However, the binary operation of subtraction on the integers does not have either of these properties. For example, $5 - 7 \neq 7 - 5$ and $9 - (8 - 3) \neq (9 - 8) - 3$.   □

Suppose we consider the two binary operations given in Example 1.

EXAMPLE 4   The binary operation $*$ defined by

$$x * y = x + y - 1$$

is commutative since

$$x * y = x + y - 1 = y + x - 1 = y * x.$$

Note that $*$ is also associative since

$$
\begin{aligned}
x * (y * z) &= x * (y + z - 1) \\
&= x + (y + z - 1) - 1 \\
&= (x + y - 1) + z - 1 \\
&= (x + y - 1) * z \\
&= (x * y) * z.
\end{aligned}
$$
□

EXAMPLE 5   The binary operation $*$ defined by

$$x * y = 1 + xy$$

is commutative since

$$x * y = 1 + xy = 1 + yx = y * x.$$

Note that * is not associative, since

$$x * (y * z) = x * (1 + yz) = 1 + x(1 + yz) = 1 + x + xyz$$

is not equal to

$$(x * y) * z = (1 + xy) * z = 1 + (1 + xy)z = 1 + z + xyz. \qquad \square$$

DEFINITION 1.14   Let * be a binary operation on the nonempty set A. An element e in A is called an **identity** element with respect to the binary operation * if e has the property that

$$e * x = x * e = x$$

for all $x \in A$.

The phase "with respect to the binary operation" is critical in this definition, since the particular binary operation being considered is all-important. This is pointed out in the next example.

EXAMPLE 6   The integer 1 is an identity with respect to the operation of multiplication ($1 \cdot x = x \cdot 1 = x$), but not with respect to the operation of addition ($1 + x \neq x$). $\qquad \square$

EXAMPLE 7   The element 1 is the identity element with respect to the binary operation *, given by

$$x * y = x + y - 1, \qquad (x, y) \in \mathbf{Z} \times \mathbf{Z},$$

since

$$x * 1 = x + 1 - 1 = x \quad \text{and} \quad 1 * x = 1 + x - 1 = x. \qquad \square$$

EXAMPLE 8   There is no identity element with respect to the binary operation * defined by

$$x * y = 1 + xy, \qquad (x, y) \in \mathbf{Z} \times \mathbf{Z},$$

since there is no fixed integer z such that

$$x * z = z * x = 1 + xz = x, \qquad \text{for all } x \in \mathbf{Z}. \qquad \square$$

Whenever a set has an identity element with respect to a binary operation on the set, it is then in order to raise the question of inverses.

DEFINITION 1.15 Suppose that e is an identity element for the binary operation $*$ on the set A, and let $a \in A$. If there exists an element $b \in A$ such that $a * b = e$, then b is called a **right inverse** of a with respect to this operation. Similarly, if $b * a = e$, b is called a **left inverse** of a. If both of $a * b = e$ and $b * a = e$ hold, then b is called an **inverse** of a.

Sometimes an inverse is referred to as a two-sided inverse to emphasize that both of the required equations hold.

EXAMPLE 9 Each element $x \in \mathbf{Z}$ has a two-sided inverse $(-x + 2) \in \mathbf{Z}$ with respect to the binary operation $*$ given by

$$x * y = x + y - 1, \quad (x, y) \in \mathbf{Z} \times \mathbf{Z},$$

since

$$x * (-x + 2) = (-x + 2) * x = -x + 2 + x - 1 = 1 = e. \qquad \square$$

DEFINITION 1.16 A bijection from a set A to itself is called a **permutation** on A. For any nonempty set A, we adopt the notation $\mathscr{S}(A)$ as standard for the set of all permutations on A. The set of all mappings from A to A will be denoted by $\mathscr{M}(A)$.

From the discussion in Section 1.2, we know that composition of mappings is an associative binary operation on $\mathscr{M}(A)$. The identity mapping I_A is defined by

$$I_A(x) = x \quad \text{for all } x \in A.$$

For any f in $\mathscr{M}(A)$,

$$(I_A \circ f)(x) = I_A(f(x)) = f(x)$$

and

$$(f \circ I_A)(x) = f(I_A(x)) = f(x),$$

so $I_A \circ f = f \circ I_A = f$. That is, I_A is an identity element for map composition.

EXAMPLE 10   In Example 8 of Section 1.2, we defined the mappings $f : \mathbf{Z} \to \mathbf{Z}$ and $g : \mathbf{Z} \to \mathbf{Z}$ by

$$f(n) = 2n$$

and

$$g(n) = \begin{cases} \dfrac{n}{2} & \text{if } n \text{ is even} \\ 4 & \text{if } n \text{ is odd.} \end{cases}$$

For these mappings, $(g \circ f)(n) = n$ for all $n \in \mathbf{Z}$, so $g \circ f = I_{\mathbf{Z}}$ and g is a left inverse for f. Note, however, that

$$(f \circ g)(n) = \begin{cases} n & \text{if } n \text{ is even} \\ 8 & \text{if } n \text{ is odd,} \end{cases}$$

so $f \circ g \neq I_{\mathbf{Z}}$, and g is not a right inverse for f.   □

Example 10 furnishes some insight into the next two lemmas.[†]

LEMMA 1.17   Let A be a nonempty set, and let $f : A \to A$. Then f is injective if and only if f has a left inverse.

Proof   Assume first that f has a left inverse g, and suppose that $f(a_1) = f(a_2)$. Since $g \circ f = I_A$, we have

$$\begin{aligned} a_1 &= I_A(a_1) = (g \circ f)(a_1) = g(f(a_1)) = g(f(a_2)) \\ &= (g \circ f)(a_2) = I_A(a_2) = a_2. \end{aligned}$$

Thus $f(a_1) = f(a_2)$ implies $a_1 = a_2$, and f is injective.

Conversely, now assume that f is injective. We shall define a left inverse g of f. Let a_0 represent an arbitrarily chosen but fixed element in A. For each x in A, $g(x)$ is defined by this rule:

1.   If there is an element y in A such that $f(y) = x$, then $g(x) = y$.
2.   If no such element y exists in A, then $g(x) = a_0$.

When the first part of the rule applies, the element y is unique because f is injective $(f(y_1) = x = f(y_2) \Rightarrow y_1 = y_2 = g(x))$. Thus $g(x)$ is unique in this case. When the second part of the rule applies, $g(x) = a_0$ is surely unique, and g is a mapping from A to A. For all a in A we have

$$(g \circ f)(a) = g(f(a)) = a$$

because $x = f(a)$ requires $g(x) = a$. Thus g is a left inverse for f.   ■

[†] A lemma is a proposition the main purpose of which is to help prove another proposition.

There is a connection between surjectiveness and right inverses that is similar to the one between injectiveness and left inverses. We state this connection in the next lemma, and leave the proof as an exercise.

LEMMA 1.18 Let A be a nonempty set, and $f: A \to A$. Then f is surjective if and only if f has a right inverse.

Lemmas 1.17 and 1.18 enable us to prove the following important theorem.

THEOREM 1.19 Let $f: A \to A$. Then f has an inverse if and only if f is a permutation on A.

Proof If f has an inverse g, then $g \circ f = I_A$ and $f \circ g = I_A$. Note that $g \circ f = I_A$ implies f is injective by Lemma 1.17, and $f \circ g = I_A$ implies f is surjective by Lemma 1.18. Thus f is a permutation on A.

Now suppose that f is a permutation on A. Then f has a left inverse g by Lemma 1.17 and a right inverse h by Lemma 1.18. We have $g \circ f = I_A$ and $f \circ h = I_A$, so

$$g = g \circ I_A = g \circ (f \circ h) = (g \circ f) \circ h = I_A \circ h = h.$$

That is, $g = h$, and f has an inverse. ∎

The last theorem shows that the members of the set $\mathscr{S}(A)$ are special in that each of them has an inverse. We denote the inverse of f by f^{-1}. It is left as an exercise to prove that f^{-1} is a permutation on A.

There is one other property of the set $\mathscr{S}(A)$ that is significant. Whenever f and g are in $\mathscr{S}(A)$, then $f \circ g$ is also in $\mathscr{S}(A)$. (See Problem 12.) We say that $\mathscr{S}(A)$ is **closed** under mapping composition.

More generally, suppose that $*$ is a binary operation on a nonempty set A, and let $B \subseteq A$. If $x * y$ is in B for all $x \in B$ and $y \in B$, we say that B is **closed** with respect to $*$.

EXAMPLE 11 Consider the binary operation $*$ defined on the set of integers $\mathbf{Z}$ by

$$x * y = |x| + |y|, \qquad (x, y) \in \mathbf{Z} \times \mathbf{Z}.$$

The set B of negative integers is not closed with respect to $*$ since $x = -1 \in B$ and $y = -2 \in B$, but

$$x * y = (-1) * (-2) = |-1| + |-2| = 3 \notin B. \qquad \square$$

EXERCISES 1.3

1 Decide if the given set B is closed with respect to the binary operation defined on the set of integers $\mathbf{Z}$. If B is not closed, exhibit elements $x \in B$ and $y \in B$, such that $x * y \notin B$.
(a) $x * y = xy,$ $B = \{-1, -2, -3, \dots\}$
(b) $x * y = x - y,$ $B = \{1, 2, 3, \dots\}$
(c) $x * y = x^2 + y^2,$ $B = \{1, 2, 3, \dots\}$
(d) $x * y = \operatorname{sgn} x + \operatorname{sgn} y,$ $B = \{-1, 0, 1\}$ where

$$\operatorname{sgn} x = \begin{cases} 1 & \text{if } x > 0 \\ 0 & \text{if } x = 0 \\ -1 & \text{if } x < 0. \end{cases}$$

2 In each part below, a rule is given that determines a binary operation $*$ on the set $\mathbf{Z}$ of all integers. Determine in each case if the operation is commutative or associative, and if there is an identity element.
(a) $x * y = x + xy$
(b) $x * y = x$
(c) $x * y = x + 2y$
(d) $x * y = 3(x + y)$
(e) $x * y = x - y$

3 Let S be a set of four elements given by $S = \{A, B, C, D\}$. In the table below, all of the elements of S are listed in a row at the top, and in a column at the left. The result $x * y$ is found in the row that starts with x at the left and in the column that has y at the top. For example, $B * C = B$ and $B * D = A$.

$*$	A	B	C	D
A	B	C	A	B
B	C	D	B	A
C	A	B	C	D
D	A	B	D	D

(a) Is the binary operation $*$ commutative? Why?
(b) Determine whether there is an identity element in S for $*$.
(c) If there is an identity element, which elements have inverses?

4 Assume that $*$ is an associative binary operation on A with identity element e. Prove that if b is a right inverse of a and c is a left inverse of a, then $b = c$.

5 For each of the mappings $f : \mathbf{Z} \to \mathbf{Z}$ given below, exhibit a right inverse of f with respect to mapping composition whenever one exists.

(a) $f(x) = \begin{cases} x & \text{if } x \text{ is even} \\ 2x - 1 & \text{if } x \text{ is odd} \end{cases}$

(b) $f(x) = 2x$
(c) $f(x) = x^2$
(d) $f(x) = x^3$

(e) $f(x) = \begin{cases} x & \text{if } x \text{ is even} \\ x - 1 & \text{if } x \text{ is odd} \end{cases}$

(f) $f(x) = |x|$

(g) $f(x) = x - |x|$

(h) $f(x) = \operatorname{sgn} x = \begin{cases} 1 & \text{if } x > 0 \\ 0 & \text{if } x = 0 \\ -1 & \text{if } x < 0 \end{cases}$

6   For each of the mappings f given in Problem 5, determine if f has a left inverse. Exhibit a left inverse whenever one exists.

7   Let $f: A \to A$, where A is nonempty. Prove that f has a left inverse if and only if $f^{-1}(f(S)) = S$ for every nonempty subset S of A.

8   Let $f: A \to A$, where A is nonempty. Prove that f has a right inverse if and only if $f(f^{-1}(T)) = T$ for every nonempty subset T of A.

9   Prove Lemma 1.18: Let A be a nonempty set, and $f: A \to A$. Then f is surjective if and only if f has a right inverse.

10   Prove that if f is a permutation on A, then f^{-1} is a permutation on A.

11   Prove that if f is a permutation on A, then $(f^{-1})^{-1} = f$.

12   Let f and g be permutations on A. Prove that $f \circ g$ is a permutation on A.

13   (a) Prove that the set of all surjective mappings from A to A is closed under composition of mappings.

(b) Prove that the set of all injective mappings from A to A is closed under mapping composition.

14   Let M be the set of all rectangular arrays $\begin{bmatrix} a & b \\ c & d \end{bmatrix}$ of two rows and two columns of real numbers. (Such an array is called a **matrix**.) Let addition and multiplication be defined for elements of M by

$$\begin{bmatrix} a & b \\ c & d \end{bmatrix} + \begin{bmatrix} x & y \\ z & w \end{bmatrix} = \begin{bmatrix} a+x & b+y \\ c+z & d+w \end{bmatrix}$$

and

$$\begin{bmatrix} a & b \\ c & d \end{bmatrix} \cdot \begin{bmatrix} x & y \\ z & w \end{bmatrix} = \begin{bmatrix} ax+bz & ay+bw \\ cx+dz & cy+dw \end{bmatrix}.$$

(a) Prove that each operation is associative.

(b) Prove that addition is commutative.

(c) Find an identity element for addition.

(d) Find an identity element for multiplication.

(e) What is the additive inverse of $\begin{bmatrix} a & b \\ c & d \end{bmatrix}$ in M?

(f) If $ad - bc \neq 0$, show that the multiplicative inverse of $\begin{bmatrix} a & b \\ c & d \end{bmatrix}$ is given by

$$\begin{bmatrix} \dfrac{d}{ad-bc} & \dfrac{-b}{ad-bc} \\ \dfrac{-c}{ad-bc} & \dfrac{a}{ad-bc} \end{bmatrix}$$

15 Let M be as in Problem 14, and let G be the set of all elements of M that have one row that consists of zeros, and one row of the form $[a \quad a]$, with $a \neq 0$.
(a) Show that G is closed under multiplication.
(b) Show that for each x in G, there is an element y in G such that $xy = yx = x$.
(c) Show that G does not have an identity element with respect to multiplication.

1.4 RELATIONS

In the study of mathematics, we deal with many examples of relations between elements of various sets. In working with the integers, we encounter relations such as "x is less than y," or "x is a factor of y." In the calculus, one function may be the derivative of some other function, or perhaps an integral of another function. The property that these examples of relations have in common is that there is an association of some sort between two elements of a set, and the ordering of the elements is important. These relations can all be described by the following definition.

> **DEFINITION 1.20** A **relation** (or a **binary relation**) on a nonempty set A is a nonempty set R of ordered pairs (x, y) of elements x and y of A.

That is, a relation R is a subset of the Cartesian product $A \times A$. If the pair (a, b) is in R, we write aRb, and say that a has the relation R to b. If $(a, b) \notin R$, we write $a\not R b$. This notation agrees with the customary notations for relations, such as $a = b$ and $a < b$.

> **EXAMPLE 1** Let $A = \{-2, -5, 2, 5\}$, and let
> $$R = \{(5, -2), (5, 2), (-5, -2), (-5, 2)\}.$$
> Then $5R2$, $-5R2$, $5R(-2)$, and $(-5)R(-2)$, but $2\not R 5$, $5\not R 5$, and so on. As is frequently the case, this relation can be described by a simple rule: xRy if and only if the absolute value of x is the same as $y^2 + 1$, that is, if $|x| = y^2 + 1$. □

Any mapping from A to A is an example of a relation, but not all relations are mappings, as Example 1 illustrates. We have $(5, 2) \in R$ and $(5, -2) \in R$, and for a mapping from A to A, the second element y in $(5, y)$ would have to be unique.

Our main concern is with relations that have additional special properties. More precisely, we are interested for the most part in *equivalence relations*.

DEFINITION 1.21   A relation R on a set A is an **equivalence relation** if the following conditions are satisfied for arbitrary x, y, z in A:
(1) xRx is true for all $x \in A$.
(2)   If xRy is true, then yRx is true.
(3)   If xRy is true and yRz is true, then xRz is true.
Properties (1), (2), and (3) are called the **reflexive, symmetric,** and **transitive** properties, respectively. They are the familiar basic properties of equality. Indeed, every equivalence relation on a set is a sort of equality on the set.

EXAMPLE 2   The relation R defined on the set of integers $\mathbf{Z}$ by

xRy if and only if $|x| = |y|$

is reflexive, symmetric and transitive. For arbitrary x, y, and z in $\mathbf{Z}$,

(1) xRx since $|x| = |x|$.
(2) $xRy \Rightarrow |x| = |y|$
$\Rightarrow |y| = |x|$
$\Rightarrow yRx.$
(3) xRy and $yRz \Rightarrow |x| = |y|$ and $|y| = |z|$
$\Rightarrow |x| = |z|$
$\Rightarrow xRz.$ $\square$

EXAMPLE 3   The relation R defined on the set of integers $\mathbf{Z}$ by

xRy if and only if $x > y$

is not an equivalence relation since it is neither reflexive nor symmetric.

(1) $x \not> x$ for all $x \in \mathbf{Z}$.
(2) $x > y \not\Rightarrow y > x$.

Note that R is transitive:

(3) $x > y$ and $y > z \Rightarrow x > z$. $\square$

The following example is a special case of an equivalence relation on the integers that will be extremely important in later work.

EXAMPLE 4   The relation "congruence modulo 4" is defined on the set $\mathbf{Z}$ of all integers as follows: x is congruent to y modulo 4 if and only if $x - y$ is a multiple of 4. We write $x \equiv y \pmod 4$ as shorthand for "x is congruent to y modulo 4." Thus $x \equiv y \pmod 4$ if and only if $x - y = 4k$ for some integer k. We demonstrate that this is an equivalence relation. For arbitrary x, y, z in $\mathbf{Z}$,

(1) $x \equiv x \pmod 4$ since $x - x = (4)(0)$.

(2) $x \equiv y \pmod 4 \Rightarrow x - y = 4k$ for $k \in \mathbf{Z}$
$\Rightarrow y - x = 4(-k)$ for $-k \in \mathbf{Z}$
$\Rightarrow y \equiv x \pmod 4$.

(3) $x \equiv y \pmod 4$ and $y \equiv z \pmod 4$
$\Rightarrow x - y = 4k$ and $y - z = 4m$ for $k, m \in \mathbf{Z}$
$\Rightarrow x - z = x - y + y - z$
$= 4(k + m)$,   where $k + m \in \mathbf{Z}$
$\Rightarrow x \equiv z \pmod 4$.

Thus congruence modulo 4 has the reflexive, symmetric, and transitive properties, and is an equivalence relation on $\mathbf{Z}$.   □

DEFINITION 1.22   Let R be an equivalence relation on the nonempty set A. For each $a \in A$, the set

$$[a] = \{x \in A \mid xRa\}$$

is called the **equivalence class** containing a.

Recall from Section 1.1 that a separation of the elements of a nonempty set A into mutually disjoint nonempty subsets is called a *partition* of A. It is not difficult to show that if R is an equivalence relation on A, then the distinct equivalence classes of R form a partition of A. Conversely, if a partition of A is given, then we can find an equivalence relation R on A that has the given subsets as its equivalence classes. We simply define R by xRy if and only if x and y are in the same subset. The proofs of these statements are requested in the exercises.

The discussion in the last paragraph illustrates a situation where we are dealing with a collection of sets about which very little is explicit. For example, the collection may be finite, or it may be infinite. In such situations, it is sometimes desirable to use the notational convenience known as indexing. We assume that the sets in the collection are **labeled**, or **indexed**, by a set $\mathscr{L}$ of symbols λ. That is, a typical set in the collection is denoted by a symbol such as A_λ, and the index λ takes on values from the set $\mathscr{L}$. For such a collection $\{A_\lambda\}$ we write $\bigcup_{\lambda \in \mathscr{L}} A_\lambda$ for the union of the collection of sets and $\bigcap_{\lambda \in \mathscr{L}} A_\lambda$ for the intersection. That is,

$$\bigcup_{\lambda \in \mathscr{L}} A_\lambda = \{x \mid x \in A_\lambda \text{ for at least one } A_\lambda\}$$

and

$$\bigcap_{\lambda \in \mathscr{L}} A_\lambda = \{x \mid x \in A_\lambda \text{ for every } A_\lambda\}.$$

If the indexing set $\mathscr{L}$ is given by $\mathscr{L} = \{1, 2, \ldots, n\}$, then the union of the collection of sets $\{A_i\}$ might be written in any one of the following three ways.

$$A_1 \cup A_2 \cup \cdots \cup A_n = \bigcup_{i \in \mathscr{L}} A_i = \bigcup_{i=1}^{n} A_i$$

EXERCISES 1.4

1   For $A = \{1, 3, 5\}$, determine which of the following relations on A are mappings from A to A, and justify your answer.
   (a) $\{(1, 3), (3, 5), (5, 1)\}$
   (b) $\{(1, 1), (3, 1), (5, 1)\}$
   (c) $\{(1, 1), (1, 3), (1, 5)\}$
   (d) $\{(1, 3), (3, 1), (5, 5)\}$
   (e) $\{(1, 5), (3, 3), (5, 3)\}$

2   In each part below, a relation R is defined on the set $\mathbf{Z}$ of all integers. Determine in each case whether or not R is reflexive, symmetric, or transitive. Justify your answers.
   (a) xRy if and only if $x = 2y$.
   (b) xRy if and only if $x = -y$.
   (c) xRy if and only if $x < y$.
   (d) xRy if and only if $x \geq y$.
   (e) xRy if and only if $x - y = 5k$ for some k in $\mathbf{Z}$.
   (f) xRy if and only if the greatest common division of x and y is 1.

3   Let A be the set of all nonempty subsets of $U = \{1, 2, 3, 4, 5\}$. Determine whether or not the given relation R on A is reflexive, symmetric, or transitive. Justify your answers.
   (a) xRy if and only if x is a subset of y.
   (b) xRy if and only if x is a proper subset of y.
   (c) xRy if and only if x and y have the same number of elements.

4   In each part below, a relation is defined on the set of all human beings. Determine if the relation is reflexive, symmetric, or transitive, and justify your answers.
   (a) xRy if and only if x lives within 400 miles of y.
   (b) xRy if and only if x is the father of y.
   (c) xRy if and only if x is a first cousin of y.
   (d) xRy if and only if x and y were born in the same year.
   (e) xRy if and only if x and y have the same mother.

5   In each part below, a relation R is defined on the power set $\mathscr{P}(A)$ of the nonempty set A. Determine in each case whether or not R is reflexive, symmetric, or transitive. Justify your answers.
   (a) BRC if and only if $B \cap C \neq \varnothing$.
   (b) BRC if and only if $B \subseteq C$.

6   For each of the following relations R defined on the set A of all triangles in a plane, determine whether or not R is reflexive, symmetric, or transitive. Justify your answers.
   (a) aRb if and only if a is similar to b.
   (b) aRb if and only if a is congruent to b.

7 Let $\mathscr{L} = \{1, 2, 3\}$, $A_1 = \{a, b, c, d\}$, $A_2 = \{c, d, e, f\}$, and $A_3 = \{a, c, f, g\}$. Write out $\bigcup_{\lambda \in \mathscr{L}} A_\lambda$ and $\bigcap_{\lambda \in \mathscr{L}} A_\lambda$.

8 Let $\mathscr{L} = \{\alpha, \beta, \gamma\}$, $A_\alpha = \{1, 2, 3, \ldots\}$, $A_\beta = \{-1, -2, -3, \ldots\}$, and $A_\gamma = \{0\}$. Write out $\bigcup_{\lambda \in \mathscr{L}} A_\lambda$ and $\bigcap_{\lambda \in \mathscr{L}} A_\lambda$.

9 Suppose R is an equivalence relation on the nonempty set A. Prove that the distinct equivalence classes of R separate the elements of A into mutually disjoint subsets.

10 Let $A = \{1, 2, 3\}$, $B_1 = \{1, 2\}$, $B_2 = \{2, 3\}$. Define the relation R on A by aRb if and only if there is a set B_i ($i = 1$ or 2) such that $a \in B_i$ and $b \in B_i$. Determine which of the properties of an equivalence relation hold for R and give an example for each property that fails to hold.

11 Suppose $\{A_\lambda\}, \lambda \in \mathscr{L}$, represents a partition of the nonempty set A. That is, $A = \bigcup_{\lambda \in \mathscr{L}} A_\lambda$, and if A_α and A_β are distinct, then $A_\alpha \cap A_\beta = \varnothing$. Define R on A by xRy if and only if there is a subset A_λ such that $x \in A_\lambda$ and $y \in A_\lambda$. Prove that R is an equivalence relation on A, and that the equivalence classes of R are the subsets A_λ.

Summary of Key Words and Phrases

Subset	Injective mapping
Proper subset	Bijective mapping
Union	Onto mapping
Intersection	One-to-one mapping
Empty set	One-to-one correspondence
Disjoint sets	Composite mapping
Power set	Binary operation
Universal set	Commutative binary operation
Complement	Associative binary operation
Partition	Identity element
Cartesian product	Right inverse
Mapping	Left inverse
Domain	Inverse
Codomain	Permutation
Range	Closed
Image	Relation
Inverse image	Equivalence relation
Surjective mapping	Equivalence class

References

Ames, Dennis B. *An Introduction to Abstract Algebra.* Scranton, Pa.: International Textbook, 1969.

Bundrick, Charles M. and John J. Leeson. *Essentials of Abstract Algebra.* Monterey, Calif.: Brooks-Cole, 1972.

Durbin, John R. *Modern Algebra.* New York: Wiley, 1979.

Fraleigh, John B. *A First Course in Abstract Algebra* (2nd ed.). Reading, Mass.: Addison-Wesley, 1976.

Herstein, I. N. *Topics in Algebra* (2nd ed.). New York: Wiley, 1975.

Larsen, Max D. *Introduction to Modern Algebraic Concepts.* Reading, Mass.: Addison-Wesley, 1969.

Mitchell, A. Richard and Roger W. Mitchell. *An Introduction to Abstract Algebra.* Monterey, Calif.: Brooks-Cole, 1970.

Shapiro, Louis. *Introduction to Abstract Algebra.* New York: McGraw-Hill, 1975.

2 THE INTEGERS

2.1 POSTULATES FOR THE INTEGERS

In the nineteenth century, the mathematician G. Peano formulated a set of postulates for the positive integers. They are known as **Peano's Postulates**, and may be stated as follows:

1. There is a positive integer 1.

2. Each positive integer n has a unique successor n'.

3. 1 is not the successor of any positive integer.

4. If $m' = n'$, then $m = n$.

5. (**The induction postulate**) If S is a set of positive integers, then S contains all of the positive integers if it has the following properties.
   (a) 1 is in S.
   (b) $n \in S$ always implies $n' \in S$.

It is possible to use these five postulates as a basis from which to develop all of the familiar properties of the positive integers. That is, it is possible to define addition and multiplication of positive integers and to prove that these operations are commutative, associative, and so forth. The set of all integers may then be constructed from the set of all positive integers. However, this procedure is quite lengthy if it is undertaken carefully, and it more properly belongs in a course dealing with the theory of numbers.

As our starting point, we shall take the system of integers as given, and assume that the system of integers satisfies a certain list of basic axioms, or postulates. More precisely, we assume that there is a set **Z** of elements, called the **integers**, that satisfies the following conditions.

1. **(Addition postulates)** There is a binary operation defined in **Z** that is called *addition* and denoted by $+$, and that has the following properties:
   (a) **Z** is **closed** under addition.
   (b) Addition is **associative**.
   (c) **Z** contains an element 0 that is an **identity element for addition**.
   (d) For each $x \in$ **Z** there is an **additive inverse** of x in **Z**, denoted by $-x$, such that $x + (-x) = 0 = (-x) + x$.
   (e) Addition is **commutative**.

2. **(Multiplication postulates)** There is a binary operation defined in **Z** that is called **multiplication** and denoted by $\cdot$, and that has the following properties:
   (a) **Z** is **closed** under multiplication.
   (b) Multiplication is **associative**.
   (c) **Z** contains an element 1 that is different from 0 and that is an **identity element for multiplication**.
   (d) Multiplication is **commutative**.

3. The **distributive law**

   $$x \cdot (y + z) = x \cdot y + x \cdot z$$

   holds for all elements x, y, $z \in$ **Z**.

4. **Z** contains a subset $\mathbf{Z}^+$, called the **positive integers**, that has the following properties:
   (a) $\mathbf{Z}^+$ is **closed** under addition.
   (b) $\mathbf{Z}^+$ is **closed** under multiplication.
   (c) For each x in **Z**, one and only one of the following statements is true:
      (i) $x \in \mathbf{Z}^+$.
      (ii) $x = 0$.
      (iii) $-x \in \mathbf{Z}^+$.

5. **(Induction postulate)** If S is a subset of $\mathbf{Z}^+$ such that
   (a) $1 \in S$, and
   (b) $x \in S$ always implies $x + 1 \in S$,
   then $S = \mathbf{Z}^+$.

Note that we are taking the entire list of postulates as *assumptions* concerning **Z**. This list is our set of basic properties. In this section, we shall investigate briefly some of the consequences of this set of properties.

2.1   POSTULATES FOR THE INTEGERS

After the term "group" has been defined in Chapter 3, we shall see that the addition postulates listed above state that **Z** is a commutative group with respect to addition. Note that there is a major difference between the multiplication and the addition postulates, in that no inverses are required with respect to multiplication.

Postulate 3, the distributive law, is sometimes known as the **left distributive law**. The requirement that

$$(y + z) \cdot x = y \cdot x + z \cdot x$$

is known as the **right distributive law**. This property can be deduced from those in our list, as can all the familiar properties of addition and multiplication of integers.

Postulate 4(c) is referred to as the **law of trichotomy** because of its assertion that *exactly one of three possibilities* must hold. In case (iii), where $-x \in \mathbf{Z}^+$, we say that x is a *negative integer*, and the set $\{x \mid -x \in \mathbf{Z}^+\}$ is the *set of all negative integers*.

The induction postulate is, of course, a basis for proofs by mathematical induction. It also has other uses, as we shall see in the next section.

As stated above, the right distributive law can be shown to follow from the set of postulates for the integers. We do this formally in the following theorem.

THEOREM 2.1   The equality

$$(y + z) \cdot x = y \cdot x + z \cdot x$$

holds for all x, y, z in **Z**.

Proof   For arbitrary x, y, z in **Z**, we have

$$\begin{aligned}
(y + z) \cdot x &= x \cdot (y + z) && \text{by postulate 2(d)} \\
&= x \cdot y + x \cdot z && \text{by postulate 3} \\
&= y \cdot x + z \cdot x && \text{by postulate 2(d).}
\end{aligned}$$ ∎

The foregoing proof is admittedly trivial, but the point we wish to make is that the usual manipulations involving integers are indeed consequences of our basic set of postulates. As another example, consider the statement[†] that $(-x)y = -(xy)$. In this equation, $-(xy)$ denotes the additive inverse of xy, just as $-x$ denotes the additive inverse of x.

THEOREM 2.2   For arbitrary x and y in **Z**, $(-x)y = -(xy)$.

[†] We adopt the usual convention that the juxtaposition of x and y in xy indicates the operation of multiplication.

35

Proof   Instead of attempting to prove this statement directly, we shall first prove a lemma.

LEMMA 2.3   (**Cancellation Law for Addition**) If a, b, and c are integers and $a + b = a + c$, then $b = c$.

Proof of the lemma   Suppose $a + b = a + c$. Now $-a$ is in **Z**, and hence

$$a + b = a + c \Rightarrow (-a) + (a + b) = (-a) + (a + c)$$
$$\Rightarrow [(-a) + a] + b = [-(a) + a] + c \quad \text{by postulate 1(b)}$$
$$\Rightarrow \qquad\quad 0 + b = 0 + c \qquad\qquad \text{by postulate 1(d)}$$
$$\Rightarrow \qquad\qquad\quad b = c \qquad\qquad\quad \text{by postulate 1(c).}$$

This completes the proof of the lemma.   ∎

Proof of the theorem   Returning to the theorem, we see that we only need to show that $xy + (-x)y = xy + [-(xy)]$. That is, we only need to show that $xy + (-x)y = 0$. We have

$$
\begin{aligned}
xy + (-x)y &= yx + y(-x) & \text{by postulate 2(d)}\\
&= y[x + (-x)] & \text{by postulate 3}\\
&= y \cdot 0 & \text{by postulate 1(d)}\\
&= y \cdot 0 + 0 & \text{by postulate 1(c)}\\
&= y \cdot 0 + \{y \cdot 0 + [-(y \cdot 0)]\} & \text{by postulate 1(d)}\\
&= (y \cdot 0 + y \cdot 0) + [-(y \cdot 0)] & \text{by postulate 1(b)}\\
&= y \cdot (0 + 0) + [-(y \cdot 0)] & \text{by postulate 3}\\
&= y \cdot 0 + [-(y \cdot 0)] & \text{by postulate 1(c)}\\
&= 0 & \text{by postulate 1(d).}
\end{aligned}
$$

We have shown that $xy + (-x)y = 0$, and the theorem is proven.   ∎

The last part of the proof above would have been shorter if the fact that $y \cdot 0 = 0$ had been available. However, it is our attitude at present to use in a proof only the basic postulates for **Z** and those facts previously proven. Several statements similar to the last two theorems are given to be proved in the exercises at the end of this section. After this section, we assume the usual properties of addition and multiplication in **Z**.

Postulate 4, which asserts the existence of the set $\mathbf{Z}^+$ of positive integers, can be used to introduce the order relation "less than" on the set of integers. We make the following definition.

DEFINITION 2.4   For integers x and y,

$x < y$ if and only if $y - x \in \mathbf{Z}^+$

where $y - x = y + (-x)$.

The symbol $<$ is read "less than," as usual. Here we have defined the relation, but we have not assumed any of its usual properties. However, they can be obtained by use of this definition and the properties of $\mathbf{Z}^+$. Before illustrating this with an example, we note that $0 < y$ if and only if $y \in \mathbf{Z}^+$.

For an arbitrary $x \in \mathbf{Z}$ and a positive integer n, we define x^n as follows:

$$x^1 = x,$$
$$x^{k+1} = x^k \cdot x \quad \text{for any positive integer } k.$$

Similarly, positive multiples nx of x are defined by:

$$1x = x$$
$$(k+1)x = kx + x \quad \text{for any positive integer } k.$$

THEOREM 2.5   For any $x \neq 0$ in $\mathbf{Z}$, $x^2 \in \mathbf{Z}^+$.

Proof   Let $x \neq 0$ in $\mathbf{Z}$. By postulate 4, either $x \in \mathbf{Z}^+$ or $-x \in \mathbf{Z}^+$. If $x \in \mathbf{Z}^+$, then $x^2 = x \cdot x$ is in $\mathbf{Z}^+$ by postulate 4(b). And if $-x \in \mathbf{Z}^+$, then $(-x)^2 = (-x) \cdot (-x)$ is in $\mathbf{Z}^+$, by the same postulate. But

$$x^2 = x \cdot x$$
$$= (-x) \cdot (-x) \quad \text{by Problem 5,}$$

so x^2 is in $\mathbf{Z}^+$ if $-x \in \mathbf{Z}^+$. In each possible case, x^2 is in $\mathbf{Z}^+$, and this completes the proof.   ∎

As a particular case of this theorem, $1 \in \mathbf{Z}^+$ since $1 = (1)^2$. That is, 1 must be a positive integer, a fact that may not be immediately evident in postulate 4. This in turn implies that $2 = 1 + 1$ is in $\mathbf{Z}^+$ by postulate 4(a). Repeated application of 4(a) gives $3 = 2 + 1 \in \mathbf{Z}^+$, $4 = 3 + 1 \in \mathbf{Z}^+$, $5 = 4 + 1 \in \mathbf{Z}^+$, and so on. It turns out that $\mathbf{Z}^+$ must necessarily be the set

$$\mathbf{Z}^+ = \{1, 2, 3, \ldots, n, n+1, \ldots\}.$$

Although our discussion of order has been in terms of *less than*, the relations *greater than*, *less than or equal to*, and *greater than or equal to* can be introduced in $\mathbf{Z}$ and similarly treated. We consider this treatment to be trivial and do not bother with it. At the same time, we accept terms such as *nonnegative* and *nonpositive* with their usual meanings and without formal definitions.

As mentioned earlier, the induction postulate is a basis for proofs by mathematical induction. The reader has no doubt previously encountered the method of proof by induction. Typically, there is a statement $\mathscr{S}_n$ to be proved true for every positive integer n. The proof consists of three steps: (1) the statement is verified for $n = 1$; (2) the statement is assumed true for $n = k$; and (3) with this assumption made, the statement is then proved to be true for $n = k + 1$.

EXAMPLE 1   We shall prove that any even power of a nonzero integer is positive. That is, if $x \neq 0$ in $\mathbf{Z}$, then x^{2n} is positive for every positive integer n.

The second formulation of the statement is suitable for a proof by induction on n. For $n = 1$, $x^{2n} = x^2$, and x^2 is positive by Theorem 2.5. Assume the statement is true for $n = k$, that is, x^{2k} is positive. For $n = k + 1$, we have

$$
\begin{aligned}
x^{2n} &= x^{2(k+1)} \\
&= x^{2k+2} \\
&= x^{2k} \cdot x^2.
\end{aligned}
$$

Since x^{2k} and x^2 are positive, the product is positive by postulate 4(b). Thus the statement is true for $n = k + 1$. It follows from the induction postulate that the statement is true for all positive integers.   □

In some cases it is more convenient to use another form of the induction postulate known as the **second principle of finite induction**, or the method of proof by **complete induction**. In this form, a proof that a statement $\mathscr{S}_n$ is true for all positive integers n consists of three steps: (1) the statement is proved true for $n = 1$; (2) for a positive integer k, the statement is assumed true for all positive integers $m < k$; and (3) under this assumption, the statement is proved to be true for $m = k$. For $k = 1$, the set of all $m < k$ is empty, so it may happen that a proof that $\mathscr{S}_1$ is true is implicit in steps (2) and (3). An example of this type of proof is provided by the proof of Theorem 2.16 in Section 2.3.

EXERCISES 2.1

Prove that the equalities in Problems 1–10 hold for all x, y, and z in $\mathbf{Z}$. Assume only the basic postulates for $\mathbf{Z}$ and those properties proved in this section. **Subtraction** is defined by $x - y = x + (-y)$.

1    $x \cdot 0 = 0$

2    $-x = (-1) \cdot x$

3    $-(-x) = x$

4    $(-1)(-1) = 1$

5    $(-x)(-y) = xy$

6    $x - 0 = x$

7    $x(y - z) = xy - xz$

8    $(y - z)x = yx - zx$

9    $-(x + y) = (-x) + (-y)$

10    $(x - y) + (y - z) = x - z$

In Problems 11–18 prove the statements concerning the relation $<$ on the set $\mathbf{Z}$ of all integers.

11   If $x < y$, then $x + z < y + z$.

12   If $x < y$ and $z < w$, then $x + z < y + w$.

13   If $x < y$ and $y < z$, then $x < z$.

14   If $x < y$ and $0 < z$, then $xz < yz$.

15   If $x < y$ and $z < 0$, then $yz < xz$.

16   If $0 < x < y$, then $x^2 < y^2$.

17   If $0 < x < y$ and $0 < z < w$ then $xz < yw$.

18   If $0 < z$ and $xz < yz$, then $x < y$.

19   Prove that if x and y are integers and $xy = 0$, then either $x = 0$ or $y = 0$. [Hint: If $x \neq 0$, then either $x \in \mathbf{Z}^+$ or $-x \in \mathbf{Z}^+$, and similarly for y. Consider xy for the various cases.]

20   Prove that the cancellation law for multiplication holds in $\mathbf{Z}$. That is, if $xy = xz$ and $x \neq 0$, then $y = z$.

Let x, y be integers, and let m and n be positive integers. Use mathematical induction to prove the statements in Problems 21–23.

21 $(xy)^n = x^n y^n$

22 $x^m \cdot x^n = x^{m+n}$

23 $(x^m)^n = x^{mn}$

Prove that the statements in Problems 24–33 are true for every positive integer n.

24 $1 + 2 + 3 + \cdots + n = \dfrac{n(n + 1)}{2}$

25 $1 + 3 + 5 + \cdots + (2n - 1) = n^2$

26 $1^2 + 2^2 + 3^2 + \cdots + n^2 = \dfrac{n(n + 1)(2n + 1)}{6}$

27 $2 + 2^2 + 2^3 + \cdots + 2^n = 2(2^n - 1)$

28 $1^3 + 2^3 + 3^3 + \cdots + n^3 = \dfrac{n^2(n + 1)^2}{4}$

29 $1^3 + 3^3 + 5^3 + \cdots + (2n - 1)^3 = n^2(2n^2 - 1)$

30 $1 \cdot 2 + 2 \cdot 3 + 3 \cdot 4 + \cdots + n(n + 1) = \dfrac{n(n + 1)(n + 2)}{3}$

31 $1 \cdot 2 + 2 \cdot 2^2 + 3 \cdot 2^3 + \cdots + n \cdot 2^n = (n - 1)2^{n+1} + 2$

32 $n < 2^n$

33 $1 + 2n \leq 3^n$

For an integer x, the **absolute value** of x is denoted by $|x|$, and is defined by

$$|x| = \begin{cases} x & \text{if } 0 \leq x \\ -x & \text{if } x < 0. \end{cases}$$

Use this definition for the proofs in Problems 34–36.

34   Prove that $-x \leq |x| \leq x$ for any integer x.

35   Prove that $|xy| = |x| \cdot |y|$ for all x and y in $\mathbf{Z}$.

36   Prove that $|x + y| \leq |x| + |y|$ for all x and y in $\mathbf{Z}$.

Let x, y be integers, and let m and n be positive integers. Prove the properties of multiples given in Problems 37–39.

37 $n(x + y) = nx + ny$

38 $(m + n)x = mx + nx$

39 $m(nx) = (mn)x$

40   Prove that if a is positive and b is negative, then ab is negative.

41   Prove that if a is positive and ab is positive, then b is positive.

42   Prove that if a is positive and ab is negative, then b is negative.

43   Consider the set $\{0\}$ consisting of 0 alone, with $0 + 0 = 0$ and $0 \cdot 0 = 0$. Which of the postulates for $\mathbf{Z}$ are satisfied?

2.2   DIVISIBILITY

From this point on, full knowledge of the properties of addition, subtraction, and multiplication of integers is assumed. We turn now to a study of divisibility in the set of integers. Our main goal in this section is to obtain the *Division Algorithm* (Theorem 2.10). To achieve this goal, we need an important consequence of the induction postulate, known as the *Well-Ordering Theorem*. Our first step in this direction is to prove that there is no positive integer smaller than 1.

THEOREM 2.6   The integer 1 is the least positive integer. That is, $1 \le x$ for all $x \in \mathbf{Z}^{+}$.

Proof   The proof is by induction. Let S be the set of all positive integers x such that $1 \le x$. Then $1 \in S$. Suppose $n \in S$. Now $0 < 1$ implies $n = n + 0 < n + 1$ by Problem 11 of Exercises 2.1, so we have $1 \le n < n + 1$. Thus $n \in S$ implies $n + 1 \in S$, and $S = \mathbf{Z}^{+}$ by postulate 5.   ∎

We proceed now to the Well-Ordering Theorem.

THEOREM 2.7   **(The Well-Ordering Theorem)** Every nonempty set S of positive integers contains a least element. That is, there is an element $m \in S$ such that $m \le x$ for all $x \in S$.

Proof   Let S be a nonempty set of positive integers. If $1 \in S$, then $1 \le x$ for all $x \in S$, by Theorem 2.6. In this case $m = 1$ is the least element in S.

Consider now the case where $1 \notin S$, and let L be the set of all positive integers p such that $p < x$ for all $x \in S$. That is,

$$L = \{p \in \mathbf{Z}^{+} \mid p < x \text{ for all } x \in S\}.$$

Since $1 \notin S$, Theorem 2.6 assures us that $1 \in L$. We shall show that there is a positive integer p_0 such that $p_0 \in L$ and $p_0 + 1 \notin L$. Suppose this is not the case. Then we have that $p \in L$ implies $p + 1 \in L$, and $L = \mathbf{Z}^+$ by the induction postulate. This contradicts the fact that S is nonempty (note that $L \cap S = \varnothing$). Therefore, there is a p_0 such that $p_0 \in L$ and $p_0 + 1 \notin L$.

We must show that $p_0 + 1 \in S$. We have $p_0 < x$ for all $x \in S$, so $p_0 + 1 \leq x$ for all $x \in S$ (see Problem 14). If $p_0 + 1 < x$ were always true, then $p_0 + 1$ would be in L. Hence $p_0 + 1 = x$ for some $x \in S$, and $m = p_0 + 1$ is the required least element in S. ∎

DEFINITION 2.8 Let a and b be integers. We say that a **divides** b if there is an integer c such that $b = ac$.

If a divides b, we write $a \mid b$. Also, we say that b is a **multiple** of a, or that a is a **factor** of b, or that a is a **divisor** of b. If a does not divide b, we write $a \nmid b$.

It may come as a surprise that we can use our previous results to prove the following simple theorem.

THEOREM 2.9 The only divisors of 1 are 1 and -1.

Proof Suppose a is a divisor of 1. Then $1 = ac$ for some integer c.

The equation $1 = ac$ requires $a \neq 0$, so either $a \in \mathbf{Z}^+$ or $-a \in \mathbf{Z}^+$. Consider first the case where $a \in \mathbf{Z}^+$. This requires $c \in \mathbf{Z}^+$ (see Problem 41 of Exercises 2.1), so we have $1 \leq a$ and $1 \leq c$, by Theorem 2.6. Now

$$1 < a \Rightarrow 1 \cdot c < a \cdot c \quad \text{by Problem 14 of Exercises 2.1}$$
$$\Rightarrow c < 1 \quad \text{since } ac = 1.$$

and this is a contradiction of $1 \leq c$. Thus $1 = a$ is the only possibility when $a \in \mathbf{Z}^+$.

The case where $-a \in \mathbf{Z}^+$ can be treated similarly, and its proof is left as an exercise. ∎

Our next result is the basic theorem on divisibility.

THEOREM 2.10 (**The Division Algorithm**) Let a and b be integers with $b > 0$. Then there exist unique integers q and r such that

$$a = bq + r, \quad \text{with } 0 \leq r < b.$$

Proof Let S be the set of all integers x that can be written in the form $x = a - bn$ for $n \in \mathbf{Z}$, and let S' denote the set of all nonnegative integers in S. The set S' is clearly nonempty. If $0 \in S'$, we have $a - bq = 0$ for some q, and $a = bq + 0$. If $0 \notin S'$, then S' contains a least element $r = a - bq$ by the Well-Ordering Theorem, and

$$a = bq + r,$$

where r is positive. Now

$$a - b(q + 1) = a - bq - b = r - b,$$

so $r - b \in S$. Since r is the least element in S' and $r - b < r$, it must be that $r - b$ is negative. That is, $r - b < 0$, and $r < b$. Combining the two cases (where $0 \in S'$ and where $0 \notin S'$), we have

$$a = bq + r, \quad \text{with } 0 \leq r < b.$$

To show that q and r are unique, suppose $a = bq_1 + r_1$ and $a = bq_2 + r_2$, where $0 \leq r_1 < b$ and $0 \leq r_2 < b$. We may assume that $r_1 \leq r_2$ without loss of generality. This means that

$$0 \leq r_2 - r_1 \leq r_2 < b.$$

However, we also have

$$0 \leq r_2 - r_1 = (a - bq_2) - (a - bq_1) = b(q_1 - q_2).$$

That is, $r_2 - r_1$ is a nonnegative multiple of b that is less than b. For any positive integer n, $1 \leq n$ implies $b \leq bn$. Therefore $r_2 - r_1 = 0$, and $r_1 = r_2$. It follows that $bq_1 = bq_2$, where $b \neq 0$. This implies that $q_1 = q_2$ (see Problem 20 of Exercises 2.1). We have shown that $r_1 = r_2$ and $q_1 = q_2$, and this proves that q and r are unique. ∎

In the Division Algorithm, the integer q is called the **quotient** and r is called the **remainder** in the division of a by b.

EXAMPLE 1 When a and b are both positive integers, the quotient q and remainder r can be found by the familiar routine of long division. For instance, if $a = 357$ and $b = 13$, long division gives

$$
\begin{array}{r}
27 \\
13\,\overline{)357} \\
26 \\
\hline
97 \\
91 \\
\hline
6
\end{array}
$$

so $q = 27$ and $r = 6$ in $a = bq + r$, with $0 \le r < b$:

$$357 = (13)(27) + 6.$$

If a is negative, a minor adjustment can be made to obtain the expression in the Division Algorithm. With $a = -357$ and $b = 13$, the preceding equation can be multiplied by -1 to obtain

$$-357 = (13)(-27) + (-6).$$

To obtain an expression with a positive remainder, we need only subtract and add 13 in the right member of the equation:

$$-357 = (13)(-27) + (13)(-1) + (-6) + 13$$
$$= (13)(-28) + 7.$$

Thus $q = -28$ and $r = 7$ in the Division Algorithm, with $a = -357$ and $b = 13$.
□

EXERCISES 2.2

1   List all divisors of the following integers.
   (a) 30   (b) 42   (c) 28   (d) 45
   (e) 24   (f) 40   (g) 32   (h) 210

   With a and b as given in Problems 2–11, find the q and r that satisfy the conditions in the Division Algorithm.

2 $a = 1776, b = 1492$ **3** $a = 512, b = 15$

4 $a = -12, b = 5$ **5** $a = -27, b = 7$

6 $a = 1492, b = 1776$ **7** $a = 15, b = 512$

8 $a = -4317, b = 12$ **9** $a = -5316, b = 171$

10 $a = 0, b = 3$ **11** $a = 0, b = 5$

12   Prove that if a, b, and c are integers such that $a \mid b$ and $a \mid c$, then $a \mid (b + c)$.

13   Let a, b, and c be integers such that $a \mid b$ and $b \mid c$. Prove that $a \mid c$.

14   Let m be an arbitrary positive integer. Prove that there is no integer n such that $m < n < m + 1$,

15   Let S be as described in the proof of Theorem 2.10. Give a specific example of a positive element of S.

16   Complete the proof of Theorem 2.9.

17   Prove that if a and b are integers such that $a \mid b$ and $b \mid a$, then either $a = b$ or $a = -b$.

18   Prove that if a and b are integers such that $b \ne 0$ and $a \mid b$, then $|a| \le |b|$.

19   Use the Division Algorithm to prove that if a and b are integers with $b \ne 0$, then there exist unique integers q and r such that $a = bq + r$, with $0 \le r < |b|$.

20   Prove that the Well-Ordering Theorem implies the finite induction postulate 5.

21   Assume that the Well-Ordering Theorem holds, and prove the second principle of finite induction.

2.3   PRIME FACTORS AND GREATEST COMMON DIVISOR

In this section we establish the existence of the greatest common divisor of two integers when at least one of them is nonzero. The **Unique Factorization Theorem**, also known as the **Fundamental Theorem of Arithmetic**, is obtained.

> **DEFINITION 2.11**   An integer d is a **greatest common divisor** of a and b if these conditions are satisfied:
> (1) d is a positive integer;
> (2) $d \mid a$ and $d \mid b$;
> (3) $c \mid a$ and $c \mid b$ imply $c \mid d$.

The next theorem shows that the greatest common divisor d of a and b exists when at least one of them is not zero. Our proof also shows that d is a linear combination of a and b, that is, $d = ma + nb$ for integers m and n. The technique of proof by use of the Well-Ordering Theorem should be compared to that used in the proof of Theorem 2.10.

> **THEOREM 2.12**   Let a and b be integers, at least one of them not 0. Then there exists a greatest common divisor d of a and b. Moreover, d can be written as
>
> $$d = am + bn$$
>
> for integers m and n, and d is the smallest positive integer that can be written in this form.

Proof   Let a and b be integers, at least one of them not 0. If $b = 0$, then $a \neq 0$, so $|a| > 0$. It is easy to see that $d = |a|$ is a greatest common divisor of a and b in this case, and either $d = a \cdot (1) + b \cdot (0)$ or $d = a \cdot (-1) + b \cdot (0)$.

Suppose now that $b \neq 0$. Consider the set S of all integers that can be written in the form $ax + by$, and let S^+ be the set of all positive integers in S. The set S contains $b = a \cdot (0) + b \cdot (1)$ and $-b = a \cdot (0) + b \cdot (-1)$, so S^+ is not empty. By the Well-Ordering Theorem, S^+ has a least element d,

$$d = am + bn.$$

We have d positive, and we shall show that d is a greatest common divisor of a and b.

By the Division Algorithm, there are integers q and r such that

$$a = dq + r, \quad \text{with } 0 \leq r < d.$$

From this equation,

$$r = a - dq$$
$$= a - (am + bn)q$$
$$= a(1 - mq) + b(-nq).$$

Thus r is in $S = \{ax + by\}$, and $0 \leq r < d$. By choice of d as the least element in S^+, it must be that $r = 0$, and $d \,|\, a$. Similarly, it can be shown that $d \,|\, b$.

If $c \,|\, a$ and $c \,|\, b$, then $a = cu$ and $b = cv$ for integers u and v. Therefore

$$d = am + bn$$
$$= cum + cvn$$
$$= c(um + vn),$$

and this shows that $c \,|\, d$. By Definition 2.11, $d = am + bn$ is a greatest common divisor of a and b. It follows from the choice of d as least element of S^+ that d is the smallest positive integer that can be written in this form. ∎

Whenever the greatest common divisor of a and b exists, it is unique. For if d_1 and d_2 are both greatest common divisors of a and b, it must be true that $d_1 \,|\, d_2$ and $d_2 \,|\, d_1$. Since d_1 and d_2 are positive integers, this means that $d_1 = d_2$ (see Problem 17 of Exercises 2.2). We shall write (a, b) to indicate the unique greatest common divisor of a and b.

When at least one of a and b is not 0, the proof of the last theorem establishes the existence of (a, b), but looking for a smallest positive integer in $S = \{ax + by\}$ is not a very satisfactory method for finding this greatest common divisor. A procedure known as the **Euclidean Algorithm** furnishes a systematic method for finding (a, b). It can also be used to find integers m and n such that $(a, b) = am + bn$. This procedure consists of repeated applications of the Division Algorithm according to the following pattern:

$$a = bq_0 + r_1, \qquad 0 \leq r_1 < b$$
$$b = r_1 q_1 + r_2, \qquad 0 \leq r_2 < r_1$$
$$r_1 = r_2 q_2 + r_3, \qquad 0 \leq r_3 < r_2$$
$$\vdots \qquad\qquad\qquad \vdots$$
$$r_k = r_{k+1} q_{k+1} + r_{k+2}, \qquad 0 \leq r_{k+2} < r_{k+1}.$$

Since the integers $r_1, r_2, \ldots, r_{k+2}$ are decreasing and are all nonnegative, there is a smallest integer n such that $r_{n+1} = 0$:

$$r_{n-1} = r_n q_n + r_{n+1}, \qquad 0 = r_{n+1}.$$

If we put $r_0 = b$, this last nonzero remainder r_n is always the greatest common divisor of a and b. The proof of this statement is left as an exercise.

As an example, we shall find the greatest common divisor of 1492 and 1776.

EXAMPLE 1   Performing the arithmetic for the Euclidean Algorithm, we have

$$1776 = (1)(1492) + 284 \qquad (q_0 = 1, r_1 = 284)$$
$$1492 = (5)(284) + 72 \qquad (q_1 = 5, r_2 = 72)$$
$$284 = (3)(72) + 68 \qquad (q_2 = 3, r_3 = 68)$$
$$72 = (1)(68) + 4 \qquad (q_3 = 1, r_4 = 4)$$
$$68 = (4)(17) \qquad (q_4 = 17, r_5 = 0).$$

Thus the last nonzero remainder is $r_n = r_4 = 4$, and $(1776,1492) = 4$.   □

As mentioned above, the Euclidean Algorithm can also be used to find integers m and n such that

$$(a, b) = am + bn.$$

We can obtain these integers by solving for the last nonzero remainder and substituting the remainders from the preceding equations successively until a and b are present in the equation. For example, the remainders in Example 1 can be expressed as

$$284 = (1776)(1) + (1492)(-1)$$
$$72 = (1492)(1) + (284)(-5)$$
$$68 = (284)(1) + (72)(-3)$$
$$4 = (72)(1) + (68)(-1).$$

Substituting the remainders from the preceding equations successively, we have

$$
\begin{aligned}
4 &= (72)(1) + [(284)(1) + (72)(-3)](-1) \\
&= (72)(1) + (284)(-1) + (72)(3) \\
&= (72)(4) + (284)(-1) \quad \text{after the first substitution} \\
&= [(1492)(1) + (284)(-5)](4) + (284)(-1) \\
&= (1492)(4) + (284)(-20) + (284)(-1) \\
&= (1492)(4) + (284)(-21) \quad \text{after the second substitution} \\
&= (1492)(4) + [(1776)(1) + (1492)(-1)](-21) \\
&= (1492)(4) + (1776)(-21) + (1492)(21) \\
&= (1776)(-21) + (1492)(25) \quad \text{after the third substitution.}
\end{aligned}
$$

Thus $m = -21$ and $n = 25$ are integers such that

$$4 = 1776m + 1492n.$$

The remainders are printed in bold type in each step above, and we carefully avoided performing a multiplication that involved a remainder.

The m and n are not unique in the equation

$$(a, b) = am + bn.$$

To see this, simply add and subtract the product ab:

$$(a, b) = am + ab + bn - ab$$
$$= a(m + b) + b(n - a).$$

Thus $m' = m + b$ and $n' = n - a$ are another pair of integers such that

$$(a, b) = am' + bn'.$$

Among the integers, there are those that have the fewest number of factors possible. Some of these are the *prime integers*.

DEFINITION 2.13   An integer p is a **prime integer** if $p > 1$ and the only divisors of p are ± 1 and $\pm p$.

Notice that the condition $p > 1$ makes p positive and ensures that $p \neq 1$. The exclusion of 1 from the set of primes makes possible the statement of the Unique Factorization Theorem. Before going into that, we prove the following important property of primes.

THEOREM 2.14   If p is a prime and $p \mid ab$, then either $p \mid a$ or $p \mid b$.

Proof   Assume p is a prime and $p \mid ab$, say $ab = pq$. If $p \mid a$, the conclusion is satisfied. Suppose, then, that p does not divide a. Then $1 = (a, p)$ since the only positive divisors of p are 1 and p. By Theorem 2.12, there are integers m and n such that $1 = am + pn$. Note that

$$1 = am + pn \Rightarrow b = abm + pbn$$
$$\Rightarrow b = pqm + pbn \quad \text{since } ab = pq$$
$$\Rightarrow b = p(qm + bn)$$
$$\Rightarrow p \mid b.$$

Thus $p \mid b$ if p does not divide a, and the theorem is true in any case.   ∎

Two integers a and b are called **relatively prime** if their greatest common divisor is 1. (This has nothing at all to do with whether or not they are prime integers.) Theorem 2.14 can be generalized in this manner: if a and b are relatively prime and $a \mid bc$, then $a \mid c$. The proof of this statement is left as an exercise.

The proof of the corollary below, which also generalizes Theorem 2.14, is requested in the exercises.

COROLLARY 2.15 If p is a prime and $p \mid (a_1 a_2 \cdots a_n)$, then p divides some a_j.

This brings us to the final theorem of this section, the **Unique Factorization Theorem**. It is of such importance that it is also frequently called the **Fundamental Theorem of Arithmetic**.

THEOREM 2.16 (**Unique Factorization Theorem**) Every positive integer n is either 1 or can be expressed as a product of prime integers, and this factorization is unique except for the order of the factors.

Proof In the statement of the theorem, the word "product" is used in an extended sense: the "product" may have just one factor. That is, if n is a prime, say $n = p$, then this expression is our product of prime integers.

Let $\mathscr{S}_n$ be the statement that either $n = 1$ or n can be expressed as a product of primes. We shall prove that $\mathscr{S}_n$ is true for all $n \in \mathbf{Z}^+$ by the second principle of finite induction.

Now $\mathscr{S}_1$ is trivially true. Assume that $\mathscr{S}_m$ is true for all positive integers $m < k$. If k is a prime, say $k = p$, then $\mathscr{S}_k$ is true, as explained in our discussion of the word "product." Suppose k is not a prime. Then $k = ab$, where neither a nor b is 1. Therefore $1 < a < k$ and $1 < b < k$. By the induction hypothesis, $\mathscr{S}_a$ is true and $\mathscr{S}_b$ is true. That is,

$$a = p_1 p_2 \cdots p_r \quad \text{and} \quad b = q_1 q_2 \cdots q_s$$

for primes p_i and q_j. These factorizations give

$$k = ab$$
$$= p_1 p_2 \cdots p_r q_1 q_2 \cdots q_s,$$

and k is thereby expressed as a product of primes. Thus $\mathscr{S}_k$ is true, and therefore $\mathscr{S}_n$ is true for all positive integers n.

To prove that the factorization is unique, suppose that

$$n = p_1 p_2 \cdots p_t \quad \text{and} \quad n = q_1 q_2 \cdots q_v$$

are factorizations of n as products of prime factors p_i and q_j. Then

$$p_1 p_2 \cdots p_t = q_1 q_2 \cdots q_v,$$

so $p_1 | (q_1 q_2 \cdots q_v)$. By Corollary 2.15, $p_1 | q_j$ for some j, and there is no loss of generality if we assume $j = 1$. However, p_1 and q_1 are primes, so $p_1 | q_1$ implies $q_1 = p_1$. This gives

$$p_1 p_2 \cdots p_t = p_1 q_2 \cdots q_v,$$

and therefore

$$p_2 \cdots p_t = q_2 \cdots q_v$$

by the cancellation law. This argument can be repeated, removing one factor p_i with each application of the cancellation law, until we obtain

$$p_t = q_t \cdots q_v.$$

Since the only positive factors of p_t are 1 and p_t, and since each q_j is a prime, this means that there must be only one q_j on the right in this equation, and it is q_t. That is, $v = t$ and $q_t = p_t$. This completes the proof. ∎

The Unique Factorization Theorem can be used to describe a standard form of a positive integer n. Suppose $p_1, p_2, \ldots, p_r$ are the *distinct* prime factors of n, arranged in order of magnitude so that

$$p_1 < p_2 < \cdots < p_r.$$

Then all repeated factors may be collected together and expressed by use of exponents to yield

$$n = p_1^{m_1} p_2^{m_2} \cdots p_r^{m_r},$$

where each m_i is a positive integer. Each m_i is called the **multiplicity** of p_i, and this factorization is known as the **standard form** for n.

EXAMPLE 2 The standard forms for two positive integers a and b can be used to find their greatest common divisor (a, b) and their least common multiple (see Problems 22 and 23). For instance, if

$$a = 31{,}752 = 2^3 \cdot 3^4 \cdot 7^2 \quad \text{and} \quad b = 126{,}000 = 2^4 \cdot 3^2 \cdot 5^3 \cdot 7,$$

then (a, b) can be found by forming the product of all the common prime factors, with each common factor raised to the least power to which it appears in either factorization:

$$(a, b) = 2^3 \cdot 3^2 \cdot 7 = 504. \qquad \square$$

EXERCISES 2.3

In this set of exercises, all variables represent integers.

1 List all the primes less than 100.

2 For each of the following pairs, write a and b in standard form and use these factorizations to find (a, b).
(a) $a = 1400, b = 980$ (b) $a = 4950, b = 10,500$
(c) $a = 3780, b = 16,200$ (d) $a = 52,920, b = 25,200$

3 In each part, find the greatest common divisor (a, b) and integers m and n such that $(a, b) = am + bn$.
(a) $a = 0, b = -3$ (b) $a = 65, b = 91$
(c) $a = 414, b = 33$ (d) $a = 252, b = 180$
(e) $a = 414, b = 693$ (f) $a = 382, b = 26$
(g) $a = 5088, b = 156$ (h) $a = 8767, b = 2178$

4 Find the smallest integer in the given set.
(a) $\{x \in \mathbf{Z} \mid x > 0 \text{ and } x = 4s + 6t \text{ for some } s, t \text{ in } \mathbf{Z}\}$
(b) $\{x \in \mathbf{Z} \mid x > 0 \text{ and } x = 6s + 15t \text{ for some } s, t \text{ in } \mathbf{Z}\}$

5 If c is a divisor of a and b, prove that c is a divisor of $ax + by$ for all $x, y \in \mathbf{Z}$.

6 Prove that if p and q are distinct primes, then there exist integers m and n such that $pm + qn = 1$.

7 Prove that if $a \mid (bc)$ and $(a, b) = 1$, then $a \mid c$.

8 Give an example where $a \mid (bc)$, but $a \nmid b$ and $a \nmid c$.

9 If $a \mid c$ and $b \mid c$, and $(a, b) = 1$, prove that ab divides c.

10 If $b > 0$ and $a = bq + r$ with $0 \le r < b$, prove that $(a, b) = (b, r)$.

11 Let $r_0 = b > 0$. With the notation used in the description of the Euclidean Algorithm, use the result in Problem 10 to prove that $(a, b) = r_n$, the last nonzero remainder.

12 Prove that every remainder r_j in the Euclidean Algorithm is a "linear combination" of a and b: $r_j = s_j a + t_j b$, for integers s_j and t_j.

13 Let a and b be integers, at least one of them not 0. Prove that an integer c can be expressed as a linear combination of a and b if and only if $(a, b) \mid c$.

14 Prove Corollary 2.15: If p is a prime and $p \mid (a_1 a_2 \cdots a_n)$, then p divides some a_j.

15 Prove that if n is a positive integer greater than 1 such that n is not a prime, then n has a divisor d such that $1 < d \le \sqrt{n}$.

16 Prove that $(ab, c) = 1$ if and only if $(a, c) = 1$ and $(b, c) = 1$.

17 Prove that if $m > 0$ and (a, b) exists then $(ma, mb) = m \cdot (a, b)$.

18 Prove that if $d = (a, b)$, $a = a_0 d$, and $b = b_0 d$, then $(a_0, b_0) = 1$.

19 Prove that if $d = (a, b)$, $a \mid c$, and $b \mid c$, then $ab \mid cd$.

20 A *least common multiple* of two nonzero integers a and b is an integer m which satisfies the conditions
(1) m is a positive integer,
(2) $a \mid m$ and $b \mid m$,
(3) $a \mid c$ and $b \mid c$ imply $m \mid c$.
Prove that the least common multiple of two nonzero integers exists and is unique.

21   Let a and b be positive integers. If $d = (a, b)$ and m is the least common multiple of a and b, prove that $dm = ab$.

22   Let a and b be positive integers. Prove that if $d = (a, b)$, $a = a_0 d$, and $b = b_0 d$, then the least common multiple of a and b is $a_0 b_0 d$.

23   Describe a procedure for using the standard forms of two positive integers to find their least common multiple.

24   For each pair of integers a, b in Problem 2, find the least common multiple of a and b by using their standard forms.

25   Let a, b, and c be three nonzero integers.
   (a) Use Definition 2.11 as a pattern to define a greatest common divisor of a, b, and c.
   (b) Use Theorem 2.12 and its proof as a pattern to prove the existence of a greatest common divisor of a, b, and c.
   (c) If d is the greatest common divisor of a, b, and c, show that $d = ((a, b), c)$.

26   Use the fact that 2 is a prime to prove that there do not exist nonzero integers a and b such that $a^2 = 2b^2$. Explain how this proves that $\sqrt{2}$ is not a rational number.

27   Use the fact that 3 is a prime to prove that there do not exist nonzero integers a and b such that $a^2 = 3b^2$. Explain how this proves that $\sqrt{3}$ is not a rational number.

2.4   CONGRUENCE OF INTEGERS

In Example 4 of Section 1.4, we defined the relation "congruence modulo 4" on the set $\mathbf{Z}$ of all integers, and we proved this relation to be an equivalence relation on $\mathbf{Z}$. That example is a special case of "congruence modulo n," as defined below.

DEFINITION 2.17   Let n be a positive integer, $n > 1$. For integers x and y, x is congruent to y modulo n if and only if $x - y$ is a multiple of n. We write

$$x \equiv y \pmod{n}$$

to indicate that x is congruent to y modulo n.

Thus $x \equiv y \pmod{n}$ if and only if n divides $x - y$, and this is equivalent to $x - y = nq$, or $x = y + nq$. Another way to describe this relation is to say that x and y yield the same remainder when each is divided by n. To see that this is true, let

$$x = nq_1 + r_1, \quad \text{with } 0 \le r_1 < n$$

and

$$y = nq_2 + r_2, \quad \text{with } 0 \le r_2 < n.$$

Then

$$x - y = n(q_1 - q_2) + (r_1 - r_2), \text{ with } 0 \le |r_1 - r_2| < n.$$

Thus $x - y$ is a multiple of n if and only if $r_1 - r_2 = 0$, that is, if and only if $r_1 = r_2$. In particular, any integer x is congruent to its remainder when divided by n. This means that any x is congruent to one of

$$0, 1, 2, \ldots, n - 1.$$

Congruence modulo n is an equivalence relation on $\mathbf{Z}$, and this fact is important enough to be stated as a theorem.

THEOREM 2.18   The relation of congruence modulo n is an equivalence relation on $\mathbf{Z}$.

Proof   We shall show that congruence modulo n is (1) reflexive, (2) symmetric, and (3) transitive. Let $n > 1$, and let x, y, and z be arbitrary in $\mathbf{Z}$.

(1) $x \equiv x \pmod{n}$ since $x - x = (n)(0)$.

(2) $x \equiv y \pmod{n} \Rightarrow x - y = nq \quad \text{for } q \in \mathbf{Z}$
$\Rightarrow y - x = n(-q) \quad \text{for } -q \in \mathbf{Z}$
$\Rightarrow y \equiv x \pmod{n}.$

(3) $x \equiv y \pmod{n}$ and $y \equiv z \pmod{n}$
$\Rightarrow x - y = nq \quad \text{and} \quad y - z = nk \quad \text{for } q,k \in \mathbf{Z}$
$\Rightarrow x - z = x - y + y - z$
$\qquad\quad = n(q + k), \quad \text{where } q + k \in \mathbf{Z}$
$\Rightarrow x \equiv z \pmod{n}.$ ∎

As with any equivalence relation, the equivalence classes for congruence modulo n form a *partition* of $\mathbf{Z}$; that is, they separate $\mathbf{Z}$ into mutually disjoint subsets. These subsets are called **congruence classes**, or **residue classes**. Referring to our discussion concerning remainders, we see that there are n distinct congruence classes modulo n, given by

$$[0] = \{\ldots, -2n, -n, 0, n, 2n, \ldots\},$$
$$[1] = \{\ldots, -2n + 1, -n + 1, 1, n + 1, 2n + 1, \ldots\},$$
$$[2] = \{\ldots, -2n + 2, -n + 2, 2, n + 2, 2n + 2, \ldots\},$$
$$\vdots$$
$$[n - 1] = \{\ldots, -n - 1, -1, n - 1, 2n - 1, 3n - 1, \ldots\}.$$

When $n = 4$, these classes appear as

$$[0] = \{\ldots, -8, -4, 0, 4, 8, \ldots\},$$
$$[1] = \{\ldots, -7, -3, 1, 5, 9, \ldots\},$$
$$[2] = \{\ldots, -6, -2, 2, 6, 10, \ldots\},$$
$$[3] = \{\ldots, -5, -1, 3, 7, 11, \ldots\}.$$

Congruence classes are useful in connection with numerous examples, and we shall see more of them later.

Although $x \equiv y \pmod{n}$ is certainly not an equation, in many ways congruences can be handled in the same fashion as equations. The next theorem asserts that the same integer can be added to both members, and that both members can be multiplied by the same integer.

THEOREM 2.19   If $a \equiv b \pmod{n}$ and x is any integer, then

$$a + x \equiv b + x \pmod{n} \quad \text{and} \quad ax \equiv bx \pmod{n}.$$

Proof   Let $a \equiv b \pmod{n}$, and $x \in \mathbf{Z}$. We shall prove that $ax \equiv bx \pmod{n}$, and leave the other part as an exercise. We have

$$a \equiv b \pmod{n} \Rightarrow a - b = nq \qquad \text{for } q \in \mathbf{Z}$$
$$\Rightarrow (a - b)x = (nq)x \qquad \text{for } q, x \in \mathbf{Z}$$
$$\Rightarrow ax - bx = n(qx) \qquad \text{for } qx \in \mathbf{Z}$$
$$\Rightarrow ax \equiv bx \pmod{n}. \qquad \blacksquare$$

Congruence modulo n also has substitution properties that are analogous to those possessed by equality.

THEOREM 2.20   Suppose $a \equiv b \pmod{n}$ and $c \equiv d \pmod{n}$. Then

$$a + c \equiv b + d \pmod{n} \quad \text{and} \quad ac \equiv bd \pmod{n}.$$

Proof   Let $a \equiv b \pmod{n}$ and $c \equiv d \pmod{n}$. By Theorem 2.19,

$$a \equiv b \pmod{n} \Rightarrow ac \equiv bc \pmod{n}$$

and

$$c \equiv d \pmod{n} \Rightarrow bc \equiv bd \pmod{n}.$$

By the transitive property, $ac \equiv bc \pmod{n}$ and $bc \equiv bd \pmod{n}$ imply $ac \equiv bd \pmod{n}$. The proof that $a + c \equiv b + d \pmod{n}$ is left as an exercise. $\blacksquare$

It is easy to show that there is a "cancellation law" for addition that holds for congruences: $a + x \equiv a + y \pmod{n}$ implies $x \equiv y \pmod{n}$. This is not the case, however, with multiplication. For example,

$$(4)(6) \equiv (4)(21) \ (\mathrm{mod}\ 30) \quad \text{but} \quad 6 \not\equiv 21 \ (\mathrm{mod}\ 30).$$

Notice that 4 and 30 are not relatively prime. When this condition is imposed, we can obtain a cancellation law for multiplication.

THEOREM 2.21   If $ax \equiv ay \ (\mathrm{mod}\ n)$ and $(a, n) = 1$, then $x \equiv y \ (\mathrm{mod}\ n)$.

Proof   Assume that $ax \equiv ay \ (\mathrm{mod}\ n)$ and that a and n are relatively prime.

$$
\begin{aligned}
ax \equiv ay \ (\mathrm{mod}\ n) &\Rightarrow n \,|\, (ax - ay) \\
&\Rightarrow n \,|\, a(x - y) \\
&\Rightarrow a(x - y) = nq
\end{aligned}
$$

Since a and n are relatively prime, there are integers s and t such that

$$
\begin{aligned}
1 &= sa + tn \\
\Rightarrow \quad (1)(x - y) &= sa(x - y) + tn(x - y) \\
\Rightarrow \quad x - y &= snq + tn(x - y) \quad \text{since } a(x - y) = nq \\
&= n(sq + tx - ty) \\
\Rightarrow n \,|\, (x - y)& \\
\Rightarrow x \equiv y \ (\mathrm{mod}\ n)&.
\end{aligned}
$$

This completes the proof. ∎

We have seen that there are analogues for many of the manipulations that may be performed with equations. There are also techniques for obtaining solutions to congruence equations of certain types. The basic technique makes use of Theorem 2.20 and the Euclidean Algorithm. It is illustrated following the proof of the next theorem.

THEOREM 2.22   If a and n are relatively prime, the congruence $ax \equiv b \ (\mathrm{mod}\ n)$ has a solution x in the integers.

Proof   Since a and n are relatively prime, there exist integers s and t such that

$$
\begin{aligned}
1 &= as + nt \\
\Rightarrow \quad b &= asb + ntb \\
\Rightarrow a(sb) - b &= n(-tb) \\
\Rightarrow \quad n &\,|\, [a(sb) - b] \\
\Rightarrow \quad a(sb) &\equiv b \ (\mathrm{mod}\ n).
\end{aligned}
$$

Thus $x = sb$ is a solution to $ax \equiv b \ (\mathrm{mod}\ n)$. ∎

EXAMPLE 1   The Euclidean Algorithm can be used to find a solution x to $ax \equiv b \pmod n$ when $(a, n) = 1$, Consider the congruence

$$20x \equiv 14 \pmod{63}.$$

We first obtain s and t such that

$$1 = 20s + 63t.$$

Applying the Euclidean Algorithm, we have

$$63 = (20)(3) + 3$$
$$20 = (3)(6) + 2$$
$$3 = (2)(1) + 1$$
$$2 = (1)(2).$$

Solving for the nonzero remainders,

$$3 = 63 - (20)(3)$$
$$2 = 20 - (3)(6)$$
$$1 = 3 - (2)(1).$$

Substituting the remainders in turn, we obtain

$$1 = 3 - (2)(1)$$
$$= 3 - [20 - (3)(6)](1)$$
$$= (3)(7) + (20)(-1)$$
$$= [63 - (20)(3)][7] + (20)(-1)$$
$$= (20)(-22) + (63)(7).$$

Multiplying this equation by $b = 14$, we have

$$14 = (20)(-308) + (63)(98)$$
$$\Rightarrow 14 \equiv (20)(-308) \pmod{63}.$$

Thus $x = -308$ is a solution. However any number is congruent modulo 63 to its remainder when divided by 63, and

$$-308 = (63)(-5) + 7.$$

This means that $x = 7$ is a solution in the range $0 \le x < 63$. $\square$

The preceding example illustrates the basic technique for obtaining a solution to $ax \equiv b \pmod n$ when a and n are relatively prime, but there are other methods that are also very useful. Some of them make use of Theorems 2.20 and 2.21. Theorem 2.21 can be used to remove a factor c from both sides of the congruence provided c and n are relatively prime. That is, c may be canceled from $crx \equiv ct \pmod n$ to obtain the equivalent congruence $rx \equiv t \pmod n$.

EXAMPLE 2   Since 2 and 63 are relatively prime, the factor 2 in both sides of

$$20x \equiv 14 \ (\text{mod } 63)$$

can be removed to obtain

$$10x \equiv 7 \ (\text{mod } 63).$$

Theorem 2.18 allows us to replace an integer by any other integer that is congruent to it modulo n. Now $7 \equiv 70 \ (\text{mod } 63)$, and this substitution yields

$$10x \equiv 70 \ (\text{mod } 63).$$

Removing the factor 10 from both sides, we have

$$x \equiv 7 \ (\text{mod } 63).$$

Thus we have obtained the solution $x = 7$ much more easily than by the method of Example 1. However, this method is less systematic, and it requires more ingenuity.   ☐

EXERCISES 2.4

In this exercise set, all variables are integers.

1   List the distinct congruence classes modulo 5, exhibiting at least three elements in each class.

2   Follow the instructions in Problem 1 for the congruence classes modulo 6.

Find a solution $x \in \mathbf{Z}, 0 \le x < n$, for each of the congruences $ax \equiv b \ (\text{mod } n)$ in Problems 3–18. Note that in each case a and n are relatively prime.

3   $2x \equiv 3 \ (\text{mod } 7)$
4   $2x \equiv 3 \ (\text{mod } 5)$

5   $3x \equiv 7 \ (\text{mod } 13)$
6   $3x \equiv 4 \ (\text{mod } 13)$

7   $8x \equiv 1 \ (\text{mod } 21)$
8   $14x \equiv 8 \ (\text{mod } 15)$

9   $11x \equiv 1 \ (\text{mod } 317)$
10   $11x \equiv 3 \ (\text{mod } 138)$

11   $8x \equiv 66 \ (\text{mod } 79)$
12   $6x \equiv 14 \ (\text{mod } 55)$

13   $8x + 3 \equiv 5 \ (\text{mod } 9)$
14   $19x + 7 \equiv 27 \ (\text{mod } 18)$

15   $25x \equiv 31 \ (\text{mod } 7)$
16   $45x \equiv 17 \ (\text{mod } 313)$

17   $35x + 14 \equiv 3 \ (\text{mod } 27)$
18   $57x + 7 \equiv 78 \ (\text{mod } 53)$

19   Complete the proof of Theorem 2.19: If $a \equiv b \ (\text{mod } n)$ and x is any integer, then $a + x \equiv b + x \ (\text{mod } n)$.

20   Complete the proof of Theorem 2.20: If $a \equiv b \ (\text{mod } n)$ and $c \equiv d \ (\text{mod } n)$, then $a + c \equiv b + d \ (\text{mod } n)$.

21   Prove that if $a + x \equiv a + y \ (\text{mod } n)$, then $x \equiv y \ (\text{mod } n)$.

22   If $ca \equiv cb \ (\text{mod } n)$ and $d = (c, n)$, where $n = dm$, prove that $a \equiv b \ (\text{mod } m)$.

23   Find the least positive integer which is congruent modulo 7 to the given product.
(a) $(4)(9)(15)(59)$ (b) $(5)(11)(17)(65)$

24   If $a \equiv b \pmod{n}$, prove that $a^m \equiv b^m \pmod{n}$ for every positive integer m.

25   Prove that if p is a prime and $c \not\equiv 0 \pmod{p}$, then $cx \equiv b \pmod{p}$ has a unique solution modulo p.

26   Prove that the congruence $ax \equiv b \pmod{n}$ has a solution if and only if $d = (a, n)$ divides b. If d divides b, prove that the congruence has exactly d incongruent solutions modulo n.

In the congruences $ax \equiv b \pmod{n}$ in Problems 27–34, a and n are not relatively prime. Use the result in Problem 26 to determine if there are solutions. If there are, find d incongruent solutions modulo n.

27 $8x \equiv 66 \pmod{78}$ **28** $6x + 3 \equiv 12 \pmod{30}$

29 $24x + 5 \equiv 50 \pmod{348}$ **30** $18x \equiv 33 \pmod{15}$

31 $18x \equiv 33 \pmod{27}$ **32** $42x \equiv 15 \pmod{30}$

33 $21x \equiv 18 \pmod{20}$ **34** $35x \equiv 28 \pmod{20}$

35   Prove that if m is an integer, then either $m^2 \equiv 0 \pmod{4}$ or $m^2 \equiv 1 \pmod{4}$. [Hint: Use the Division Algorithm, and consider the possible remainders in $m = 4q + r$.]

36   If m is an integer, show that m^2 is congruent modulo 8 to one of the integers 0, 1, or 4. (See the hint in Problem 35.)

37   Let x and y be integers. Prove that if there is an equivalence class $[a]$ modulo n such that $x \in [a]$ and $y \in [a]$, then $(x, n) = (y, n)$.

38   Let p be a prime integer. Prove **Fermat's Theorem**: For any positive integer a, $a^p \equiv a \pmod{p}$. [Hint: Use induction on a, with p held fixed.]

39   Assume that m and n are relatively prime, and let a and b be integers. Prove that there exists an integer x such that $x \equiv a \pmod{m}$ and $x \equiv b \pmod{n}$.

40   In Problem 39, prove that any two solutions are congruent modulo mn.

41   Solve the following simultaneous congruences. (See Problem 39.)
(a) $x \equiv 2 \pmod{5}$ and $x \equiv 3 \pmod{8}$
(b) $x \equiv 4 \pmod{5}$ and $x \equiv 2 \pmod{3}$

2.5   CONGRUENCE CLASSES

In connection with the relation of congruence modulo n, we have observed that there are n distinct congruence classes. Let $\mathbf{Z}_n$ denote this set of classes:

$$\mathbf{Z}_n = \{[0], [1], [2], \ldots, [n-1]\}.$$

When addition and multiplication are defined in a natural and appropriate manner in $\mathbf{Z}_n$, these sets provide useful examples for our work in later chapters.

THEOREM 2.23   Consider the rule given by

$$[a] + [b] = [a + b].$$

(a)   This rule defines an addition which is a binary operation on $\mathbf{Z}_n$.

(b)   Addition is associative in $\mathbf{Z}_n$:

$$[a] + ([b] + [c]) = ([a] + [b]) + [c].$$

(c) $\mathbf{Z}_n$ has the additive identity $[0]$.

(d)   Each $[a]$ in $\mathbf{Z}_n$ has an additive inverse $[-a]$ in $\mathbf{Z}_n$.

(e)   Addition is commutative in $\mathbf{Z}_n$:

$$[a] + [b] = [b] + [a].$$

Proof

(a)   It is clear that the rule $[a] + [b] = [a + b]$ yields an element of $\mathbf{Z}_n$, but the uniqueness of this result needs to be verified. In other words, closure is obvious, but we need to show that the operation is well-defined. To do this, suppose that $[a] = [x]$ and $[b] = [y]$. Then

$$[a] = [x] \Rightarrow a \equiv x \pmod{n}$$

and

$$[b] = [y] \Rightarrow b \equiv y \pmod{n}.$$

By Theorem 2.20,

$$a + b \equiv x + y \pmod{n},$$

and therefore $[a + b] = [x + y]$.

(b)   The associative property follows from

$$
\begin{aligned}
[a] + ([b] + [c]) &= [a] + [b + c] \\
&= [a + (b + c)] \\
&= [(a + b) + c] \\
&= [a + b] + [c] \\
&= ([a] + [b]) + [c].
\end{aligned}
$$

Notice that the key step here is the fact that addition is associative in $\mathbf{Z}$: $a + (b + c) = (a + b) + c$.

(c) $[0]$ is an additive identity since

$$[a] + [0] = [a + 0] = [a]$$

and

$$[0] + [a] = [0 + a] = [a].$$

(d) $[-a] = [n - a]$ is the additive inverse of $[a]$ since

$$[a] + [-a] = [a + (-a)] = [0]$$

and

$$[-a] + [a] = [-a + a] = [0].$$

(e)   The commutative property follows from

$$[a] + [b] = [a + b]$$
$$= [b + a]$$
$$= [b] + [a].$$ ∎

EXAMPLE 1   Following the procedure described in Problem 3 of Exercises 1.3, we can construct an addition table for $\mathbf{Z}_4 = \{[0], [1], [2], [3]\}$. In computing the entries for this table, $[a] + [b]$ is entered in the row with $[a]$ at the left and in the column with $[b]$ at the top. For instance,

$$[3] + [2] = [5] = [1]$$

is entered in the row with $[3]$ at the left and in the column with $[2]$ at the top. The complete addition table is shown below.

+	[0]	[1]	[2]	[3]
[0]	[0]	[1]	[2]	[3]
[1]	[1]	[2]	[3]	[0]
[2]	[2]	[3]	[0]	[1]
[3]	[3]	[0]	[1]	[2]

□

In the following theorem, multiplication in $\mathbf{Z}_n$ is defined in a natural way, and the basic properties for this operation are stated. The proofs of the various parts of the theorem are quite similar to those for the corresponding parts of Theorem 2.23, and are left as exercises.

THEOREM 2.24   Consider the rule for multiplication in $\mathbf{Z}_n$ given by

$$[a][b] = [ab].$$

(a)   Multiplication as defined by this rule is a binary operation on $\mathbf{Z}_n$.

(b)   Multiplication is associative in $\mathbf{Z}_n$:

$$[a]([b][c]) = ([a][b])[c].$$

(c) $\mathbf{Z}_n$ has the multiplicative identity $[1]$.

(d)   Multiplication is commutative in $\mathbf{Z}_n$:

$$[a][b] = [b][a].$$

When we compare the properties listed in Theorems 2.23 and 2.24, we see that the existence of multiplicative inverses, even for the nonzero elements, is conspicuously missing. The following example shows that this is appropriate, because it illustrates a case where some of the nonzero elements of $\mathbf{Z}_n$ do not have multiplicative inverses.

EXAMPLE 2   A multiplication table for $\mathbf{Z}_4$ is as follows.

×	[0]	[1]	[2]	[3]
[0]	[0]	[0]	[0]	[0]
[1]	[0]	[1]	[2]	[3]
[2]	[0]	[2]	[0]	[2]
[3]	[0]	[3]	[2]	[1]

The third row of the table shows that [2] is a nonzero element of $\mathbf{Z}_4$ that has no multiplicative inverse; there is no $[x]$ in $\mathbf{Z}_4$ such that $[2][x] = [1]$. Another interesting point in connection with this table is that the equality $[2][2] = [0]$ shows that in $\mathbf{Z}_n$ the product of nonzero factors may be zero.   □

The next theorem characterizes those elements of $\mathbf{Z}_n$ that have multiplicative inverses.

THEOREM 2.25   An element $[a]$ of $\mathbf{Z}_n$ has a multiplicative inverse in $\mathbf{Z}_n$ if and only if a and n are relatively prime.

Proof   Suppose first that $[a]$ has a multiplicative inverse $[b]$ in $\mathbf{Z}_n$. Then

$[a][b] = [1]$.

This means that

$[ab] = [1]$ and $ab \equiv 1 \pmod{n}$.

Therefore

$ab - 1 = nq$

for some integer q, and

$a(b) + n(-q) = 1$.

By Theorem 2.12, we have $(a, n) = 1$.
    Conversely, if $(a, n) = 1$, then there exist integers s and t such that

$as + nt = 1$

$\Rightarrow as - 1 = n(-t)$

$\Rightarrow \quad as \equiv 1 \pmod{n}$

$\Rightarrow [a][s] = [1]$.

Thus $[a]$ has a multiplicative inverse $[s]$ in $\mathbf{Z}_n$.   ■

COROLLARY 2.26 Every nonzero element of Z_n has a multiplicative inverse if and only if n is a prime.

Proof The corollary follows from the fact that n is a prime if and only if every integer a such that $1 \leq a < n$ is relatively prime to n.

EXAMPLE 3 The elements of Z_{15} that have multiplicative inverses can be listed by writing down those $[a]$ that are such that $(a, 15) = 1$. These elements are

$$[1], [2], [4], [7], [8], [11], [13], [14].$$ □

EXAMPLE 4 Suppose we wish to find the multiplicative inverse of $[13]$ in Z_{191}. The modulus $n = 191$ is so large that it is not practical to test all of the elements in Z_{191}, so we utilize the Euclidean Algorithm and proceed according to the last part of the proof of Theorem 2.25:

$$191 = (13)(14) + 9$$
$$13 = (9)(1) + 4$$
$$9 = (4)(2) + 1.$$

Substituting the remainders in turn, we have

$$1 = 9 - (4)(2)$$
$$= 9 - [13 - (9)(1)](2)$$
$$= (9)(3) - (13)(2)$$
$$= [191 - (13)(14)](3) - (13)(2)$$
$$= (191)(3) + (13)(-44).$$

Thus

$$(13)(-44) \equiv 1 \pmod{191}$$

or

$$[13][-44] = [1].$$

The desired inverse is

$$[13]^{-1} = [-44] = [147].$$ □

EXERCISES 2.5

1 Perform the following computations in Z_{12}.
   (a) $[8] + [7]$
   (b) $[10] + [9]$
   (c) $[8][11]$
   (d) $[6][9]$

(e) $[6]([9] + [7])$ (f) $[5]([8] + [11])$

(g) $[6][9] + [6][7]$ (h) $[5][8] + [5][11]$

2   (a) Verify that $[1][2][3][4] = [4]$ in $\mathbf{Z}_5$.

      (b) Verify that $[1][2][3][4][5][6] = [6]$ in $\mathbf{Z}_7$.

      (c) Evaluate $[1][2][3]$ in $\mathbf{Z}_4$.

      (d) Evaluate $[1][2][3][4][5]$ in $\mathbf{Z}_6$.

3   Make addition tables for each of the following.

      (a) $\mathbf{Z}_2$ (b) $\mathbf{Z}_3$ (c) $\mathbf{Z}_5$

      (d) $\mathbf{Z}_6$ (e) $\mathbf{Z}_7$ (f) $\mathbf{Z}_8$

4   Make multiplication tables for each of the following.

      (a) $\mathbf{Z}_2$ (b) $\mathbf{Z}_3$ (c) $\mathbf{Z}_5$

      (d) $\mathbf{Z}_6$ (e) $\mathbf{Z}_7$ (f) $\mathbf{Z}_8$

5   Find the multiplicative inverse of each given element.

      (a) $[3]$ in $\mathbf{Z}_{13}$ (b) $[7]$ in $\mathbf{Z}_{11}$

      (c) $[17]$ in $\mathbf{Z}_{20}$ (d) $[16]$ in $\mathbf{Z}_{27}$

      (e) $[11]$ in $\mathbf{Z}_{317}$ (f) $[9]$ in $\mathbf{Z}_{128}$

6   For each of the following $\mathbf{Z}_n$, list all the elements of $\mathbf{Z}_n$ which have multiplicative inverses in $\mathbf{Z}_n$.

      (a) $\mathbf{Z}_6$ (b) $\mathbf{Z}_8$ (c) $\mathbf{Z}_{10}$

      (d) $\mathbf{Z}_{12}$ (e) $\mathbf{Z}_{18}$ (f) $\mathbf{Z}_{20}$

7   For each of the following $\mathbf{Z}_n$, find all the elements $[a]$ in $\mathbf{Z}_n$ for which the equation $[a][x] = [0]$ has a nonzero solution $[x] \neq [0]$ in $\mathbf{Z}_n$.

      (a) $\mathbf{Z}_6$ (b) $\mathbf{Z}_8$ (c) $\mathbf{Z}_{10}$

      (d) $\mathbf{Z}_{12}$ (e) $\mathbf{Z}_{18}$ (f) $\mathbf{Z}_{20}$

8   Whenever possible, find a solution for each of the following equations in the given $\mathbf{Z}_n$.

      (a) $[3][x] = [2]$ in $\mathbf{Z}_6$ (b) $[6][x] = [4]$ in $\mathbf{Z}_8$

      (c) $[4][x] = [6]$ in $\mathbf{Z}_8$ (d) $[8][x] = [6]$ in $\mathbf{Z}_{12}$

9   Let $[a]$ be an element of $\mathbf{Z}_n$ that has a multiplicative inverse $[a]^{-1}$ in $\mathbf{Z}_n$. Prove that $[x] = [a]^{-1}[b]$ is the unique solution in $\mathbf{Z}_n$ to the equation $[a][x] = [b]$.

10   Solve each of the following equations by finding $[a]^{-1}$ and using the result in Problem 9.

      (a) $[4][x] = [5]$ in $\mathbf{Z}_{13}$ (b) $[8][x] = [7]$ in $\mathbf{Z}_{11}$

      (c) $[6][x] = [5]$ in $\mathbf{Z}_{319}$ (d) $[9][x] = [8]$ in $\mathbf{Z}_{242}$

In Problems 11 and 12 solve the systems of equations in $\mathbf{Z}_7$.

11 $[3][x] + [2][y] = [1]$

        $[5][x] + [6][y] = [5]$

12 $[2][x] + [5][y] = [6]$

        $[4][x] + [6][y] = [6]$

13   Prove Theorem 2.24.

14   Prove the following distributive property in $\mathbf{Z}_n$:

      $[a]([b] + [c]) = [a][b] + [a][c]$.

15   Let p be a prime integer. Prove that if $[a][b] = [0]$ in $\mathbf{Z}_p$, then either $[a] = [0]$ or $[b] = [0]$.

16 Show that if n is not a prime, then there exist $[a]$ and $[b]$ in $\mathbf{Z}_n$ such that $[a] \neq [0]$ and $[b] \neq [0]$, but $[a][b] = [0]$.

17 Let p be a prime integer. Prove the following cancellation law in $\mathbf{Z}_p$: If $[a][x] = [a][y]$ and $[a] \neq [0]$, then $[x] = [y]$.

18 Show that if n is not a prime, the cancellation law stated in Problem 17 does not hold in $\mathbf{Z}_n$.

Key Words and Phrases

Induction postulate	Prime integer
Law of trichotomy	Relatively prime integers
Cancellation law for addition	Unique Factorization Theorem
Less than, greater than	Standard form of a positive integer
Second principle of finite induction	Least common multiple
Well-Ordering Theorem	Congruence modulo n
Division Algorithm	Congruence classes
Greatest common divisor	Properties of addition in $\mathbf{Z}_n$
Euclidean Algorithm	Properties of multiplication in $\mathbf{Z}_n$

References

Ames, Dennis B. *An Introduction to Abstract Algebra*. Scranton, Pa.: International Textbook, 1969.

Ball, Richard W. *Principles of Abstract Algebra*. New York: Holt, Rinehart and Winston, 1963.

Birkhoff, Garrett and Saunders MacLane. *A Survey of Modern Algebra* (4th ed.). New York: Macmillan, 1977.

Durbin, John, R. *Modern Algebra*. New York: Wiley, 1979.

Hillman, Abraham P. and Gerald L. Alexanderson. *A First Undergraduate Course in Abstract Algebra* (2nd ed.). Belmont, Calif.: Wadsworth, 1978.

Keesee, John W. *Elementary Abstract Algebra*. Boston: Heath, 1965.

Larsen, Max D. *Introduction to Modern Algebraic Concepts*. Reading, Mass.: Addison-Wesley, 1969.

McCoy, Neal H. *Introduction to Modern Algebra* (3rd ed.). Boston: Allyn and Bacon, 1975.

Mitchell, A. Richard, and Roger W. Mitchell. *An Introduction to Abstract Algebra*. Monterey, Calif.: Brooks-Cole, 1970.

Mostow, George D., Joseph H. Sampson, and Jean-Pierre Meyer. *Fundamental Structures of Algebra*. New York: McGraw-Hill, 1963.

Weiss, Marie J., and Roy Dubisch. *Higher Algebra for the Undergraduate* (2nd ed.). New York: Wiley, 1962.

3 GROUPS

3.1 DEFINITION OF A GROUP

The fundamental notions of set, mapping, binary operation, and binary relation were presented in Chapter 1. These notions are essential for a study of an algebraic system. An *algebraic system* is a nonempty set in which at least one equivalence relation (equality) and one or more binary operations are defined. The simplest cases occur when there is only one binary operation, as is the case with the algebraic system known as a *group*.

An introduction to the theory of groups is presented in this chapter, and it is appropriate to point out that this is only an introduction. Entire books have been devoted to the theory of groups, as is consistent with the usefulness of the group concept in both pure and applied mathematics.

A group may be defined as follows.

DEFINITION 3.1   Let $G \subseteq S$, and suppose that a binary operation $*$ is defined in S. Then G is a **group** (with respect to $*$) provided these conditions hold:

(1) G is **closed** under $*$. That is, $x \in G$ and $y \in G$ imply $x * y$ is in G.

(2) $*$ is **associative**. For all x, y, z in G, $x * (y * z) = (x * y) * z$.

(3) G has an **identity element** e. There is an e in G such that $x * e = e * x = x$ for all $x \in G$.

(4) G contains **inverses**. For each $a \in G$, there exists $b \in G$ such that $a * b = b * a = e$.

If G is a group, then G is a **commutative group**, or an **abelian group** if
(5) $*$ is **commutative**. For all x, y in G, $x * y = y * x$.

The phrase "with respect to $*$" should be noted. For example, the set **Z** of all integers is a group with respect to addition, but not with respect to multiplication (it has no inverses for elements other than ± 1). Similarly, the set $G = \{1, -1\}$ is a group with respect to multiplication but not with respect to addition. In most instances, however, only one binary operation is under consideration, and we say simply that "G is a group." If the binary operation is unspecified, we adopt the multiplicative notation and use the juxtaposition xy to indicate the result of combining x and y. It should be kept in mind, though, that the binary operation is not necessarily multiplication.

EXAMPLE 1   We can obtain some simple examples of groups by considering appropriate subsets of the familiar number systems.

(a)   The set of all *complex numbers* is an abelian group with respect to addition.

(b)   The set of all *nonzero rational numbers* is an abelian group with respect to multiplication.

(c)   The set of all *positive real numbers* is an abelian group with respect to multiplication, but is not a group with respect to addition (it has no additive identity and no additive inverses).   □

The following examples give some indication of the great variety there is in groups.

EXAMPLE 2   Recall from Chapter 1 that a permutation on a set A is a bijection from A to A, and that $\mathcal{S}(A)$ denotes the set of all permutation on A. We have seen that $\mathcal{S}(A)$ is closed with respect to the binary operation $\circ$ of mapping composition, and that the operation $\circ$ is associative. The identity mapping I_A is an identity element:

$$f \circ I_A = f = I_A \circ f$$

for all $f \in \mathcal{S}(A)$, and each $f \in \mathcal{S}(A)$ has an inverse in $\mathcal{S}(A)$. Thus we have, in fact, established in Chapter 1 that $\mathcal{S}(A)$ is a group with respect to map composition.   □

EXAMPLE 3   We shall take $A = \{1, 2, 3\}$ and obtain an explicit example of $\mathcal{S}(A)$. In order to define an element f of $\mathcal{S}(A)$, we need to specify $f(1)$,

65

$f(2)$, and $f(3)$. There are three possible choices for $f(1)$. Since f is to be bijective, there are two choices for $f(2)$ after $f(1)$ has been designated, and then only one choice for $f(3)$. Hence there are $3! = 3 \cdot 2 \cdot 1$ different mappings f in $\mathcal{S}(A)$. These are given by:

$$e = I_A: \begin{cases} e(1) = 1 \\ e(2) = 2 \\ e(3) = 3 \end{cases} \qquad \sigma: \begin{cases} \sigma(1) = 2 \\ \sigma(2) = 1 \\ \sigma(3) = 3 \end{cases}$$

$$\rho: \begin{cases} \rho(1) = 2 \\ \rho(2) = 3 \\ \rho(3) = 1 \end{cases} \qquad \gamma: \begin{cases} \gamma(1) = 3 \\ \gamma(2) = 2 \\ \gamma(3) = 1 \end{cases}$$

$$\tau: \begin{cases} \tau(1) = 3 \\ \tau(2) = 1 \\ \tau(3) = 2 \end{cases} \qquad \delta: \begin{cases} \delta(1) = 1 \\ \delta(2) = 3 \\ \delta(3) = 2. \end{cases}$$

Thus $\mathcal{S}(A) = \{e, \rho, \tau, \sigma, \gamma, \delta\}$. Following the same convention as in Problem 3 of Exercises 1.3, we shall construct a "multiplication" table for $\mathcal{S}(A)$ with the elements of $\mathcal{S}(A)$ listed in a row at the top and in a column at the left as shown in Figure 3.1. The result $f \circ g$ will be entered in the row with f at the left and in the column with g at the top. When the product $\rho^2 = \rho \circ \rho$ is computed, we have

$$\rho^2(1) = \rho(\rho(1)) = \rho(2) = 3$$
$$\rho^2(2) = \rho(\rho(2)) = \rho(3) = 1$$
$$\rho^2(3) = \rho(\rho(3)) = \rho(1) = 2$$

so $\rho^2 = \tau$. Similarly, $\rho \circ \sigma = \gamma$, $\sigma \circ \rho = \delta$, and so on.

$\circ$	e	ρ	ρ^2	σ	γ	δ
e	e	ρ	ρ^2	σ	γ	δ
ρ	ρ	ρ^2	e	γ	δ	σ
ρ^2	ρ^2	e	ρ	δ	σ	γ
σ	σ	δ	γ	e	ρ^2	ρ
γ	γ	σ	δ	ρ	e	ρ^2
δ	δ	γ	σ	ρ^2	ρ	e

FIGURE 3.1

Whenever a group table such as this is constructed, it is easy to locate the identity and inverses of elements. If the elements are listed on the left in the same order from top to bottom as the order from left to right across the top of the table, it is also possible to use the table to check for commutativity. The operation is commutative if and only if equal elements appear in all positions that are symmetrically placed relative to the diagonal from upper left to lower right. In the case above, the group is not abelian since the table is not symmetric. For example $\gamma \circ \rho^2 = \delta$ is in row 5, column 3, and $\rho^2 \circ \gamma = \sigma$ is in row 3, column 5.

EXAMPLE 4   Let G be the set of complex numbers given by $G = \{1, -1, i, -i\}$, where $i = \sqrt{-1}$, and consider the operation of multiplication of complex numbers in G. The table in Figure 3.2 shows that G is closed with respect to multiplication.

$\times$	1	-1	i	$-i$
1	1	-1	i	$-i$
-1	-1	1	$-i$	i
i	i	$-i$	-1	1
$-i$	$-i$	i	1	-1

FIGURE 3.2

Multiplication in G is associative and commutative since multiplication has these properties in the set of all complex numbers. We can observe from Figure 3.2 that 1 is the identity element, and that all elements have inverses. Each of 1 and -1 is its own inverse, and i and $-i$ are inverses of each other. Thus G is a group with respect to multiplication.   □

EXAMPLE 5   It is an immediate corollary of Theorem 2.23 that the set

$$\mathbf{Z}_n = \{[0], [1], [2], \ldots, [n-1]\}$$

of congruence classes modulo n forms a group with respect to addition.   □

EXAMPLE 6   Let $G = \{e, a, b, c\}$ with multiplication as defined by the table in Figure 3.3.

$\cdot$	e	a	b	c
e	e	a	b	c
a	a	b	c	e
b	b	c	e	a
c	c	e	a	b

FIGURE 3.3

From the table, we observe that

1. G is closed under this multiplication.
2. e is the identity element.
3. Each of e and b is its own inverse, and c and a are inverses of each other.

This multiplication is associative, but we shall not verify it here because it is a laborious task. It follows that G is a group. $\square$

EXAMPLE 7   The table in Figure 3.4 defines a binary operation $*$ on the set $S = \{A, B, C, D\}$.

$*$	A	B	C	D
A	B	C	A	B
B	C	D	B	A
C	A	B	C	D
D	A	B	D	D

FIGURE 3.4

From the table, we see that

1. G is closed under $*$.
2. C is an identity element.
3. D does not have an inverse, since $DX = C$ has no solution.

Thus S is not a group with respect to $*$. $\square$

DEFINITION 3.2   If a group G has a finite number of elements, G is called a **finite group**, or a **group of finite order**. The number of elements in G is called the **order** of G, and is denoted by $o(G)$. If G does not have a finite number of elements, G is called an **infinite group**, or a **group of infinite order**.

EXAMPLE 8   In Example 3, the group

$$G = \{e, \rho, \rho^2, \sigma, \gamma, \delta\}$$

has order $o(G) = 6$. In Example 5, $o(\mathbf{Z}_n) = n$. The set $\mathbf{Z}$ of all integers is a group under addition, and this is an example of an infinite group. If A is an infinite set, then $\mathscr{S}(A)$ furnishes an example of an infinite group. $\square$

Several simple consequences of the definition of a group are recorded in the theorem below.

THEOREM 3.3   Let G be a group with respect to a binary operation that is written as multiplication.

(a)   The identity element e in G is unique.
(b)   For each $x \in G$, the inverse x^{-1} in G is unique.
(c)   For each $x \in G$, $(x^{-1})^{-1} = x$.
(d)   For any x and y in G, $(xy)^{-1} = y^{-1}x^{-1}$.
(e)   **(Cancellation Law)** If a, x, and y are in G and $ax = ay$, then $x = y$.

Proof   We prove parts (b) and (d) and leave the others as exercises.
To prove part (b), let $x \in G$, and suppose that each of y and z are inverses of x. That is,

$$xy = e = yx \quad \text{and} \quad xz = e = zx.$$

Then

$$
\begin{aligned}
y &= ey & &\text{since } e \text{ is an identity} \\
&= (zx)y & &\text{since } zx = e \\
&= z(xy) & &\text{by associativity} \\
&= z(e) & &\text{since } xy = e \\
&= z & &\text{since } e \text{ is an identity.}
\end{aligned}
$$

Thus $y = z$, and this justifies the notation x^{-1} as the unique inverse of x in G.

We shall use part (b) in the proof of part (d). Specifically, we shall use the fact that the inverse $(xy)^{-1}$ is unique. This means that in order to show that $y^{-1}x^{-1} = (xy)^{-1}$, we only need to verify that $(xy)(y^{-1}x^{-1}) = e = (y^{-1}x^{-1})(xy)$. These calculations are straightforward:

$$(xy)(y^{-1}x^{-1}) = x(yy^{-1})x^{-1} = xex^{-1} = xx^{-1} = e$$

and

$$(y^{-1}x^{-1})(xy) = y^{-1}(x^{-1}x)y = y^{-1}ey = y^{-1}y = e. \qquad \blacksquare$$

Although our definition of a group is the standard one, alternate forms can be made. One of these is given in the next theorem.

THEOREM 3.4   Let G be a nonempty set that is closed under an associative binary operation called multiplication. Then G is a group if and only if the equations $ax = b$ and $ya = b$ have solutions x and y in G for all choices of a and b in G.

Proof   Assume first that G is a group, and let a and b represent arbitrary elements of G. Now a^{-1} is in G, and so are $x = a^{-1}b$ and

$y = ba^{-1}$. With these choices for x and y, we have

$$ax = a(a^{-1}b) = (aa^{-1})b = eb = b$$

and

$$ya = (ba^{-1})a = b(a^{-1}a) = be = b.$$

Thus G contains solutions x and y to $ax = b$ and $ya = b$.

Suppose now that the equations always have solutions in G. We first show that G has an identity element. Let a represent an arbitrary but fixed element in G. The equation $ax = a$ has a solution in G, say $x = u$. We shall show that u is a right identity for every element in G. To do this, let b be arbitrary in G. With z a solution to $ya = b$, we have $za = b$ and

$$bu = (za)u = z(au) = za = b.$$

Thus u is a right identity for every element in G. In a similar fashion, there exists an element v in G such that $vb = b$ for all b in G. Then $vu = v$ since u is a right identity, and $vu = u$ since v is a left identity. That is, the element $e = u = v$ is an identity element for G.

Now for any a in G, let x be a solution to $ax = e$, and let y be a solution to $ya = e$. Combining these equations, we have

$$
\begin{aligned}
x &= ex \\
&= yax \\
&= ye \\
&= y,
\end{aligned}
$$

and $x = y$ is an inverse for a. This proves that G is a group. ∎

In a group G, the associative property can be extended to products involving more than three factors. For example, if a_1, a_2, a_3, and a_4 are elements of G, then applications of condition (2) in Definition 3.1 yield

$$[a_1(a_2a_3)]a_4 = [(a_1a_2)a_3]a_4$$

and

$$(a_1a_2)(a_3a_4) = [(a_1a_2)a_3]a_4.$$

These equalities suggest (but do not completely prove) that, regardless of how symbols of grouping are introduced in a product $a_1a_2a_3a_4$, the resulting expression can be reduced to

$$[(a_1a_2)a_3]a_4.$$

With these observations in mind, we make the following definition.

DEFINITION 3.5 Let n be a positive integer, $n \geq 2$. For elements $a_1, a_2, \ldots, a_n$ in a group G, the expression $a_1 a_2 \cdots a_n$ is defined recursively by

$$a_1 a_2 \cdots a_k a_{k+1} = (a_1 a_2 \cdots a_k) a_{k+1}, \quad \text{for } k \geq 1.$$

We can now prove the following generalization of the associative property.

THEOREM 3.6 (**Generalized Associative Law**) Let $n \geq 2$ be a positive integer, and let $a_1, a_2, \ldots, a_n$ denote elements of a group G. For any positive integer m such that $1 \leq m < n$,

$$(a_1 a_2 \cdots a_m)(a_{m+1} \cdots a_n) = a_1 a_2 \cdots a_n.$$

Proof For $n \geq 2$, let $\mathscr{S}_n$ denote the statement of the theorem. With $n = 2$, the only possible value for m is $m = 1$, and $\mathscr{S}_2$ asserts the trivial equality

$$(a_1)(a_2) = a_1 a_2.$$

Assume now that $\mathscr{S}_k$ is true: for any positive integer m such that $1 \leq m < k$,

$$(a_1 a_2 \cdots a_m)(a_{m+1} \cdots a_k) = a_1 a_2 \cdots a_k.$$

Consider the statement $\mathscr{S}_{k+1}$, and let m be a positive integer such that $1 \leq m < k + 1$. We treat separately the cases where $m = k$ and where $1 \leq m < k$. If $m = k$, the desired equality is true at once from Definition 3.5, as follows:

$$(a_1 a_2 \cdots a_m)(a_{m+1} \cdots a_{k+1}) = (a_1 a_2 \cdots a_k) a_{k+1}.$$

If $1 \leq m < k$, then

$$a_{m+1} \cdots a_k a_{k+1} = (a_{m+1} \cdots a_k) a_{k+1}$$

by Definition 3.5, and consequently

$$
\begin{aligned}
(a_1 a_2 \cdots a_m)&(a_{m+1} \cdots a_k a_{k+1}) \\
&= (a_1 a_2 \cdots a_m)[(a_{m+1} \cdots a_k) a_{k+1}] \\
&= [(a_1 a_2 \cdots a_m)(a_{m+1} \cdots a_k)] a_{k+1} \quad \text{by the associative property} \\
&= [a_1 a_2 \cdots a_k] a_{k+1} \quad \text{by } \mathscr{S}_k \\
&= a_1 a_2 \cdots a_{k+1} \quad \text{by Definition 3.5.}
\end{aligned}
$$

Thus $\mathscr{S}_{k+1}$ is true whenever $\mathscr{S}_k$ is true, and the proof of the theorem is complete. ∎

EXERCISES 3.1

In Problems 1–10 decide if each of the given sets is a group with respect to the indicated operation. If it is not a group, state all of the conditions in Definition 3.1 which fail to hold.

1 The set of all rational numbers with operation addition.

2 The set of all irrational numbers with operation addition.

3 The set of all positive irrational numbers with operation multiplication.

4 The set of all positive rational numbers with operation multiplication.

5 The set of all real numbers x such that $0 < x \leq 1$, with operation multiplication.

6 For a fixed positive integer n, the set of all complex numbers x such that $x^n = 1$ (that is, the set of all n^{th} roots of 1), with operation multiplication.

7 The set of all complex numbers x that have absolute value 1, with operation multiplication.

8 The set in Problem 7 with operation addition.

9 The set $\mathbf{E}$ of all even integers with operation addition.

10 The set $\mathbf{E}$ of all even integers with operation multiplication.

In Problems 11 and 12, the given table defines an operation of multiplication on the set $S = \{e, a, b, c\}$. In each case, find a condition in Definition 3.1 that fails to hold, and thereby show that S is not a group.

11 See Figure 3.5.

12 See Figure 3.6.

×	e	a	b	c
e	e	a	b	c
a	a	b	a	b
b	b	c	b	c
c	c	e	c	e

FIGURE 3.5

×	e	a	b	c
e	e	a	b	c
a	e	a	b	c
b	e	a	b	c
c	e	a	b	c

FIGURE 3.6

13 In each of the following parts, let the binary operation * be defined on $\mathbf{Z}$ by the given rule. Determine in each case if $\mathbf{Z}$ is a group with respect to *, and if it is an abelian group. State which, if any, conditions fail to hold.

(a) $x * y = x + y + 1$ (b) $x * y = x + xy + y$
(c) $x * y = x + xy$ (d) $x * y = x - y$

In Problems 14–19 decide if each of the given sets is a group with respect to the indicated operation. If it is not a group, state all of the conditions in Definition 3.1 that fail to hold.

14 The set $\{[1], [3]\} \subseteq \mathbf{Z}_8$ with operation multiplication.

15 The set $\{[1], [2], [3], [4]\} \subseteq \mathbf{Z}_5$ with operation multiplication.

16 The set $\{[0], [2], [4]\} \subseteq \mathbf{Z}_8$ with operation multiplication.

17 The set $\{[0], [2], [4], [6], [8]\} \subseteq \mathbf{Z}_{10}$ with operation multiplication.

18 The set $\{[0],[2],[4],[6],[8]\} \subseteq \mathbf{Z}_{10}$ with operation addition.

19 The set $\{[0],[2],[4],[6]\} \subseteq \mathbf{Z}_8$ with operation addition.

20 (a) Let $G = \{[a] \,|\, [a] \neq [0]\} \subseteq \mathbf{Z}_n$. Show that G is a group with respect to multiplication in $\mathbf{Z}_n$ if and only if n is a prime.

   (b) Construct a multiplication table for the group G of all nonzero elements in $\mathbf{Z}_7$, and identify the inverse of each element.

21 Let G be the set of eight elements $G = \{\pm 1, \pm i, \pm j, \pm k\}$ with multiplication given by[†]

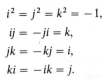

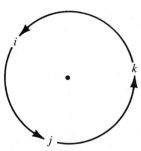

FIGURE 3.7

(The cyclic order of multiplication is indicated by the diagram in Figure 3.7.) Given that G is a group, write out the multiplication table for G. This group is known as the **quaternion group**.

22 Prove parts (a), (c), and (e) of Theorem 3.3.

23 Prove that if a, x, and y are elements of a group G such that $xa = ya$, then $x = y$.

24 In Example 3, find elements a and b of $\mathcal{S}(A)$ such that $(ab)^{-1} \neq a^{-1}b^{-1}$.

25 In Example 3, find elements a, b, and c of $\mathcal{S}(A)$ such that $ab = bc$ but $a \neq c$.

26 In Example 3, find elements a and b of $\mathcal{S}(A)$ such that $(ab)^2 \neq a^2b^2$.

27 An element x in a multiplicative group G is called **idempotent** if $x^2 = x$. Prove that the identity element e is the only idempotent element in a group G.

28 Prove that if $x = x^{-1}$ for all x in the group G, then G is abelian.

29 Prove that if $(ab)^{-1} = a^{-1}b^{-1}$ for all a and b in the group G, then G is abelian.

30 Let a, b, and c be elements of a group G. Prove that $((ab)c)^{-1} = (c^{-1}b^{-1})a^{-1}$.

31 Let a, b, c, and d be elements of a group G. Find an expression for $(abcd)^{-1}$ in terms of a^{-1}, b^{-1}, c^{-1}, and d^{-1}.

32 For an arbitrary set A, the power $\mathcal{P}(A)$ was defined in Section 1.1 by $\mathcal{P}(A) = \{X \,|\, X \subseteq A\}$, and addition in $\mathcal{P}(A)$ was defined by

$$X + Y = (X \cup Y) - (X \cap Y)$$
$$= (X - Y) \cup (Y - X).$$

Prove that $\mathcal{P}(A)$ is a group with respect to this operation of addition.

33 Write out the elements of $\mathcal{P}(A)$ for the set $A = \{a, b, c\}$ and construct an addition table for $\mathcal{P}(A)$.

[†] In a multiplicative group, a^2 is defined by $a^2 = a \cdot a$.

34   Let $A = \{a, b, c\}$. Show that $\mathscr{P}(A)$ is not a group with respect to the operation of union.

35   Let $A = \{a, b, c\}$. Show that $\mathscr{P}(A)$ is not a group with respect to the operation of intersection.

36   Reword Definition 3.5 for a group G with respect to addition.

37   State and prove Theorem 3.6 for an additive group.

3.2   SUBGROUPS

Among the nonempty subsets of a group G, there are some that themselves form a group with respect to the binary operation $*$ in G. That is, a subset $H \subseteq G$ may be such that H is also a group with respect to $*$. Such a subset H is called a *subgroup* of G.

> **DEFINITION 3.7**   Let G be a group with respect to the binary operation $*$. A subset H of G is called a **subgroup** of G if H forms a group with respect to the binary operation $*$ that is defined in G.
>
> The subsets $H = \{e\}$ and $H = G$ are always subgroups of the group G. They are referred to as **trivial subgroups** and all other subgroups of G are called **non-trivial**.

> **EXAMPLE 1**   The set $\mathbf{Z}$ of all integers is a group with respect to addition, and the set $\mathbf{E}$ of all even integers is a nontrivial subgroup of $\mathbf{Z}$ (see Problem 9 of Exercises 3.1). $\qquad\square$

> **EXAMPLE 2**   The set of all nonzero complex numbers is a group under multiplication, and $G = \{1, -1, i, -i\}$ is a nontrivial subgroup of this group (see Example 4 of Section 3.1). $\qquad\square$

If G is a group with respect to $*$, then $*$ is an associative operation on any nonempty subset of G. A subset H of G is a subgroup, provided

(i) H contains the identity,

(ii) H is closed under $*$, and

(iii) H contains an inverse for each of its elements.

The following theorem gives a slightly different set of conditions.

> **THEOREM 3.8**   A subset H of the group G is a subgroup of G if and only if these conditions are satisfied:
>
> (a) H is nonempty,
>
> (b) $x \in H$ and $y \in H$ imply $xy \in H$, and
>
> (c) $x \in H$ implies $x^{-1} \in H$.

Proof If H is a subgroup of G, the conditions follow at once from Definitions 3.7 and 3.1.

Suppose that H is a subset of G that satisfies the conditions. Since H is nonempty, there is at least one $a \in H$. By condition (c), $a^{-1} \in H$. But $a \in H$ and $a^{-1} \in H$ imply $aa^{-1} = e \in H$, by condition (b). Thus H contains e, is closed, and contains inverses. Hence H is a subgroup. ∎

EXAMPLE 3 It follows from Example 5 of Section 3.1 that

$$G = \mathbf{Z}_8 = \{[0],[1],[2],[3],[4],[5],[6],[7]\}$$

forms an abelian group with respect to addition $[a] + [b] = [a + b]$. Consider the subset

$$H = \{[0],[2],[4],[6]\}$$

of G. An addition table for H is given in Figure 3.8. The subset H is nonempty, and it is evident from the table that H is closed and contains the inverse of each of its elements. Hence H is a nontrivial abelian subgroup of $\mathbf{Z}_8$ under addition.

$+$	$[0]$	$[2]$	$[4]$	$[6]$
$[0]$	$[0]$	$[2]$	$[4]$	$[6]$
$[2]$	$[2]$	$[4]$	$[6]$	$[0]$
$[4]$	$[4]$	$[6]$	$[0]$	$[2]$
$[6]$	$[6]$	$[0]$	$[2]$	$[4]$

FIGURE 3.8

EXAMPLE 4 In Problem 20 of Exercises 3.1, it was shown that

$$G = \{[1],[2],[3],[4],[5],[6]\} \subseteq \mathbf{Z}_7$$

is a group with respect to multiplication in $\mathbf{Z}_7$. The multiplication table in Figure 3.9 shows that the nonempty subset

$$H = \{[1],[2],[4]\}$$

is closed and contains inverses, and therefore is an abelian subgroup of G.

$\cdot$	$[1]$	$[2]$	$[4]$
$[1]$	$[1]$	$[2]$	$[4]$
$[2]$	$[2]$	$[4]$	$[1]$
$[4]$	$[4]$	$[1]$	$[2]$

FIGURE 3.9

An even shorter set of conditions for a subgroup is given in the next theorem.

THEOREM 3.9 A subset H of the group G is a subgroup of G if and only if

(a) H is nonempty, and
(b) $a \in H$ and $b \in H$ imply $ab^{-1} \in H$.

Proof Assume H is a subgroup of G. Then H is nonempty since $e \in H$. Let $a \in H$ and $b \in H$. Then $b^{-1} \in H$ since H contains inverses, and $ab^{-1} \in H$ since H is closed. Thus conditions (a) and (b) are satisfied.

Suppose conversely that conditions (a) and (b) hold for H. There is at least one $a \in H$, and condition (b) implies that $aa^{-1} = e \in H$. For an arbitrary $x \in H$, we have $e \in H$ and $x \in H$, which implies that $ex^{-1} = x^{-1} \in H$. Thus H contains inverses. To show closure, let $x \in H$ and $y \in H$. Since H contains inverses, $y^{-1} \in H$. But $x \in H$ and $y^{-1} \in H$ imply $x(y^{-1})^{-1} = xy \in H$, by condition (b). Hence H is closed; therefore H is a subgroup of G. ∎

When the phrase "H is a subgroup of G" is used, it indicates that H is a group with respect to the group operation in G. Consider the following example.

EXAMPLE 5 The operation of multiplication is defined in $\mathbf{Z}_{10}$ by

$$[a][b] = [ab].$$

This rule defines a binary operation that is associative, and $\mathbf{Z}_{10}$ is closed under this multiplication. Also $[1]$ is an identity element. However, $\mathbf{Z}_{10}$ is *not* a group with respect to multiplication, since some of its elements do not have inverses. For example, the products

$$[2][0] = [0] \qquad [2][5] = [0]$$
$$[2][1] = [2] \qquad [2][6] = [2]$$
$$[2][2] = [4] \qquad [2][7] = [4]$$
$$[2][3] = [6] \qquad [2][8] = [6]$$
$$[2][4] = [8] \qquad [2][9] = [8]$$

show that $[2][x] = [1]$ has no solution in $\mathbf{Z}_{10}$.

Now let us examine the multiplication table for the subset $H = \{[2],[4],[6],[8]\}$ of $\mathbf{Z}_{10}$ (see Figure 3.10). It is surprising, perhaps, but the table shows that $[6]$ is an identity element for H, and that H actually forms a group with respect to multiplication. However, H is *not* a subgroup of Z_{10}, since Z_{10} is not a group with respect to multiplication.

×	[2]	[4]	[6]	[8]
[2]	[4]	[8]	[2]	[6]
[4]	[8]	[6]	[4]	[2]
[6]	[2]	[4]	[6]	[8]
[8]	[6]	[2]	[8]	[4]

FIGURE 3.10   □

Integral *exponents* can be defined for elements of a group as follows.

DEFINITION 3.10   Let G be a group with the binary operation written as multiplication. For any $a \in G$, we define **nonnegative integral exponents** by

$$a^0 = e, \qquad a^1 = a,$$

and

$$a^{k+1} = a^k \cdot a \quad \text{for any positive integer } k.$$

Negative integral exponents are defined by

$$a^{-k} = (a^{-1})^k \quad \text{for any positive integer } k.$$

It is common practice to write the binary operation as addition in the case of abelian groups. When the operation is addition, the corresponding **multiples** of a are given by

$$0a = 0$$
$$1a = a$$
$$(k + 1)a = ka + a$$
$$(-k)a = k(-a)$$

for any positive integer k. The notation ka here does not present a *product* of k and a, but rather a *sum*

$$ka = a + a + \cdots + a$$

with k terms. In $0a = 0$, the zero on the left is the zero integer, and the zero on the right represents the additive identity in the group.

The familiar laws of exponents hold in a group.

THEOREM 3.11   Let x and y be elements of the group G, and let m and n denote integers. Then

(a) $x^n \cdot x^{-n} = e.$
(b) $x^m \cdot x^n = x^{m+n}.$
(c) $(x^m)^n = x^{mn}.$
(d) If G is abelian, $(xy)^n = x^n y^n.$

Proof The proof of each statement involves the use of mathematical induction. It would be redundant, and even boring, to include a complete proof of the theorem, so we shall assume statement (a) and prove (b) for the case where m is a positive integer. Even then, the argument is tedious. The proofs of statements (a), (c), and (d) are left as exercises.

Let m be an arbitrary, but fixed, positive integer. There are three cases to consider for n:

(i) $n = 0.$
(ii) n a positive integer.
(iii) n a negative integer.

First, let $n = 0$ for case (i). Then

$$x^m \cdot x^n = x^m \cdot x^0 = x^m \cdot e = x^m \quad \text{and} \quad x^{m+n} = x^{m+0} = x^m.$$

Thus $x^m \cdot x^n = x^{m+n}$ in the case where $n = 0.$

Second, we shall use induction on n for case (ii) where n is a positive integer. If $n = 1$, we have

$$x^m \cdot x^n = x^m \cdot x = x^{m+1} = x^{m+n},$$

and statement (b) of the theorem holds when $n = 1$. Assume that (b) is true for $n = k$. That is, assume that

$$x^m \cdot x^k = x^{m+k}.$$

Then, for $n = k + 1$, we have

$$
\begin{aligned}
x^m \cdot x^n &= x^m \cdot x^{k+1} \\
&= x^m \cdot (x^k \cdot x) && \text{by definition of } x^{k+1} \\
&= (x^m \cdot x^k) \cdot x && \text{by associativity} \\
&= x^{m+k} \cdot x && \text{by the induction hypothesis} \\
&= x^{m+k+1} && \text{by definition of } x^{(m+k)+1} \\
&= x^{m+n} && \text{since } n = k + 1.
\end{aligned}
$$

Thus (b) is true for $n = k + 1$, and it follows that it is true for all positive integers n.

Third, consider case (iii) where n is a negative integer. This means that $n = -p$, where p is a positive integer. We consider three possibilities for p: $p = m$, $p < m$, and $m < p$.

If $p = m$, then $n = -p = -m$, and we have

$$x^m \cdot x^n = x^m \cdot x^{-m} = e$$

by statement (a) of the theorem, and

$$x^{m+n} = x^{m-m} = x^0 = e.$$

We have $x^m \cdot x^n = x^{m+n}$ when $p = m$.

If $p < m$, let $m - p = q$, so that $m = q + p$, where q and p are positive integers. We have already proved statement (b) when m and n are positive integers, so we may use $x^{q+p} = x^q \cdot x^p$. This gives

$$
\begin{aligned}
x^m \cdot x^n &= x^{q+p} \cdot x^{-p} \\
&= x^q \cdot x^p \cdot x^{-p} \\
&= x^q \cdot e \qquad \text{by statement (a)} \\
&= x^q \\
&= x^{q+p-p} \\
&= x^{m+n}.
\end{aligned}
$$

That is, $x^m \cdot x^n = x^{m+n}$ for the case where $p < m$.

Finally, suppose that $m < p$. Let $r = p - m$, so that r is a positive integer and $p = m + r$. By the definition of x^{-p},

$$
\begin{aligned}
x^{-p} &= (x^{-1})^p \\
&= (x^{-1})^{m+r} \\
&= (x^{-1})^m \cdot (x^{-1})^r \quad \text{since } m \text{ and } r \text{ are positive integers} \\
&= x^{-m} \cdot x^{-r}.
\end{aligned}
$$

Substituting this value for x^{-p} in $x^m \cdot x^n = x^m \cdot x^{-p}$, we have

$$
\begin{aligned}
x^m \cdot x^n &= x^m \cdot (x^{-m} \cdot x^{-r}) \\
&= (x^m \cdot x^{-m}) \cdot x^{-r} \\
&= e \cdot x^{-r} \\
&= x^{-r}.
\end{aligned}
$$

We also have

$$
\begin{aligned}
x^{m+n} &= x^{m-p} \\
&= x^{m-(m+r)} \\
&= x^{-r},
\end{aligned}
$$

so $x^m \cdot x^n = x^{m+n}$ when $m < p$.

We have proved that $x^m \cdot x^n = x^{m+n}$ in the cases where m is a positive integer, and n is any integer (zero, positive, or negative). Of course, this is not a complete proof of statement (b) of the theorem. A complete proof would require considering cases where $m = 0$ or where m is a negative integer. The proofs for these cases are similar to those given above, and we omit them entirely. ∎

Let G be a group, let a be an element of G, and let H be the set of all elements of the form a^n, where n is an integer. That is,

$$H = \{x \in G \mid x = a^n \text{ for } n \in \mathbf{Z}\}.$$

Then H is nonempty and actually forms a subgroup of G. For if $x = a^m \in H$ and $y = a^n \in H$, then $xy = a^{m+n} \in H$ and $x^{-1} = a^{-m} \in H$. It follows from Theorem 3.8 that H is a subgroup.

DEFINITION 3.12 Let G be a group. For any $a \in G$, the subgroup

$$H = \{x \in G \mid x = a^n \text{ for } n \in \mathbf{Z}\}$$

is the **subgroup generated by** a and is denoted by $\langle a \rangle$. A given subgroup K of G is a **cyclic subgroup** if there exists an element b in G such that

$$K = \langle b \rangle = \{y \in G \mid y = b^n \text{ for some } n \in \mathbf{Z}\}.$$

In particular, G is a **cyclic group** if there is an element $a \in G$ such that $G = \langle a \rangle$.

EXAMPLE 6

(a) The set $\mathbf{Z}$ of integers is a cyclic group under addition. We have $\mathbf{Z} = \langle 1 \rangle$ and $\mathbf{Z} = \langle -1 \rangle$.

(b) The subgroup $\mathbf{E} \subseteq \mathbf{Z}$ of all even integers is a cyclic subgroup of the additive group $\mathbf{Z}$, generated by 2. Hence $\mathbf{E} = \langle 2 \rangle$.

(c) In Example 5, we saw that

$$H = \{[2], [4], [6], [8]\} \subseteq \mathbf{Z}_{10}$$

is an abelian group with respect to multiplication. Since

$$[2]^2 = [4], \qquad [2]^3 = [8], \qquad [2]^4 = [6],$$

$$H = \langle [2] \rangle.$$

(d) The group $\mathcal{S}(A) = \{e, \rho, \tau, \sigma, \gamma, \delta\}$ of Example 3 in Section 3.1 is not a cyclic group. This can be verified by considering $\langle a \rangle$ for all possible choices of a in $\mathcal{S}(A)$. ◻

EXERCISES 3.2

1 Let $\mathcal{S}(A) = \{e, \rho, \tau, \sigma, \gamma, \delta\}$ be as in Example 3 in Section 3.1. Decide whether or not each of the following subsets is a subgroup of $\mathcal{S}(A)$. If a set is not a subgroup, give a reason. [Hint: Construct a multiplication table for each subset.]
(a) $\{e, \sigma\}$ (b) $\{e, \delta\}$
(c) $\{e, \rho\}$ (d) $\{e, \tau\}$

(e) $\{e, \rho, \tau\}$ (f) $\{e, \rho, \sigma\}$

(g) $\{e, \sigma, \gamma\}$ (h) $\{e, \sigma, \gamma, \delta\}$

2   Decide whether or not each of the following sets is a subgroup of the group $G = \{1, -1, i, -i\}$ under multiplication. If a set is not a subgroup, give a reason why it is not.

(a) $\{1, -1\}$ (b) $\{1, i\}$

(c) $\{i, -i\}$ (d) $\{1, -i\}$

3   Consider the group $\mathbf{Z}_{16}$ under addition. List all the elements of the subgroup $\langle [6] \rangle$.

4   Assume that the nonzero elements of $\mathbf{Z}_{13}$ form a group G under multiplication $[a][b] = [ab]$.

(a) List the elements of the subgroup $\langle [4] \rangle$ of G.

(b) List the elements of the subgroup $\langle [8] \rangle$ of G.

5   Find a subset of $\mathbf{Z}$ that is closed under addition but is not a subgroup of the additive group $\mathbf{Z}$.

6   Let G be the group of all nonzero real numbers under multiplication. Find a subset of G that is closed under multiplication but is not a subgroup of G.

7   Let $n > 1$ be an integer, and let a be a fixed integer. Prove that the set

$$H = \{x \in \mathbf{Z} \mid ax \equiv 0 \pmod{n}\}$$

is a subgroup of $\mathbf{Z}$ under addition.

8   Let H be a subgroup of G, let a be a fixed element of G, and let K be the set of all elements of the form aha^{-1}, where $h \in H$. That is,

$$K = \{x \in G \mid x = aha^{-1} \text{ for some } h \in H\}.$$

Prove that K is a subgroup of G.

9   Prove that $H = \{h \in G \mid h^{-1} = h\}$ is a subgroup of the group G if G is abelian.

10 For any group G, the set of all elements that commute with every element of G is called the **center** of G, and is denoted by $\mathscr{C}(G)$:

$$\mathscr{C}(G) = \{a \in G \mid ax = xa \text{ for every } x \in G\}.$$

Prove that $\mathscr{C}(G)$ is a subgroup of G.

11 For a fixed element a of a group G, the set $\mathscr{C}_a = \{x \in G \mid ax = xa\}$ is the **centralizer** of a in G. Prove that for any $a \in G$, $\mathscr{C}_a$ is a subgroup of G.

12 Suppose that H_1 and H_2 are subgroups of the group G. Prove that $H_1 \cap H_2$ is a subgroup of G.

13 Let $\{H_\lambda\}, \lambda \in \mathscr{L}$, be an arbitrary, nonempty collection of subgroups H_λ of the group G, and let $K = \bigcap_{\lambda \in \mathscr{L}} H_\lambda$. Prove that K is a subgroup of G.

14 Find subgroups H_1 and H_2 of the group $\mathscr{S}(A)$ in Example 3 of Section 3.1 such that $H_1 \cup H_2$ is *not* a subgroup of $\mathscr{S}(A)$.

15 Assume H_1 and H_2 are subgroups of the abelian group G. Prove that $H_1 H_2 = \{g \in G \mid g = h_1 h_2 \text{ for } h_1 \in H_1 \text{ and } h_2 \in H_2\}$ is a subgroup of G.

16 Find subgroups H_1 and H_2 of the group $\mathscr{S}(A)$ in Example 3 of Section 3.1 such that the set $H_1 H_2$ defined in Problem 15 is not a subgroup of $\mathscr{S}(A)$.

17   Let G be a cyclic group, $G = \langle a \rangle$. Prove that G is abelian.

18   Let H be a subgroup of the group G. If e_1 is the identity element in H and e_2 is the identity element in G, prove that $e_1 = e_2$. (Example 5 of this section should be noted in connection with this result.)

19   For an arbitrary n in $\mathbf{Z}$, the cyclic subgroup $\langle n \rangle$ of $\mathbf{Z}$, generated by n under addition, is the set of all multiples of n. Describe the subgroup $\langle m \rangle \cap \langle n \rangle$ for arbitrary m and n in $\mathbf{Z}$.

20   Prove statement (a) of Theorem 3.11: $x^n \cdot x^{-n} = e$ for all integers n.

21   Prove statement (c) of Theorem 3.11: $(x^m)^n = x^{mn}$ for all integers m and n.

22   Prove statement (d) of Theorem 3.11: if G is abelian, $(xy)^n = x^n y^n$ for all integers n.

23   Suppose H is a nonempty subset of a group G. Prove that H is a subgroup of G if and only if $a^{-1}b \in H$ for all $a \in H$ and $b \in H$.

24   Assume that G is a finite group, and let H be a nonempty subset of G. Prove that H is closed if and only if H is a subgroup of G.

3.3   CYCLIC GROUPS

In the last section a group G was defined to be *cyclic* if there exists an element $a \in G$ such that $G = \langle a \rangle$. It may happen that there is more than one element $a \in G$ such that $G = \langle a \rangle$. For the additive group $\mathbf{Z}$, we have $\mathbf{Z} = \langle 1 \rangle$ and also $\mathbf{Z} = \langle -1 \rangle$, since any $n \in \mathbf{Z}$ can be written as $n = (-n)(-1)$. (Hence $(-n)(-1)$ does not indicate a product, but rather a multiple of -1 as described in Section 3.2.)

DEFINITION 3.13   Any element a of the group G such that $G = \langle a \rangle$ is a **generator** of G.

If a is a generator of G, then a^{-1} is also, since any element $x \in G$ can be written as

$$x = a^n = (a^{-1})^{-n}$$

for some integer n.

EXAMPLE 1   The additive group

$$\mathbf{Z}_n = \{[0], [1], \ldots, [n-1]\}$$

is a cyclic group with generator $[1]$ since any $[k]$ in $\mathbf{Z}_n$ can be written as

$$[k] = k[1]$$

where $k[1]$ indicates a multiple of $[1]$ as described in Section 3.2. Elements other than [1] may also be generators. To illustrate this, consider the particular case

$$\mathbf{Z}_6 = \{[0],[1],[2],[3],[4],[5]\}.$$

The element [5] is also a generator of $\mathbf{Z}_6$ under addition since

$$1[5] = [5]$$
$$2[5] = [5] + [5] = [4]$$
$$3[5] = [5] + [5] + [5] = [3]$$
$$4[5] = [2]$$
$$5[5] = [1]$$
$$6[5] = [0].$$

The cyclic subgroups generated by the other elements of $\mathbf{Z}_6$ under addition are as follows:

$$\langle[0]\rangle = \{[0]\}$$
$$\langle[2]\rangle = \{[2],[4],[0]\}$$
$$\langle[3]\rangle = \{[3],[0]\}$$
$$\langle[4]\rangle = \{[4],[2],[0]\} = \langle[2]\rangle.$$

Thus [1] and [5] are the only elements that are generators of the entire group.

$\square$

EXAMPLE 2   We saw in Example 6 of Section 3.2 that

$$H = \{[2],[4],[6],[8]\} \subseteq \mathbf{Z}_{10}$$

forms a cyclic group with respect to multiplication, and that [2] is a generator of H. The element $[8] = [2]^{-1}$ is also a generator of H as the following computations confirm:

$$[8]^2 = [4], \qquad [8]^3 = [2], \qquad [8]^4 = [6].$$

$\square$

EXAMPLE 3   (See Problem 21 of Exercises 3.1.) In the quaternion group $G = \{\pm 1, \pm i, \pm j, \pm k\}$, we have

$$i^2 = -1$$
$$i^3 = i^2 \cdot i = -i$$
$$i^4 = i^3 \cdot i = -i^2 = 1:$$

Thus i generates the cyclic subgroup of order 4 given by

$$\langle i \rangle = \{i, -1, -i, 1\}.$$

$\square$

Whether a group G is cyclic or not, each element a of G generates the cyclic subgroup $\langle a \rangle$, and

$$\langle a \rangle = \{ x \in G \mid x = a^n \text{ for } n \in \mathbf{Z} \}.$$

Suppose it happens that $a^n \neq e$ for every positive integer n. Then all the powers of a are distinct, and $\langle a \rangle$ is an infinite group. To see that this is true, suppose that

$$a^p = a^q$$

where $p \neq q$. We may assume that $p > q$. Then

$$a^p = a^q \Rightarrow a^p \cdot a^{-q} = a^q \cdot a^{-q}$$
$$\Rightarrow a^{p-q} = e.$$

Since $p - q$ is a positive integer, this result contradicts $a^n \neq e$ for every positive integer n. Therefore, it must be that $a^p \neq a^q$ whenever $p \neq q$.

Of course, it may happen that $a^n = e$ for some positive integers n. In this case, also, $\langle a \rangle$ can be described completely.

THEOREM 3.14   Let a be an element in a group G, and suppose $a^n = e$ for some positive integer n. If m is the least positive integer such that $a^m = e$, then

(a) $\langle a \rangle$ has order m, and
(b) $a^s = a^t$ if and only if $s \equiv t \pmod{m}$.

Proof   Assume that m is the least positive integer such that $a^m = e$. We first show that the elements

$$a^0 = e, a, a^2, \ldots, a^{m-1}$$

are all distinct. Suppose

$$a^i = a^j, \quad \text{where } 0 \leq i < m \quad \text{and} \quad 0 \leq j < m.$$

There is no loss of generality in assuming $i \geq j$. Then $a^i = a^j$ implies

$$a^{i-j} = a^i \cdot a^{-j} = e, \quad \text{where } 0 \leq i - j < m.$$

Since m is the least positive integer such that $a^m = e$ and since $i - j < m$, it must be that $i - j = 0$, and therefore $i = j$. Thus $\langle a \rangle$ contains the m distinct elements $a^0 = e, a, a^2, \ldots, a^{m-1}$. The proof of part (a) will be complete if we can show that any power of a is equal to one of these elements. Consider an arbitrary a^k. By the Division Algorithm, there exist integers q and r such that

$$k = mq + r, \quad \text{with } 0 \leq r < m.$$

Thus

$$a^k = a^{mq+r}$$
$$= a^{mq} \cdot a^r \quad \text{by part (b) of Theorem 3.11}$$
$$= (a^m)^q \cdot a^r \quad \text{by part (c) of Theorem 3.11}$$
$$= e^q \cdot a^r$$
$$= a^r,$$

where r is in the set $\{0, 1, 2, \ldots, m-1\}$, and it follows that

$$\langle a \rangle = \{e, a, a^2, \ldots, a^{m-1}\}.$$

To obtain part (b), we first observe that if $k = mq + r$, with $0 \le r < m$, then $a^k = a^r$, where r is in the set $\{0, 1, 2, \ldots, m-1\}$. In particular, $a^k = e$ if and only if $r = 0$; that is, if and only if $k \equiv 0 \pmod{m}$. Thus

$$a^s = a^t \Leftrightarrow a^{s-t} = e$$
$$\Leftrightarrow s - t \equiv 0 \pmod{m}$$
$$\Leftrightarrow s \equiv t \pmod{m},$$

and the proof is complete. ∎

We have defined the order $o(G)$ of a group G to be the number of elements in the group.

DEFINITION 3.15 The **order** $o(a)$ of an element a of the group G is the order of the subgroup generated by a. That is, $o(a) = o(\langle a \rangle)$.

Part (a) of Theorem 3.14 immediately translates into the following corollary.

COROLLARY 3.16 If $o(a)$ is finite, then $m = o(a)$ is the least positive integer such that $a^m = e$.

As might be expected, every subgroup of a cyclic group is also a cyclic group. It is even possible to predict a generator of the subgroup.

THEOREM 3.17 Let G be a cyclic group with $a \in G$ as a generator, and let H be a subgroup of G. If k is the least positive integer such that $a^k \in H$, then H is the cyclic subgroup generated by a^k.

Proof Let G, H, a, and k be as described in the theorem. Since H is closed and $a^k \in H$, all powers $(a^k)^t = a^{kt}$ are in H. We need to show that any element of H is a power of a^k. Let $a^n \in H$. There are integers q and r such that

$n = kq + r,$ with $0 \le r < k.$

Now $a^{-kq} = (a^k)^{-q} \in H$ and $a^n \in H$ imply that

$a^n \cdot a^{-kq} = a^{kq+r} \cdot a^{-kq} = a^r$

is in H. Since $0 \le r < k$ and k is the least positive integer such that $a^k \in H$, r must be zero and $a^n = a^{kq}$. Thus $H = \langle a^k \rangle$.   ∎

Notice that Theorem 3.17 applies to infinite cyclic groups as well as finite ones. The next theorem, however, applies only to finite groups.

THEOREM 3.18   Let G be a finite cyclic group of order n with $a \in G$ as a generator. For any integer m, the subgroup generated by a^m is the same as the subgroup generated by a^d, where $d = (m, n)$.

Proof   Let $d = (m, n)$, and let $m = dp, n = dq$. Since $a^m = a^{dp} = (a^d)^p$, a^m is in $\langle a^d \rangle$, and therefore $\langle a^m \rangle \subseteq \langle a^d \rangle$.

In order to show that $\langle a^d \rangle \subseteq \langle a^m \rangle$, it is sufficient to show that a^d is in $\langle a^m \rangle$. By Theorem 2.12, there exists integers x and y such that

$d = mx + ny.$

Since a is a generator of G and $o(G) = n$, $a^n = e$. Using this fact, we have

$$
\begin{aligned}
a^d &= a^{mx+ny} \\
&= a^{mx} \cdot a^{ny} \\
&= (a^m)^x \cdot (a^n)^y \\
&= (a^m)^x \cdot (e)^y \\
&= (a^m)^x.
\end{aligned}
$$

Thus a^d is in $\langle a^m \rangle$, and the proof of the theorem is complete.   ∎

As an immediate corollary to Theorem 3.18 we have the following result.

COROLLARY 3.19   Let G be a finite cyclic group of order n with $a \in G$ as a generator. The distinct subgroups of G are those subgroups $\langle a^d \rangle$, where d is a divisor of n.

The corollary provides a systematic way to obtain all the subgroups of a cyclic group of order n.

EXAMPLE 4   Let $G = \langle a \rangle$ be a cyclic group of order 12. The divisors of 12 are 1, 2, 3, 4, 6, and 12, so the distinct subgroups of G are

$$\langle a \rangle = G,$$
$$\langle a^2 \rangle = \{a^2, a^4, a^6, a^8, a^{10}, a^{12} = e\},$$
$$\langle a^3 \rangle = \{a^3, a^6, a^9, a^{12} = e\},$$
$$\langle a^4 \rangle = \{a^4, a^8, a^{12} = e\},$$
$$\langle a^6 \rangle = \{a^6, a^{12} = e\},$$
$$\langle a^{12} \rangle = \langle e \rangle = \{e\}. \qquad \square$$

A method for finding all generators of a finite cyclic group is described in the next theorem.

THEOREM 3.20   Let $G = \langle a \rangle$ be a cyclic group of order n. Then a^m is a generator of G if and only if m and n are relatively prime.

Proof   On the one hand, if m is such that m and n are relatively prime, then $d = (m, n) = 1$, and a^m is a generator of G by Theorem 3.18. On the other hand, if a^m is a generator of G, then $a = (a^m)^p$ for some integer p. By part (b) of Theorem 3.14, this implies that $1 \equiv mp \pmod{n}$. That is,

$$1 - mp = nq$$

for some integer q. This gives

$$1 = mp + nq$$

and it follows from Theorem 2.12 that $(m, n) = 1$.   ■

EXAMPLE 5   Let $G = \langle a \rangle$ be a cyclic group of order 10. The positive integers less than 10 and relatively prime to 10 are 1, 3, 7, and 9. Therefore all generators of G are included in the list

$$a, \quad a^3, \quad a^7, \quad \text{and} \quad a^9. \qquad \square$$

EXERCISES 3.3

1   List all cyclic subgroups of the group $\mathcal{S}(A)$ in Example 3 of Section 3.1. (We shall see later that these are all of the nontrivial subgroups of $\mathcal{S}(A)$.)

2   Let $G = \{\pm 1, \pm i, \pm j, \pm k\}$ be the quaternion group given in Problem 21 of Exercises 3.1. List all cyclic subgroups of G.

3   Find the order of each element of the group $\mathcal{S}(A)$ in Example 3 of Section 3.1.

4   Find the order of each element of the group G in Problem 2.

5   For each of the following values of n, find all generators of the cyclic group Z_n under addition.

(a) $n = 8$ (b) $n = 10$
(c) $n = 12$ (d) $n = 15$

6   For each of the following values of n, find all subgroups of the cyclic group Z_n under addition.

(a) $n = 8$ (b) $n = 10$
(c) $n = 12$ (d) $n = 15$

7   According to Problem 20 of Exercises 3.1, the nonzero elements of Z_n form a group G with respect to multiplication if n is a prime. For each of the following values of n, show that this group G is cyclic.

(a) $n = 5$ (b) $n = 7$
(c) $n = 11$ (d) $n = 13$

8   For each of the following values of n, find all generators of the group G described in Problem 7.

(a) $n = 5$ (b) $n = 7$
(c) $n = 11$ (d) $n = 13$

9   For each of the following values of n, find all subgroups of the group G described in Problem 7.

(a) $n = 5$ (b) $n = 7$
(c) $n = 11$ (d) $n = 13$

10   For an integer $n > 1$, let G be the set of all $[a]$ in Z_n that have multiplicative inverses. Prove that G is a group with respect to multiplication.

11   Let G be as described in Problem 10. Prove that $[a] \in G$ if and only if a and n are relatively prime.

12   Let G be as described in Problem 10. For each value of n, write out the elements of G and construct a multiplication table for G.

(a) $n = 8$ (b) $n = 20$
(c) $n = 24$ (d) $n = 30$

13   Which of the groups in Problem 12 are cyclic?

14   Consider the set G of all $[a]$ in Z_9 that have multiplicative inverses. Given that G is a cyclic group under multiplication, find all subgroups of G.

15   Suppose $G = \langle a \rangle$ is a cyclic group of order 24. List all generators of G.

16   List all distinct subgroups of the group in Problem 15.

17   Describe all subgroups of the group Z under addition.

18   Find all generators of an infinite cyclic group $G = \langle a \rangle$.

19   Assume that G is a cyclic group with a as a generator. Prove that for every $b \in G$, bab^{-1} is also a generator of G.

20   Let a and b be elements of a finite group G.
(a) Prove that a and bab^{-1} have the same order.
(b) Prove that ab and ba have the same order.

21   If G is a cyclic group of order p and p is a prime, how many elements in G are generators of G?

22   If a is an element of order m in a group G and $a^k = e$, prove that m divides k.

23   Suppose a is an element of order m in a group G, and let k be an integer. If $d = (k, m)$, prove that a^k has order m/d.

24   Assume that $G = \langle a \rangle$ is a cyclic group of order n. Prove that if r divides n, then G has a subgroup of order r.

25   Suppose a is an element of order mn in a group G, where m and n are relatively prime. Prove that a is the product of an element of order m and an element of order n.

3.4   PERMUTATION GROUPS AND ISOMORPHISMS

It turns out that the permutation groups can serve as models for all groups. For this reason, we examine permutation groups in great detail later in this chapter. In order to describe their relation to groups in general, we need the concept of an *isomorphism*. Before formally introducing this concept, however, we consider some examples.

EXAMPLE 1   Consider a cyclic group of order 4. If G is a cyclic group of order 4, it must contain an identity element e and a generator $a \neq e$ in G. The proof of Theorem 3.14 shows that

$$G = \{e, a, a^2, a^3\},$$

where $a^4 = e$. A multiplication table for G would have the form shown in Figure 3.11.

$\cdot$	e	a	a^2	a^3
e	e	a	a^2	a^3
a	a	a^2	a^3	e
a^2	a^2	a^3	e	a
a^3	a^3	e	a	a^2

FIGURE 3.11

In a very definite way, then, the structure of G is determined. The details as to what the element a might be and what the operation in G might be may vary, but the basic structure of G fits the pattern in the table.   □

EXAMPLE 2   Let us consider a group related to geometry. We begin with an equilateral triangle T with center at point O and vertices labeled V_1, V_2, and V_3 (see Figure 3.12).

The equilateral triangle, of course, consists of the set of all points on the three sides of the triangle. By a **rigid motion** of the triangle we mean a bijection of

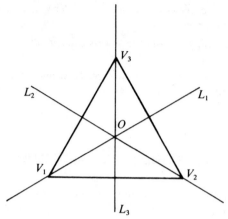

FIGURE 3.12

the set of points of the triangle onto itself that leaves the distance between any two points unchanged. In other words, a rigid motion of the triangle is a bijection that preserves distances. Such a rigid motion must map a vertex onto a vertex, and the entire mapping is determined by the images of the vertices V_1, V_2, and V_3. These rigid motions (or **symmetries** as they are often called) form a group with respect to mapping composition. (Verify this.) There are a total of six elements in the group, and they may be described as follows:

1. e, the identity mapping, that leaves all points unchanged.
2. r, a counterclockwise rotation through $120°$ about O in the plane of the triangle.
3. $r^2 = r \circ r$, a counterclockwise rotation through $240°$ about O in the plane of the triangle.
4. A reflection f about the line L_1 through V_1 and O.
5. A reflection g about the line L_2 through V_2 and O.
6. A reflection h about the line L_3 through V_3 and O.

These rigid motions can be described by indicating their values at the vertices as follows:

$$e: \begin{cases} e(V_1) = V_1 \\ e(V_2) = V_2 \\ e(V_3) = V_3 \end{cases} \qquad h: \begin{cases} h(V_1) = V_2 \\ h(V_2) = V_1 \\ h(V_3) = V_3 \end{cases}$$

$$r: \begin{cases} r(V_1) = V_2 \\ r(V_2) = V_3 \\ r(V_3) = V_1 \end{cases} \qquad g: \begin{cases} g(V_1) = V_3 \\ g(V_2) = V_2 \\ g(V_3) = V_1 \end{cases}$$

$$r^2: \begin{cases} r^2(V_1) = V_3 \\ r^2(V_2) = V_1 \\ r^2(V_3) = V_2 \end{cases} \qquad f: \begin{cases} f(V_1) = V_1 \\ f(V_2) = V_3 \\ f(V_3) = V_2. \end{cases}$$

We have a group

$$G = \{e, r, r^2, h, g, f\},$$

and G has the multiplication table shown in Figure 3.13.

$\circ$	e	r	r^2	h	g	f
e	e	r	r^2	h	g	f
r	r	r^2	e	g	f	h
r^2	r^2	e	r	f	h	g
h	h	f	g	e	r^2	r
g	g	h	f	r	e	r^2
f	f	g	h	r^2	r	e

FIGURE 3.13

We shall compare this group G with the group $\mathscr{S}(A)$ in Example 3 of Section 3.1, and we shall see that they are the same except for notation. Let the elements of G correspond to those of $\mathscr{S}(A)$ according to the mapping $\phi: G \to \mathscr{S}(A)$ given by

$$\phi(e) = I_A \qquad \phi(h) = \sigma$$
$$\phi(r) = \rho \qquad \phi(g) = \gamma$$
$$\phi(r^2) = \rho^2 \qquad \phi(f) = \delta.$$

This mapping is a bijection from G to $\mathscr{S}(A)$. Moreover, ϕ has the property that

$$\phi(xy) = \phi(x) \cdot \phi(y)$$

for all x and y in G. This statement can be verified by using the multiplication tables for G and $\mathscr{S}(A)$ in the following manner: in the entire multiplication table for G, we replace each element $x \in G$ by its image $\phi(x)$ in $\mathscr{S}(A)$. This yields the table in Figure 3.14 that has $\phi(xy)$ in the row with $\phi(x)$ at the left and in the column with $\phi(y)$ at the top.

	I_A	ρ	ρ^2	σ	γ	δ
I_A	I_A	ρ	ρ^2	σ	γ	δ
ρ	ρ	ρ^2	I_A	γ	δ	σ
ρ^2	ρ^2	I_A	ρ	δ	σ	γ
σ	σ	δ	γ	I_A	ρ^2	ρ
γ	γ	σ	δ	ρ	I_A	ρ^2
δ	δ	γ	σ	ρ^2	ρ	I_A

FIGURE 3.14

The multiplication table for $\mathscr{S}(A)$ given in Example 3 of Section 3.1 furnishes a table of values for $\phi(x) \cdot \phi(y)$, and the two tables agree in every position.[†] This means that $\phi(xy) = \phi(x) \cdot \phi(y)$ for all x and y in G. Thus G and $\mathcal{S}(A)$ are the same except for notation. ☐

A mapping such as ϕ in the preceding example is called an *isomorphism*.

DEFINITION 3.21 Let G be a group with respect to ⊛, and let G' be a group with respect to ⊡. A mapping $\phi : G \to G'$ is an **isomorphism** from G to G' if
(1) ϕ is a bijection from G to G', and
(2) $\phi(x \circledast y) = \phi(x) \boxast \phi(y)$ for all x and y in G.
If an isomorphism from G to G' exists, we say that G is **isomorphic** to G'. An isomorphism from a group G to G itself is called an **automorphism** of G.

The use of ⊛ and ⊡ in Definition 3.21 is intended to emphasize the fact that the group operations may be different. Now that this point has been made, we revert to our convention of using the multiplicative notation for the group operation. An isomorphism is said to "preserve the operation" since condition (2) of Definition 3.21 requires that the result be the same if the group operation is performed before or after the mapping.

If e and e' are the respective identity elements of isomorphic groups G and G', then e and e' must correspond under any isomorphism ϕ from G to G'. Also, $\phi(x^{-1}) = [\phi(x)]^{-1}$ must hold for all x in G. The proofs of these statements are requested in the exercises.

The concept of isomorphism introduces the relation of being isomorphic on the set of all groups, and this relation is an equivalence relation as the following statements show:

1.   Any group G is isomorphic to itself. The identity mapping I_G is an automorphism of G.

2.   If G is isomorphic to G', then G' is isomorphic to G. In fact, if ϕ is an isomorphism from G to G', then ϕ^{-1} is an isomorphism from G' to G. (See Problem 2 of Exercises 3.4.)

3.   If G_1 is isomorphic to G_2 and G_2 is isomorphic to G_3, then G_1 is isomorphic to G_3. It is left as an exercise to show that if ϕ_1 is an isomorphism from G_1 to G_2 and ϕ_2 is an isomorphism from G_2 to G_3, then $\phi_2\phi_1$ is an isomorphism from G_1 to G_3.

The fundamental idea behind isomorphisms is this: groups that are isomorphic have the same structure relative to their respective group operation.

[†] Note that the e in Example 3 of Section 3.1 stands for I_A.

They are algebraically the same, although details such as the appearance of the elements or the rule defining the operation may vary.

From our discussion at the beginning of this section, we see that any two cyclic groups of order 4 are isomorphic. In fact, any two cyclic groups of the same order are isomorphic (see Problems 18 and 19).

The next example emphasizes the fact that the elements of two isomorphic groups and their group operations may be quite different from each other.

EXAMPLE 3 Consider $G = \{1, i, -1, -i\}$ under multiplication and $G' = \mathbf{Z}_4 = \{[0], [1], [2], [3]\}$ under addition. Let $\phi : G \to G'$ be defined by

$$\phi(1) = [0], \qquad \phi(i) = [1], \qquad \phi(-1) = [2], \qquad \phi(-i) = [3].$$

This defines a bijection ϕ from G to G'. To see that ϕ is an isomorphism from G to G', we use the group tables for G and G' in the same way as in Example 2 of this section. Beginning with the multiplication table for G, we replace each x in the table by $\phi(x)$ (see Figures 3.15 and 3.16). Since the resulting table (Figure 3.16) agrees completely with the addition table for $\mathbf{Z}_4$, we conclude that

$$\phi(xy) = \phi(x) + \phi(y)$$

for all $x \in G$, $y \in G$, and therefore that ϕ is an isomorphism from G to G'.

$\cdot$	1	i	-1	$-i$
1	1	i	-1	$-i$
i	i	-1	$-i$	1
-1	-1	$-i$	1	i
$-i$	$-i$	1	i	-1

$+$	[0]	[1]	[2]	[3]
[0]	[0]	[1]	[2]	[3]
[1]	[1]	[2]	[3]	[0]
[2]	[2]	[3]	[0]	[1]
[3]	[3]	[0]	[1]	[2]

$\xrightarrow{\ \phi\ }$

Multiplication table for G
FIGURE 3.15

Table of $\phi(xy)$
FIGURE 3.16   □

The fundamental importance of permutation groups is given in the following theorem.

THEOREM 3.22 (**Cayley's Theorem**) Every group is isomorphic to a group of permutations.

Proof Let G be a given group. The permutations that we use in the proof will be mappings defined on the set of all elements in G.

For each element a in G, we define a mapping $f_a : G \to G$ by

$$f_a(x) = ax \quad \text{for all } x \text{ in } G.$$

That is, the image of each x in G is obtained by multiplying x on the left by a. Now f_a is an injective mapping since

$$f_a(x) = f_a(y) \Rightarrow ax = ay$$
$$\Rightarrow x = y.$$

To see that f_a is surjective, let b be arbitrary in G. Then $x = a^{-1}b$ is in G, and

$$f_a(x) = ax$$
$$= a(a^{-1}b) = b.$$

Thus f_a is a permutation on the set of elements of G.

We shall show that the set

$$G' = \{f_a \mid a \in G\}$$

actually forms a group of permutations. Since mapping composition is always associative, we only need to show that G' is closed, has an identity, and contains inverses.

For any f_a and f_b in G', we have

$$f_a f_b(x) = f_a(f_b(x))$$
$$= f_a(bx) = a(bx)$$
$$= (ab)(x) = f_{ab}(x)$$

for all x in G. Thus $f_a f_b = f_{ab}$, and G' is closed. Since

$$f_e(x) = ex = x$$

for all x in G, f_e is the identity permutation. Using the result $f_a f_b = f_{ab}$, we have

$$f_a f_{a^{-1}} = f_{aa^{-1}} = f_e$$

and

$$f_{a^{-1}} f_a = f_{a^{-1}a} = f_e.$$

Thus $(f_a)^{-1} = f_{a^{-1}}$ is in G', and G' is a group of permutations.

All that remains is to show that G is isomorphic to G'. The mapping $\phi : G \rightarrow G'$ defined by

$$\phi(a) = f_a$$

is clearly surjective. It is injective since

$$\phi(a) = \phi(b) \Rightarrow f_a = f_b$$
$$\Rightarrow f_a(x) = f_b(x) \quad \text{for all } x \in G$$
$$\Rightarrow ax = bx \quad\quad\ \text{for all } x \in G$$
$$\Rightarrow a = b.$$

Finally, ϕ is an isomorphism since

$$\phi(a)\phi(b) = f_a f_b$$
$$= f_{ab}$$
$$= \phi(ab)$$

for all a, b in G.   ∎

Notice that the group $G' = \{f_a \mid a \in G\}$ is a subgroup of the group $\mathcal{S}(G)$ of all permutations on G, and $G' \neq \mathcal{S}(G)$ in most cases.

EXAMPLE 4   We shall follow the proof of Cayley's Theorem with the group $G = \{1, i, -1, -i\}$ to obtain a group of permutations G' that is isomorphic to G, and an isomorphism from G to G'.

With $f_a : G \to G$ defined by $f_a(x) = ax$ for each $a \in G$, we obtain the following permutations on the set of elements of G:

$$f_1 : \begin{cases} f_1(1) = 1 \\ f_1(i) = i \\ f_1(-1) = -1 \\ f_1(-i) = -i \end{cases} \qquad f_i : \begin{cases} f_i(1) = i \\ f_i(i) = -1 \\ f_i(-1) = -i \\ f_i(-i) = 1 \end{cases}$$

$$f_{-1} : \begin{cases} f_{-1}(1) = -1 \\ f_{-1}(i) = -i \\ f_{-1}(-1) = 1 \\ f_{-1}(-i) = i \end{cases} \qquad f_{-i} : \begin{cases} f_{-i}(1) = -i \\ f_{-i}(i) = 1 \\ f_{-i}(-1) = i \\ f_{-i}(-i) = -1. \end{cases}$$

According to the proof of Cayley's Theorem, the set

$$G' = \{f_1, f_i, f_{-1}, f_{-i}\}$$

is a group of permutations, and the mapping $\phi : G \to G'$ defined by

$$\phi : \begin{cases} \phi(1) = f_1 \\ \phi(i) = f_i \\ \phi(-1) = f_{-1} \\ \phi(-i) = f_{-i} \end{cases}$$

is an isomorphism from G to G'.   □

EXERCISES 3.4

1   Suppose ϕ is an isomorphism from the group G to the group G'.
(a) If e and e' are the identity elements of G and G', respectively, prove that $\phi(e) = e'$.
(b) Prove that $\phi(x^{-1}) = [\phi(x)]^{-1}$ for all x in G.

2 Prove that if ϕ is an isomorphism from the group G to the group G', then ϕ^{-1} is an isomorphism from G' to G.

3 Let G_1, G_2, and G_3 be groups. Prove that if ϕ_1 is an isomorphism from G_1 to G_2 and ϕ_2 is an isomorphism from G_2 to G_3, then $\phi_2\phi_1$ is an isomorphism from G_1 to G_3.

4 Let $G = \{1, i, -1, -i\}$ under multiplication, and let $G' = \mathbf{Z}_4 = \{[0], [1], [2], [3]\}$ under addition. Find an isomorphism from G to G' that is different from the one given in Example 3 of this section.

5 Find an isomorphism from the additive group[†] $\mathbf{Z}_4 = \{[0]_4, [1]_4, [2]_4, [3]_4\}$ to the multiplicative group $G = \{[1]_5, [2]_5, [3]_5, [4]_5\} \subseteq \mathbf{Z}_5$.

6 Find an isomorphism from the additive group $\mathbf{Z}_6 = \{[a]_6\}$ to the multiplicative group $G = \{[a]_7 \in \mathbf{Z}_7 \mid [a]_7 \neq [0]_7\}$.

7 Consider the group G of four elements $G = \{e, a, b, ab\}$ with the multiplication table in Figure 3.17. This group is known as the **four group**. Write out the elements of a group of permutations that is isomorphic to G, and exhibit an isomorphism from G to this group.

	e	a	b	ab
e	e	a	b	ab
a	a	e	ab	b
b	b	ab	e	a
ab	ab	b	a	e

FIGURE 3.17

8 Let G be the multiplicative group $G = \{[1], [2], [3], [4]\} \subseteq \mathbf{Z}_5$. Write out the elements of a group of permutations that is isomorphic to G, and exhibit an isomorphism from G to this group.

9 Let $G = \{[2], [4], [6], [8]\} \subseteq \mathbf{Z}_{10}$. It is given that G forms a group with respect to multiplication. Write out the elements of a group of permutations that is isomorphic to G, and exhibit an isomorphism from G to this group.

10 Let G be the additive group of all real numbers, and let G' be the group of all positive real numbers under multiplication. Verify that the mapping $\phi: G \to G'$ defined by $\phi(x) = 10^x$ is an isomorphism from G to G'.

11 For each a in the group G, define a mapping $h_a: G \to G$ by $h_a(x) = xa$ for all x in G.
   (a) Prove that each h_a is a permutation on the set of elements in G.
   (b) Prove that $H = \{h_a \mid a \in G\}$ is a group with respect to mapping composition.
   (c) Define $\phi: G \to H$ by $\phi(a) = h_a$. Determine whether or not ϕ is always an isomorphism.

12 For each element a in the group G, define a mapping $k_a: G \to G$ by $k_a(x) = xa^{-1}$ for all x in G.
   (a) Prove that each k_a is a permutation on the set of elements of G.

[†] For clarity, we are temporarily writing $[a]_n$ for $[a] \in \mathbf{Z}_n$.

(b) Prove that $K = \{k_a \mid a \in G\}$ is a group with respect to mapping composition.

(c) Define $\phi : G \to K$ by $\phi(a) = k_a$ for each a in G. Determine whether or not ϕ is always an isomorphism.

13 Let G be an arbitrary group. Prove or disprove that the mapping $\phi(a) = a^{-1}$ is an isomorphism from G to G.

14 For each a in the group G, define a mapping $t_a : G \to G$ by $t_a(x) = axa^{-1}$. Prove that t_a is an isomorphism from G to G.

15 Assume G is a (not necessarily finite) cyclic group generated by a in G, and let ϕ be an automorphism of G. Prove that each element of G is equal to a power of $\phi(a)$, that is, $\phi(a)$ is a generator of G.

16 Let G be as in Problem 15. Suppose also that a^r is a generator of G. Define f on G by $f(a) = a^r$, $f(a^i) = (a^r)^i = a^{ri}$. Prove that f is an automorphism of G.

17 Let G be the group of all nonzero elements of $\mathbf{Z}_7$ under multiplication. Use the results of Problems 15 and 16 to list all the automorphisms of G. For each automorphism ϕ write out the images $\phi(x)$ for all x in G.

18 For an arbitrary positive integer n, prove that any two cyclic groups of order n are isomorphic.

19 Prove that any infinite cyclic group is isomorphic to $\mathbf{Z}$ under addition.

20 Suppose that G and G' are isomorphic groups. Prove that if G is abelian, then G' is abelian.

21 Prove that if G and G' are two groups that contain exactly two elements each, then G and G' are isomorphic.

22 Prove that any two groups of order 3 are isomorphic.

23 If G and G' are groups and $\phi : G \to G'$ is an isomorphism, prove that a and $\phi(a)$ have the same order, for any $a \in G$.

24 Find two groups of order 4 that are not isomorphic.

25 Suppose that ϕ is an isomorphism from the group G to the group G'.
(a) Prove that if H is any subgroup of G, then $\phi(H)$ is a subgroup of G'.
(b) Prove that if K is any subgroup of G', then $\phi^{-1}(K)$ is a subgroup of G.

3.5 FINITE PERMUTATION GROUPS

Suppose A is a finite set of n elements, say

$$A = \{a_1, a_2, \ldots, a_n\}.$$

Any permutation f on A is determined by the n choices for the values of

$$f(a_1), f(a_2), \ldots, f(a_n).$$

In assigning these values, there are n choices for $f(a_1)$, then $n - 1$ choices of $f(a_2)$, then $n - 2$ choices of $f(a_3)$, and so on. Thus there are $n(n - 1)\cdots(2)(1) = n!$ different ways in which f can defined, and $\mathscr{S}(A)$ has $n!$ elements. Each element f in $\mathscr{S}(A)$ can be represented by a matrix (rectangular array) in which the image of a_i is written under a_i:

$$f = \begin{bmatrix} a_1 & a_2 & \cdots & a_n \\ f(a_1) & f(a_2) & \cdots & f(a_n) \end{bmatrix}.$$

Each permutation f on A can be made to correspond to a permutation f' on $B = \{1, 2, \ldots, n\}$ by replacing a_k by k for $k = 1, 2, \ldots, n$:

$$f' = \begin{bmatrix} 1 & 2 & \cdots & n \\ f'(1) & f'(2) & \cdots & f'(n) \end{bmatrix}.$$

The mapping $f \to f'$ is an isomorphism from $\mathscr{S}(A)$ to $\mathscr{S}(B)$, and the groups are the same except for notation. For this reason, we will henceforth consider a permutation on a set of n elements as being written on the set $B = \{1, 2, \ldots, n\}$. The group $\mathscr{S}(B)$ is known as the **symmetric group** on n elements, and is denoted by S_n.

EXAMPLE 1   As an illustration of the matrix representation, the notation

$$f = \begin{bmatrix} 1 & 2 & 3 & 4 & 5 \\ 3 & 5 & 1 & 4 & 2 \end{bmatrix}$$

indicates that f is an element of S_5, and that $f(1) = 3$, $f(2) = 5$, $f(3) = 1$, $f(4) = 4$, and $f(5) = 2$.   □

DEFINITION 3.23   An element f of S_n is a **cycle** if there exists a set $\{i_1, i_2, \ldots, i_r\}$ of distinct integers such that

$$f(i_1) = i_2, \quad f(i_2) = i_3, \quad \ldots, \quad f(i_{r-1}) = i_r, \quad f(i_r) = i_1,$$

and f leaves all other elements fixed.

By this definition, f is a cycle if there are distinct integers $i_1, i_2, \ldots, i_r$ such that f maps these elements according to the cyclic pattern

$$i_1 \to i_2 \to i_3 \to \cdots \to i_{r-1} \to i_r$$

and f leaves all other elements fixed. A cycle such as this can be written in the form

$$f = (i_1, i_2, \ldots, i_r),$$

where it is understood that $f(i_k) = i_{k+1}$ for $1 \le k < r$, and $f(i_r) = i_1$.

EXAMPLE 2   The permutation

$$f = \begin{bmatrix} 1 & 2 & 3 & 4 & 5 & 6 & 7 \\ 1 & 6 & 3 & 7 & 5 & 4 & 2 \end{bmatrix}$$

can be written simply as

$$f = (2, 6, 4, 7).$$

This expression is not unique, for

$$\begin{aligned} f &= (2, 6, 4, 7) \\ &= (6, 4, 7, 2) \\ &= (4, 7, 2, 6) \\ &= (7, 2, 6, 4). \end{aligned}$$

$\square$

EXAMPLE 3   The positive integral powers[†] of a cycle f are easy to compute since f^m will map each integer in the cycle onto the integer located m places farther along in the cycle. For instance, if

$$f = (1, 2, 3, 4, 5, 6, 7, 8, 9),$$

then f^2 maps each element onto the element 2 places farther along, according to the pattern

$$\overgroup{1, 2, 3, 4, 5, 6, 7}, \ldots$$
$$f^2 = (1, 3, 5, 7, 9, 2, 4, 6, 8).$$

Similarly, f^3 maps each element onto the element 3 places farther along, and so on for higher powers:

$$\begin{aligned} f^3 &= (1, 4, 7)(2, 5, 8)(3, 6, 9), \\ f^4 &= (1, 5, 9, 4, 8, 3, 7, 2, 6), \end{aligned}$$

and so on.

$\square$

In connection with Example 3, we note that the order of an r-cycle (a cycle with r elements) is r.

EXAMPLE 4   It is easy to write the inverse of a cycle. Since $f(i_k) = i_{k+1}$ implies $f^{-1}(i_{k+1}) = i_k$, we only need to reverse the order of the cyclic pattern. For

$$f = (1, 2, 3, 4, 5, 6, 7, 8, 9)$$

we have

$$f^{-1} = (1, 9, 8, 7, 6, 5, 4, 3, 2).$$

$\square$

[†] $f^2 = f \circ f, f^3 = f \circ f^2 = f \circ f \circ f$, and so on.

Not all elements of S_n are cycles, but every permutation can be written as a product of mutually disjoint cycles. As an example, consider the permutation

$$f = \begin{bmatrix} 1 & 2 & 3 & 4 & 5 & 6 & 7 & 8 & 9 \\ 3 & 8 & 2 & 6 & 7 & 4 & 9 & 1 & 5 \end{bmatrix}.$$

Using the same representation scheme with $f(k)$ written beneath k, the result of a rearrangement of the columns in the matrix still represents f:

$$f = \begin{bmatrix} 1 & 3 & 2 & 8 & 4 & 6 & 5 & 7 & 9 \\ 3 & 2 & 8 & 1 & 6 & 4 & 7 & 9 & 5 \end{bmatrix}.$$

The columns have been arranged in a special way: if $f(p) = q$, the column with q at the top has been written next after the column with p at the top. This arranges the elements in the first row so that f maps them according to the following pattern:

$$1 \rightarrow 3 \rightarrow 2 \rightarrow 8 \rightarrow 1$$
$$4 \rightarrow 6 \rightarrow 4$$
$$5 \rightarrow 7 \rightarrow 9 \rightarrow 5.$$

Thus 1, 3, 2, and 8 are mapped in a circular pattern, and so are 4 and 6, or 5, 7, and 9. This procedure has led to a separation of the elements of $\{1, 2, 3, 4, 5, 6, 7, 8, 9\}$ into disjoint subsets $\{1, 3, 2, 8\}$, $\{4, 6\}$, and $\{5, 7, 9\}$ according to the pattern determined by the following computations:

$f(1) = 3$	$f(4) = 6$	$f(5) = 7$
$f^2(1) = f(3) = 2$	$f^2(4) = f(6) = 4$	$f^2(5) = f(7) = 9$
$f^3(1) = f(2) = 8$		$f^3(5) = f(9) = 5.$
$f^4(1) = f(8) = 1$		

The disjoint subsets $\{1, 3, 2, 8\}$, $\{4, 6\}$, and $\{5, 7, 9\}$ are called the **orbits** of f.

For each orbit of f, we define a cycle that maps the elements in that orbit in the same way as does f:

$$g_1 = (1, 3, 2, 8)$$
$$g_2 = (4, 6)$$
$$g_3 = (5, 7, 9).$$

These cycles are automatically on disjoint sets of elements since the orbits are disjoint, and their product is f:

$$f = g_1 g_2 g_3$$
$$= (1, 3, 2, 8)(4, 6)(5, 7, 9).$$

Note that these cycles commute with each other since they are on disjoint sets of elements.

Ordinarily, cycles that are not on disjoint sets of elements will not commute, but their product is defined using mapping composition. For example, suppose $f = (1, 3, 2, 4)$ and $g = (1, 7, 6, 2)$. Then[†]

$$fg = (1, 7, 6, 4)(2, 3)$$

since

$$
\begin{array}{ccc}
1 \xrightarrow{\;g\;} 7 \xrightarrow{\;f\;} 7 \\
7 \xrightarrow{\;g\;} 6 \xrightarrow{\;f\;} 6 \\
6 \xrightarrow{\;g\;} 2 \xrightarrow{\;f\;} 4 \\
4 \xrightarrow{\;g\;} 4 \xrightarrow{\;f\;} 1 \\
2 \xrightarrow{\;g\;} 1 \xrightarrow{\;f\;} 3 \\
3 \xrightarrow{\;g\;} 3 \xrightarrow{\;f\;} 2 \\
5 \xrightarrow{\;g\;} 5 \xrightarrow{\;f\;} 5.
\end{array}
$$

Similarly,

$$gf = (1, 3)(2, 4, 7, 6)$$

and $gf \neq fg$. We adopt the notation that a 1-cycle such as (5) indicates that the element is left fixed. For example, gf could also be written as

$$gf = (1, 3)(2, 4, 7, 6)(5).$$

This allows expressions such as $e = (1)$ or $e = (1)(2)$ for the identity permutation.

EXAMPLE 5 The expression of permutations as products of cycles enables us to write the elements of S_n in a very compact form. The elements of S_3 are given by

$$
\begin{array}{ll}
e = (1) & \sigma = (1, 2) \\
\rho = (1, 2, 3) & \gamma = (1, 3) \\
\rho^2 = (1, 3, 2) & \delta = (2, 3).
\end{array}
$$
$\square$

A 2-cycle such as $(3, 7)$ is called a **transposition**. Every permutation can be written as a product of transpositions, for every permutation can be written as a product of cycles, and any cycle $(i_1, i_2, \ldots, i_r)$ can be written as

$$(i_1, i_2, \ldots, i_r) = (i_1, i_r)(i_1, i_{r-1}) \cdots (i_1, i_2).$$

[†] The product fg is computed from right to left, according to $f(g(x))$. Some texts multiply permutations from left to right.

For example,

$$(1,3,2,4) = (1,4)(1,2)(1,3).$$

The factorization into a product of transpositions is not unique, as the next example shows.

EXAMPLE 6   Consider the product fg, where $f = (1,3,2,4)$ and $g = (1,7,6,2)$. This product can be written as

$$(1,3,2,4)(1,7,6,2) = (1,4)(1,2)(1,3)(1,2)(1,6)(1,7)$$

and also as

$$\begin{aligned}(1,3,2,4)(1,7,6,2) &= (1,7,6,4)(2,3)\\ &= (1,4)(1,6)(1,7)(2,3).\end{aligned}$$ □

Although the expression of a permutation as a product of transpositions is not unique, the number of transpositions used for a certain permutation is either *always odd*, or else it is *always even*. Our proof of this fact takes us somewhat astray from our main course in this chapter. It involves consideration of a polynomial P in n variables $x_1, x_2, \ldots, x_n$ that is the product of all factors of the form $(x_i - x_j)$ with $1 \le i < j \le n$:

$$P = \prod_{i<j}^{n} (x_i - x_j).$$

(The symbol $\prod$ indicates a product in the same way that $\sum$ is used to indicate sums in the calculus.) For example, if $n = 3$, then

$$\begin{aligned}P &= \prod_{i<j}^{3} (x_i - x_j)\\ &= (x_1 - x_2)(x_1 - x_3)(x_2 - x_3).\end{aligned}$$

For $n = 4$, P is given by

$$\begin{aligned}P &= \prod_{i<j}^{4} (x_i - x_j)\\ &= (x_1 - x_2)(x_1 - x_3)(x_1 - x_4)(x_2 - x_3)(x_2 - x_4)(x_3 - x_4),\end{aligned}$$

and similarly for larger values of n.

If f is any permutation on $\{1, 2, \ldots, n\}$, then f is applied to P by the rule

$$f(P) = \prod_{i<j}^{n} (x_{f(i)} - x_{f(j)}).$$

As an illustration, let us apply the transposition $t = (2, 4)$ to the polynomial

$$P = \prod_{i<j}^{4} (x_i - x_j)$$

$$= (x_1 - x_2)(x_1 - x_3)(x_1 - x_4)(x_2 - x_3)(x_2 - x_4)(x_3 - x_4).$$

We have

$$t(P) = (x_1 - x_4)(x_1 - x_3)(x_1 - x_2)(x_4 - x_3)(x_4 - x_2)(x_3 - x_2)$$

since 2 and 4 are interchanged by t. Analyzing this result, we see that

1. The factor $(x_2 - x_4)$ in P is changed to $(x_4 - x_2)$ in $t(P)$, so this factor changes sign.
2. The factor $(x_1 - x_3)$ is unchanged.
3. The remaining factors in $t(P)$ may be grouped in pairs as

   $$(x_1 - x_4)(x_1 - x_2) \text{ and } (x_4 - x_3)(x_3 - x_2) = -(x_3 - x_4)(x_3 - x_2).$$

   The products of these pairs are unchanged by t.

Thus $t(P) = (-1)P$ in this particular case. The sort of analysis we have used here can be used to prove the following lemma.

LEMMA 3.24   If $t = (r, s)$ is any transposition on $\{1, 2, \ldots, n\}$ and $P = \prod_{i<j}^{n} (x_i - x_j)$, then

$$t(P) = (-1)P.$$

Proof   Since $t = (r, s) = (s, r)$, we may assume that $r < s$. We have

$$t(P) = \prod_{i<j}^{n} (x_{t(i)} - x_{t(j)}).$$

The factors of $t(P)$ may be analyzed as follows:

1. The factor $(x_r - x_s)$ in P is changed to $(x_s - x_r)$ in $t(P)$, so this factor changes sign.
2. The factors $(x_i - x_j)$ in P with both subscripts different from r and s are unchanged by t.
3. The remaining factors in P have exactly one subscript different from r and s, and may be grouped into pairs as

   $$(x_k - x_r)(x_k - x_s) \quad \text{or} \quad -(x_k - x_r)(x_k - x_s).$$

   The products of these pairs are unchanged by the interchange of r and s.

Thus $t(P) = (-1)P$, and the proof of the lemma is complete.   ∎

THEOREM 3.25 If a certain permutation f is expressed as a product of p transpositions and also as a product of q transpositions, then either p and q are both even, or else p and q are both odd.

Proof Suppose

$$f = t_1 t_2 \cdots t_p \quad \text{and} \quad f = t_1' t_2' \cdots t_q',$$

where each t_i and each t_j' are transpositions. With the first factorization, the result of applying f to

$$P = \prod_{i<j}^n (x_i - x_j)$$

can be obtained by successive application of the transpositions $t_p, t_{p-1}, \ldots, t_2, t_1$. By Lemma 3.24, each t_i changes the sign of P, so

$$f(P) = (-1)^p P.$$

Repeating this same line of reasoning with the second factorization, we obtain

$$f(P) = (-1)^q P.$$

This means that

$$(-1)^p P = (-1)^q P,$$

and consequently

$$(-1)^p = (-1)^q.$$

Therefore, either p and q are both even, or p and q are both odd. ∎

Permutations that can be expressed as a product of an even number of transpositions are called **even permutations**, and those that require an odd number of transpositions are called **odd permutations**. The product fg in Example 6 was written as a product of six transpositions and then as a product of four transpositions, and fg is an even permutation.

The factorization of an r-cycle $(i_1, i_2, \ldots, i_r)$ as

$$(i_1, i_2, \ldots, i_r) = (i_1, i_r)(i_1, i_{r-1}) \cdots (i_1, i_2)$$

uses $r - 1$ transpositions. This shows that *an r-cycle is an even permutation if r is odd and an odd permutation if r is even.* The identity is an even permutation since $e = (1, 2)(1, 2)$. The product of two even permutations is clearly an even permutation. Since any permutation can be written as a product of disjoint cycles, and since the inverse of an r-cycle is an r-cycle, the inverse of an even permutation is an even permutation. These remarks show that the set A_n of all even permutations in S_n is a subgroup of S_n. It is called the *alternating group* on n elements.

DEFINITION 3.26   The **alternating group** A_n is the subgroup of S_n that consists of all even permutations in S_n.

EXAMPLE 7   The elements of the group A_4 are as follows:

$$(1) \qquad (1,2,4) \qquad (1,4,2) \qquad (1,2)(3,4)$$
$$(1,2,3) \qquad (1,4,3) \qquad (2,3,4) \qquad (1,3)(2,4)$$
$$(1,3,2) \qquad (1,3,4) \qquad (2,4,3) \qquad (1,4)(2,3).$$ $\square$

The concept of conjugate elements in a group is basic to the next chapter, which is devoted to certain aspects of the structure of groups. This concept is defined as follows.

DEFINITION 3.27   If a and b are elements of the group G, the **conjugate** of a by b is the element bab^{-1}.

We should point out that this concept is trivial in an abelian group G because $bab^{-1} = bb^{-1}a = ea = a$ for all $b \in G$.

There is a procedure by which conjugates of elements in a permutation group may be computed with ease. To see how this works, suppose that f and g are permutations on $\{1, 2, \ldots, n\}$ that have been written as products of disjoint cycles, and consider gfg^{-1}. If i_1 and i_2 are integers such that $f(i_1) = i_2$, then gfg^{-1} maps $g(i_1)$ to $g(i_2)$, as the following diagram shows:

$$g(i_1) \xrightarrow{\;g^{-1}\;} i_1 \xrightarrow{\;f\;} i_2 \xrightarrow{\;g\;} g(i_2).$$

This means that if

$$(i_1, i_2, \ldots, i_r)$$

is one of the disjoint cycles in f, then

$$\bigl(g(i_1), g(i_2), \ldots, g(i_r)\bigr)$$

is a corresponding cycle in gfg^{-1}. Thus, if

$$f = (i_1, i_2, \ldots, i_r)(j_1, j_2, \ldots, j_s) \cdots (k_1, k_2, \ldots, k_t),$$

then

$$gfg^{-1} = \bigl(g(i_1), g(i_2), \ldots, g(i_r)\bigr)\bigl(g(j_1), \ldots, g(j_s)\bigr) \cdots \bigl(g(k_1), \ldots, g(k_t)\bigr).$$

EXAMPLE 8   If

$$f = (1,3,6,9,5)(2,4,7)$$

and

$$g = (1, 2, 8)(3, 6)(4, 5, 7),$$

then gfg^{-1} may be obtained from f as follows:

$$f = (1, 3, 6, 9, 5)(2, 4, 7)$$
$$\downarrow \downarrow \downarrow \downarrow \downarrow \quad \downarrow \downarrow \downarrow$$
$$gfg^{-1} = (2, 6, 3, 9, 7)(8, 5, 4)$$
$$= (2, 6, 3, 9, 7)(4, 8, 5),$$

where the arrows indicate replacement of i by $g(i)$. This result may be verified by direct computation of g^{-1} and the product gfg^{-1}. □

EXERCISES 3.5

1 Express each permutation as a product of disjoint cycles.

(a) $\begin{bmatrix} 1 & 2 & 3 & 4 & 5 \\ 4 & 5 & 3 & 1 & 2 \end{bmatrix}$

(b) $\begin{bmatrix} 1 & 2 & 3 & 4 & 5 \\ 1 & 3 & 2 & 5 & 4 \end{bmatrix}$

(c) $\begin{bmatrix} 1 & 2 & 3 & 4 & 5 \\ 4 & 1 & 3 & 5 & 2 \end{bmatrix}$

(d) $\begin{bmatrix} 1 & 2 & 3 & 4 & 5 \\ 3 & 5 & 2 & 4 & 1 \end{bmatrix}$

(e) $\begin{bmatrix} 1 & 2 & 3 & 4 & 5 \\ 3 & 2 & 4 & 1 & 5 \end{bmatrix}$

(f) $\begin{bmatrix} 1 & 2 & 3 & 4 & 5 \\ 2 & 5 & 3 & 4 & 1 \end{bmatrix}$

(g) $\begin{bmatrix} 1 & 2 & 3 & 4 & 5 \\ 1 & 3 & 4 & 5 & 2 \end{bmatrix}\begin{bmatrix} 1 & 2 & 3 & 4 & 5 \\ 3 & 2 & 4 & 1 & 5 \end{bmatrix}$

(h) $\begin{bmatrix} 1 & 2 & 3 & 4 & 5 \\ 2 & 3 & 4 & 1 & 5 \end{bmatrix}\begin{bmatrix} 1 & 2 & 3 & 4 & 5 \\ 1 & 3 & 5 & 4 & 2 \end{bmatrix}$

2 Express each permutation as a product of disjoint cycles.
   (a) $(1, 3, 4, 2)(1, 7, 6, 2)(2, 5, 9, 8)$
   (b) $(1, 2, 3)(4, 5)(1, 6, 7, 8, 9)(1, 5)$
   (c) $(2, 5, 9, 8)(1, 7, 6, 2)(1, 2, 6, 4, 5)$
   (d) $(1, 4, 2, 3, 5)(1, 3, 4, 5)(2, 3, 4)$
   (e) $(1, 5, 3, 2, 4)(1, 2, 5, 4)$
   (f) $(1, 3, 5, 2)(1, 2, 8, 7, 6)(2, 3, 4, 9)$
   (g) $(1, 3, 4)(2, 3)(1, 7, 5, 4, 6)(2, 5)$
   (h) $(1, 7, 8, 9, 5)(7, 5, 3)(2, 9, 8)(4, 6)$

3 In each part of Problem 1, decide whether the permutation is even or odd.

4 In each part of Problem 2, decide whether the permutation is even or odd.

5 Find the order of each permutation in Problem 1.

6 Find the order of each permutation in Problem 2.

7 Express each permutation in Problem 1 as a product of transpositions.

8 Express each permutation in Problem 2 as a product of transpositions.

9 Compute gfg^{-1} for each pair f, g.
   (a) $f = (1, 4, 2, 3)$; $g = (1, 3, 2)$
   (b) $f = (1, 2, 4, 6)$; $g = (2, 5, 4, 6)$
   (c) $f = (1, 5, 3, 4)$; $g = (1, 3, 2)(4, 5)$
   (d) $f = (1, 3)(2, 4)$; $g = (1, 2, 3)$
   (e) $f = (1, 3, 5)(2, 4)$; $g = (2, 5)(3, 4)$
   (f) $f = (1, 3, 5, 2)(4, 6)$; $g = (1, 3, 6)(2, 4, 5)$

10 For the given permutations f and h, find a permutation g such that $gfg^{-1} = h$.
   (a) $f = (1, 5, 3)$; $h = (2, 6, 4)$
   (b) $f = (1, 2, 5, 7)$; $h = (3, 4, 6, 8)$
   (c) $f = (1, 3, 4)(2, 5)$; $h = (2, 4, 3)(1, 5)$
   (d) $f = (1, 2, 3)(4, 5)$; $h = (2, 3, 4)(1, 6)$
   (e) $f = (1, 3, 6)(2, 4, 5)$; $h = (1, 5, 4)(2, 3, 6)$
   (f) $f = (1, 3, 5)(2, 4, 6)$; $h = (1, 2, 4)(3, 5, 6)$

11 Write the permutation $f = (1, 2, 3, 4, 5, 6)$ as a product of a permutation g of order 2 and a permutation h of order 3. (See Problem 25 of Exercises 3.3.)

12 List all the elements of the alternating group A_3, written in cyclic notation.

13 List all the elements of S_4, written in cyclic notation.

14 Find all the distinct cyclic subgroups of A_4.

15 Find cyclic subgroups of S_4 that have three different orders.

16 The following eight permutations form a group G known as the **octic group**. Construct a multiplication table for the group $G = \{e, \alpha, \alpha^2, \alpha^3, \beta, \gamma, \Delta, \theta\}$.

   $e = (1)$ $\alpha = (1, 2, 3, 4)$ $\alpha^2 = (1, 3)(2, 4)$

   $\alpha^3 = (1, 4, 3, 2)$ $\beta = (1, 4)(2, 3)$ $\gamma = (1, 2)(3, 4)$

   $\Delta = (1, 3)$ $\theta = (2, 4)$

17 Find all the distinct cyclic subgroups of the octic group in Problem 16.

18 Prove that in any group the relation "x is a conjugate of y" is an equivalence relation.

19 For any element a of a group G, the **normalizer** of a in G is the set

$$\mathcal{N}(a) = \{x \in G \mid xax^{-1} = a\} = \{x \in G \mid xa = ax\}.$$

Prove that $\mathcal{N}(a)$ is a subgroup of G.

20 Use the multiplication table constructed in Problem 16 to find the normalizer $\mathcal{N}(a)$ for each element a of the octic group.

Key Words and Phrases

Group	Subgroup
Identity element	Nontrivial subgroup
Inverse	Integral exponents
Abelian group	Integral multiples
Finite group	Subgroup generated by a
Infinite group	Cyclic group
Order of a group	Generator of a group
Generalized associative law	Order of an element

Isomorphism

Automorphism

Cayley's Theorem

Symmetric group

Cycle

Orbit

Transposition

Even permutation

Odd permutation

Alternating group

Conjugate of *a* by *b*

References

Ames, Dennis B. *An Introduction to Abstract Algebra*. Scranton, Pa.: International Textbook, 1969.

Ball, Richard W. *Principles of Abstract Algebra*. New York: Holt, Rinehart and Winston, 1963.

Birkhoff, Garrett and Saunders MacLane. *A Survey of Modern Algebra* (4th ed.). New York: Macmillan, 1977.

Bundrick, Charles M. and John J. Leeson. *Essentials of Abstract Algebra*. Monterey, Calif.: Brooks-Cole, 1972.

Durbin, John R. *Modern Algebra*. New York: Wiley, 1979.

Fraleigh, John B. *A First Course in Abstract Algebra* (2nd ed.). Reading, Mass.: Addison-Wesley, 1976.

Hall, Marshall, Jr. *The Theory of Groups* (2nd ed.). New York: Chelsea, 1961.

Herstein, I. N. *Topics in Algebra* (2nd ed.). New York: Wiley, 1975.

Hillman, Abraham P. and Gerald L. Alexanderson. *A First Undergraduate Course in Abstract Algebra* (2nd ed.). Belmont, Calif.: Wadsworth, 1978.

Jones, Burton W. *An Introduction to Modern Algebra*. New York: Macmillan, 1975.

Larsen, Max D. *Introduction to Modern Algebraic Concepts*. Reading, Mass.: Addison-Wesley, 1969.

McCoy, Neal H. *Introduction to Modern Algebra* (3rd ed.). Boston: Allyn and Bacon, 1975.

Maxfield, John E. and Margaret W. Maxfield. *Abstract Algebra and Solution by Radicals*. New York: Dover, 1983.

Mitchell, A. Richard and Roger W. Mitchell. *An Introduction to Abstract Algebra*. Monterey, Calif.: Brooks-Cole, 1970.

Schilling, Otto F. G. and W. Stephen Piper. *Basic Abstract Algebra*. Boston: Allyn and Bacon, 1975.

Shapiro, Louis. *Introduction to Abstract Algebra*. New York: McGraw-Hill, 1975.

4 MORE ON GROUPS

4.1 NORMAL SUBGROUPS

The binary operation in a given group can be used in a natural way to define a product between subsets of the group. This product of subsets is a very useful concept.

DEFINITION 4.1 Let A and B be nonempty subsets of the group G. The **product** AB is defined by

$$AB = \{x \in G \mid x = ab \text{ for some } a \in A, b \in B\}.$$

This product is formed by using the group operation in G. A more precise formulation would be

$$A * B = \{x \in G \mid x = a * b \text{ for some } a \in A, b \in B\}$$

where $*$ is the group operation in G.

Several properties of this product are worth mentioning. For nonempty subsets A, B, and C of G,

$$\begin{aligned}
A(BC) &= \{a(bc) \mid a \in A, b \in B, c \in C\} \\
&= \{(ab)c \mid a \in A, b \in B, c \in C\} \\
&= (AB)C.
\end{aligned}$$

It is obvious that

$$B = C \Rightarrow AB = AC \quad \text{and} \quad BA = CA,$$

but we must be careful about the order, because AB and BA may be different sets.

EXAMPLE 1 Consider the subsets $A = \{(1,2,3),(1,2)\}$ and $B = \{(1,3),(2,3)\}$ in $G = S_3$. We have

$$\begin{aligned} AB &= \{(1,2,3)(1,3),\, (1,2)(1,3),\, (1,2,3)(2,3),\, (1,2)(2,3)\} \\ &= \{(2,3),\, (1,3,2),\, (1,2),\, (1,2,3)\} \end{aligned}$$

and

$$\begin{aligned} BA &= \{(1,3)(1,2,3),\, (2,3)(1,2,3),\, (1,3)(1,2),\, (2,3)(1,2)\} \\ &= \{(1,2),\, (1,3),\, (1,2,3),\, (1,3,2)\}, \end{aligned}$$

so $AB \neq BA$. $\square$

For a nonabelian group G, we would probably expect AB and BA to be different. A fact that is not quite so natural is that

$$AB = AC \nRightarrow B = C.$$

EXAMPLE 2 An example where $AB = AC$ but $B \neq C$ is provided by $A = \{(1,2,3),(1,3,2)\}$, $B = \{(1,3),(2,3)\}$, and $C = \{(1,2),(1,3)\}$. Straightforward calculations show that

$$AB = \{(2,3),(1,2),(1,3)\} = AC,$$

but $B \neq C$. $\square$

If B consists of a single element of a group G, say $B = \{g\}$, then AB is written simply as Ag instead of as $A\{g\}$:

$$Ag = \{x \in G \mid x = ag \text{ for some } a \in A\}.$$

Similarly,

$$gA = \{x \in G \mid x = ga \text{ for some } a \in A\}.$$

This is one instance in which a cancellation law does hold:

$$Ag = Bg \Rightarrow A = B.$$

This is true because

$$Ag = Bg \Rightarrow (Ag)g^{-1} = (Bg)g^{-1}$$
$$\Rightarrow A(gg^{-1}) = B(gg^{-1})$$
$$\Rightarrow Ae = Be$$
$$\Rightarrow A = B.$$

For convenience of reference, we summarize these results in a theorem.

THEOREM 4.2 Let A, B, and C denote nonempty subsets of the group G, and let g denote an element of G. Then the following statements hold:

(a) $A(BC) = (AB)C$.
(b) $B = C$ implies $AB = AC$ and $BA = CA$.
(c)   The product AB is not commutative.
(d) $AB = AC$ does not imply $B = C$.
(e) $Ag = Bg$ implies $A = B$.

Statements (d) and (e) have obvious duals in which the common factor is on the other side.

We shall be concerned mainly with products of subsets in which one of the factors is a subgroup. The cosets of a subgroup are of special importance.

DEFINITION 4.3 Let H be a subgroup of the group G. For any a in G,

$$aH = \{x \in G \,|\, x = ah \text{ for some } h \in H\}$$

is a **left coset** of H in G. Similarly, Ha is called a **right coset** of H in G.

The left coset aH and the right coset Ha are never disjoint since $a = ae = ea$ is in both sets. In spite of this, aH and Ha may happen to be different sets, as the next example shows.

EXAMPLE 3 Consider the subgroup

$$K = \{(1), (1, 2)\}$$

of

$$G = S_3 = \{(1), (1, 2, 3), (1, 3, 2), (1, 2), (1, 3), (2, 3)\}.$$

For $a = (1, 2, 3)$ we have

$$aK = \{(1, 2, 3), (1, 2, 3)(1, 2)\}$$
$$= \{(1, 2, 3), (1, 3)\}$$

and

$$Ka = \{(1, 2, 3), (1, 2)(1, 2, 3)\}$$
$$= \{(1, 2, 3), (2, 3)\}.$$

In this case, $aK \neq Ka$. $\qquad \square$

Although a left coset of H and a right coset of H may be neither equal nor disjoint, this cannot happen with two left cosets of H. This fact is fundamental to the proof of Lagrange's Theorem (Theorem 4.6), so we designate it as a lemma.

LEMMA 4.4 Let H be a subgroup of the group G. The distinct left cosets of H in G form a partition of G; that is, they separate the elements of G into mutually disjoint subsets.

Proof It is sufficient to show that any two left cosets of H that are not disjoint must be the same left coset.

Suppose that aH and bH have at least one element in common, say $z \in aH \cap bH$. Then $z = ah_1$ for some $h_1 \in H$ and $z = bh_2$ for some $h_2 \in H$. This means that $ah_1 = bh_2$ and $a = bh_2 h_1^{-1}$. We have that $h_2 h_1^{-1}$ is in H since H is a subgroup, so $a = bh_3$ where $h_3 = h_2 h_1^{-1} \in H$. Now, for every $h \in H$,

$$ah = bh_3 h$$
$$= bh_4$$

where $h_4 = h_3 \cdot h$ is in H. That is, $ah \in bH$ for all $h \in H$. This proves that $aH \subseteq bH$. A similar argument shows that $bH \subseteq aH$ and thus $aH = bH$. $\qquad \blacksquare$

EXAMPLE 4 Consider again the subgroup

$$K = \{(1), (1, 2)\}$$

of

$$G = S_3 = \{(1), (1, 2, 3), (1, 3, 2), (1, 2), (1, 3), (2, 3)\}.$$

In Example 3 of this section we saw that

$$(1, 2, 3)K = \{(1, 2, 3), (1, 3)\}.$$

Since $(1, 3)$ is in this left coset, it follows from Lemma 4.4 that

$$(1, 3)K = (1, 2, 3)K = \{(1, 2, 3), (1, 3)\}.$$

Straightforward computations show that

$$(1)K = (1, 2)K = \{(1), (1, 2)\}.$$

and

$$(2, 3)K = (1, 3, 2)K = \{(1, 3, 2), (2, 3)\}.$$

Thus the distinct left cosets of K in G are given by

$$K, \quad (1, 2, 3)K, \quad (1, 3, 2)K. \qquad \qquad \square$$

DEFINITION 4.5 Let H be a subgroup of G. The number of distinct left cosets of H in G is called the **index** of H in G.

In the proof of the next theorem we show that if $o(G)$ is finite, then

$o(G) = $ (order of H) · (index of H in G).

THEOREM 4.6 (**Lagrange's Theorem**) The order of a subgroup of a finite group must divide the order of the group.

Proof Let G be a finite group of order n and let H be a subgroup of G with order k. We shall show that k divides n.

From Lemma 4.4 we know that the left cosets of H in G separate the elements of G into mutually disjoint subsets. Let m be the index of H in G; that is, there are m distinct left cosets of H in G. We shall show that each left coset has exactly k elements.

Let aH represent an arbitrary left coset of H. The mapping $\phi : H \to aH$ defined by

$\phi(h) = ah$

is clearly injective. It is also surjective, since any x in aH can be written $x = ah$ for $h \in H$. Thus ϕ is a bijection from H to aH, and this means that aH has the same number of elements as does H.

We have the n elements of G separated into m disjoint subsets, and each subset has k elements. Therefore $n = mk$, and k divides n. ∎

Lagrange's Theorem is of great value if we are interested in finding all the subgroups of a finite group. In connection with this task, it is worthwhile to record this immediate corollary.

COROLLARY 4.7   The order of an element of a finite group must divide the order of the group.

EXAMPLE 5   To illustrate the usefulness of the foregoing results, we shall exhibit all of the subgroups of S_3. Any subgroup of S_3 must be of order 1, 2, 3, or 6 since $o(S_3) = 6$. An element in a subgroup of order 3 must have order dividing 3, and therefore any subgroup of order 3 is cyclic. Similarly, any subgroup of order 2 is cyclic. The following list is thus a complete list of the subgroups of S_3:

$$H_1 = \{(1)\} \qquad H_4 = \{(1), (2, 3)\}$$
$$H_2 = \{(1), (1, 2)\} \qquad H_5 = \{(1), (1, 2, 3), (1, 3, 2)\}$$
$$H_3 = \{(1), (1, 3)\} \qquad H_6 = S_3.$$

$\square$

It is easy to see that if p is a prime, any group G of order p must by cyclic (any a in G such that $a \neq e$ must be a generator). This means that, up to an isomorphism, there is only one group of order p, if p is a prime. In particular, the only groups of order 2, 3 or 5 are the cyclic groups.

By examination of the possible orders of the elements and the possible multiplication tables, it can be shown that a group of order 4 is either cyclic, or is isomorphic to the four group

$$G = \{e, a, b, ab = ba\}$$

of Problem 7 in Exercises 3.4.

Among the subgroups of a group are those known as the *normal subgroups*. The significance of the normal subgroups is revealed in the next section.

DEFINITION 4.8   Let H be a subgroup of G. Then H is a **normal** (or **invariant**) subgroup of G if $xH = Hx$ for all $x \in G$.

Note that the condition $xH = Hx$ is an equality of sets, and it does not require that $xh = hx$ for all h in H.

EXAMPLE 6   Let

$$H = A_3 = \{(1), (1, 2, 3), (1, 3, 2)\}$$

and

$$G = S_3 = \{(1), (1, 2, 3), (1, 3, 2), (1, 2), (1, 3), (2, 3)\}.$$

For $x = (1, 2)$ we have

$$xH = \{(1,2)(1), (1,2)(1,2,3), (1,2)(1,3,2)\}$$
$$= \{(1,2), (2,3), (1,3)\}$$

and

$$Hx = \{(1)(1,2), (1,2,3)(1,2), (1,3,2)(1,2)\}$$
$$= \{(1,2), (1,3), (2,3)\}.$$

We have $xH = Hx$, but $xh \neq hx$ when $h = (1,2,3) \in H$. Similar computations show that

$$(1)H = (1,2,3)H = (1,3,2)H = \{(1), (1,2,3), (1,3,2)\}$$
$$H(1) = H(1,2,3) = H(1,3,2) = \{(1), (1,2,3), (1,3,2)\}$$
$$(1,2)H = (1,3)H = (2,3)H = \{(1,2), (1,3), (2,3)\}$$
$$H(1,2) = H(1,3) = H(2,3) = \{(1,2), (1,3), (2,3)\}.$$

Thus H is a normal subgroup of G. In this example, we have $hH = H = Hh$ for all $h \in H$. The proof that this is always the case is left as an exercise. $\qquad\square$

EXAMPLE 7 As an example of a subgroup that is *not* normal, consider $K = \{(1), (1,2)\}$ in S_3. With $x = (1,2,3)$, we have

$$xK = \{(1,2,3), (1,2,3)(1,2)\}$$
$$= \{(1,2,3), (1,3)\}$$
$$Kx = \{(1,2,3), (1,2)(1,2,3)\}$$
$$= \{(1,2,3), (2,3)\}.$$

Thus $xK \neq Kx$, and K is not a normal subgroup of S_3. $\qquad\square$

The definition of a normal subgroup can be formulated in several different ways. For instance, we can write

$$xH = Hx \quad \text{for all } x \in G \Leftrightarrow xHx^{-1} = H \quad \text{for all } x \in G$$
$$\Leftrightarrow x^{-1}Hx = H \quad \text{for all } x \in G.$$

Other formulations can be found in the exercises at the end of this section. One that is frequently taken as the definition is given in Theorem 4.9.

THEOREM 4.9 Let H be a subgroup of G. Then H is a normal subgroup of G if and only if $xhx^{-1} \in H$ for every $h \in H$ and every $x \in G$.

Proof   If H is a normal subgroup of G, then the condition follows easily since H normal requires

$$xHx^{-1} = H \quad \text{for all } x \in G \Rightarrow xHx^{-1} \subseteq H \quad \text{for all } x \in G$$
$$\Rightarrow xhx^{-1} \in H \quad \text{for all } h \in H \text{ and all } x \in G.$$

Suppose now that the condition holds. For any $x \in G$, $xHx^{-1} \subseteq H$ follows immediately, and we only need to show that $H \subseteq xHx^{-1}$. Let h be arbitrary in H, and let $x \in G$. Now x^{-1} is an element in G, and the condition implies that

$$(x^{-1})(h)(x^{-1})^{-1} = x^{-1}hx$$

is in H; that is,

$$x^{-1}hx = h_1 \quad \text{for some } h_1 \in H \Rightarrow h = xh_1x^{-1} \quad \text{for some } h_1 \in H$$
$$\Rightarrow h \in xHx^{-1}.$$

Thus $H \subseteq xHx^{-1}$, and we have $xHx^{-1} = H$ for all $x \in G$. It follows that H is a normal subgroup of G.   ∎

The concept of generators can be extended from cyclic subgroups $\langle a \rangle$ to more complicated situations where a subgroup is generated by more than one element. We only touch on this topic here, but it is a fundamental idea in more advanced study of groups.

DEFINITION 4.10   If A is a nonempty subset of the group G, then $\langle A \rangle$ is the set defined by

$$\langle A \rangle = \{x \in G \mid x = a_1 a_2 \cdots a_n \text{ with either } a_i \in A \text{ or } a_i^{-1} \in A\}.$$

In other words, $\langle A \rangle$ is the set of all products that can be formed with a finite number of factors, each of which is either an element of A or has an inverse that is an element of A.

THEOREM 4.11   For any nonempty subset A of a group G, the set $\langle A \rangle$ is a subgroup of G called the **subgroup of G generated by A**.

Proof   There exists at least one $a \in A$ since $A \neq \emptyset$. Then $e = aa^{-1} \in \langle A \rangle$, so $\langle A \rangle$ is nonempty.

If $x \in \langle A \rangle$ and $y \in \langle A \rangle$, then

$$x = x_1 x_2 \cdots x_n \quad \text{with either } x_i \in A \text{ or } x_i^{-1} \in A$$

and

$$y = y_1 y_2 \cdots y_k \quad \text{with either } y_j \in A \text{ or } y_j^{-1} \in A.$$

Thus

$$xy = x_1 x_2 \cdots x_n y_1 y_2 \cdots y_k,$$

where each factor on the right is either in A or has an inverse that is an element of A. Also,

$$x^{-1} = x_n^{-1} \cdots x_2^{-1} x_1^{-1} \quad \text{with either } x_i^{-1} \in A \text{ or } x_i \in A.$$

The set $\langle A \rangle$ is closed and contains inverses, and therefore it is a subgroup of G. ∎

In work with *finite groups*, the result in Problem 24 of Exercises 3.2 is extremely helpful in finding $\langle A \rangle$ since it implies that $\langle A \rangle$ is the smallest subset of G that contains A and is closed under the operation. (This is true *only for finite groups*.) The subgroup $\langle A \rangle$ can be constructed systematically by starting a multiplication table using the elements of A and enlarging the table by adjoining additional elements until closure is obtained. A practical first step in this direction is to begin the table using all the elements of A and all their distinct powers. This is illustrated in the next example.

EXAMPLE 8   Consider the problem of finding $\langle A \rangle$ in S_4 for the set $A = \{(1,2,3,4), (1,4)(2,3)\}$. We begin by computing the distinct powers of the elements of A:

$$\alpha = (1, 2, 3, 4) \qquad\qquad \alpha^2 = (1, 3)(2, 4)$$
$$\alpha^3 = \alpha^{-1} = (1, 4, 3, 2) \qquad \alpha^4 = e = (1)$$
$$\beta = (1, 4)(2, 3) \qquad\qquad \beta^2 = e.$$

Starting a multiplication table using $e, \alpha, \alpha^2, \alpha^3, \beta$, we find the following new elements of $\langle A \rangle$:

$$\alpha\beta = (1, 2, 3, 4)(1, 4)(2, 3) = (2, 4) = \gamma$$
$$\alpha^2\beta = (1, 3)(2, 4)(1, 4)(2, 3) = (1, 2)(3, 4) = \Delta$$
$$\alpha^3\beta = (1, 4, 3, 2)(1, 4)(2, 3) = (1, 3) = \theta.$$

We then enlarge the table so as to use all eight elements

$$e, \alpha, \alpha^2, \alpha^3, \beta, \gamma, \Delta, \theta.$$

Proceeding to fill out the enlarged table, we obtain the table in Figure 4.1, which shows that the set

$$G = \{e, \alpha, \alpha^2, \alpha^3, \beta, \gamma, \Delta, \theta\}$$

is the subgroup of S_4 generated by $A = \{\alpha, \beta\}$. This group G is frequently useful to provide examples involving subgroups. It is known as the **octic group**.

117

	e	α	α^2	α^3	β	γ	Δ	θ
e	e	α	α^2	α^3	β	γ	Δ	θ
α	α	α^2	α^3	e	γ	Δ	θ	β
α^2	α^2	α^3	e	α	Δ	θ	β	γ
α^3	α^3	e	α	α^2	θ	β	γ	Δ
β	β	θ	Δ	γ	e	α^3	α^2	α
γ	γ	β	θ	Δ	α	e	α^3	α^2
Δ	Δ	γ	β	θ	α^2	α	e	α^3
θ	θ	Δ	γ	β	α^3	α^2	α	e

FIGURE 4.1   □

EXERCISES 4.1

1   Let H be a subgroup of the group G. Prove that if two right cosets Ha and Hb are not disjoint, then $Ha = Hb$.

2   If H is any subgroup of the group G, prove that $hH = H = Hh$ for all $h \in H$.

3   If H is a subgroup of the group G, prove that gHg^{-1} is a subgroup of G, for any $g \in G$. We say that gHg^{-1} is a **conjugate** of H and that H and gHg^{-1} are **conjugate subgroups**.

4   Show that every subgroup of an abelian group is normal.

5   For an arbitrary subgroup H of the group G, define the mapping θ from the set of left cosets of H in G to the set of right cosets of H in G by $\theta(aH) = Ha^{-1}$. Prove that θ is a bijection.

6   Consider the octic group G of Example 8.
    (a) Find a subgroup of G that has order 2 and is a normal subgroup of G.
    (b) Find a subgroup of G that has order 2 and is *not* a normal subgroup of G.

7   Find all subgroups of the octic group in Example 8.

8   Find all subgroups of the alternating group A_4.

9   Find all subgroups of the quaternion group. (See Problem 21 of Exercises 3.1.)

10   Find all normal subgroups of the octic group in Example 8.

11   Find all normal subgroups of the alternating group A_4.

12   Find all normal subgroups of the quaternion group. (See Problem 21 of Exercises 3.1.)

13   Find groups H and G such that $H \subseteq A_4 \subseteq G$ and the following conditions are satisfied:
    (a) H is a normal subgroup of A_4;
    (b) A_4 is a normal subgroup of G;
    (c) H is not a normal subgroup of G.

14   Find two groups of order 6 that are not isomorphic.

15   Let H be a subgroup of G, and assume that every left coset aH of H in G is equal to a right coset Hb of H in G. Prove that H is a normal subgroup of G.

16   If $\{H_\lambda\}, \lambda \in \mathscr{L}$, is a collection of normal subgroups H_λ of G, prove that $\bigcap_{\lambda \in \mathscr{L}} H_\lambda$ is a normal subgroup of G.

17 If H is a subgroup of G and K is a normal subgroup of G, prove that $HK = KH$.

18 With H and K as in Problem 17, prove that HK is a subgroup of G.

19 With H and K as in Problem 17, prove that $H \cap K$ is a normal subgroup of H.

20 With H and K as in Problem 17, prove that K is a normal subgroup of HK.

21 If H and K are arbitrary subgroups of G, prove that $HK = KH$ if and only if HK is a subgroup of G.

22 If H and K are both normal subgroups of G, prove that HK is a normal subgroup of G.

23 Prove that if H and K are normal subgroups of G such that $H \cap K = \{e\}$, then $hk = kh$ for all $h \in H, k \in K$.

24 The **center** $\mathscr{C}(G)$ of the group G is defined by

$$\mathscr{C}(G) = \{a \in G \mid ax = xa \text{ for all } x \in G\}.$$

Prove that $\mathscr{C}(G)$ is a normal subgroup of G.

25 Find the center of the octic group. (See Example 8.)

26 Find the center of A_4.

27 For an arbitrary subgroup H of the group G, the **normalizer** of H in G is the set $\mathscr{N}(H) = \{x \in G \mid xHx^{-1} = H\}$.
(a) Prove that $\mathscr{N}(H)$ is a subgroup of G.
(b) Prove that H is a normal subgroup of $\mathscr{N}(H)$.
(c) Prove that if K is a subgroup of G that contains H as a normal subgroup, then $K \subseteq \mathscr{N}(H)$.

28 Find the normalizer of the subgroup $\{(1), (1, 3)(2, 4)\}$ of the octic group.

29 Show that the octic group in Example 8 is isomorphic to the group of rigid motions of a square. [Hint: Label the vertices of the square as $1, 2, 3, 4$.]

30 Let H be a subgroup of G. Define the relation "congruence modulo H" on G by

$a \equiv b \pmod{H}$ if and only if $a^{-1}b \in H$.

Prove that congruence modulo H is an equivalence relation on G.

31 Describe the equivalence classes in Problem 30.

32 Let $n > 1$ in the group of integers under addition, and let $H = \langle n \rangle$. Prove that

$a \equiv b \pmod{H}$ if and only if $a \equiv b \pmod{n}$.

33 Prove that any subgroup H of G that has index 2 in G is a normal subgroup of G.

34 Show that A_n has index 2 in S_n, and thereby conclude that A_n is always a normal subgroup of S_n.

35 Find the subgroup of S_4 that is generated by the given set.
(a) $\{(2, 3), (1, 3, 4)\}$
(b) $\{(1, 2, 4), (2, 3, 4)\}$
(c) $\{(1, 2), (1, 3), (1, 4)\}$

36 Let n be a positive integer, $n > 1$. Prove by induction that the set of transpositions $\{(1, 2), (1, 3), \ldots, (1, n)\}$ generates the entire group S_n.

4.2   QUOTIENT GROUPS AND HOMOMORPHISMS

If H is a normal subgroup of G, then $xH = Hx$ for all x in G, so there is no distinction between left and right cosets of H in G. In this case, we refer simply to the cosets of H in G.

If H is any subgroup of G, then $hH = H = Hh$ for all h in H. It follows easily from this that $H^2 = H \cdot H = H$ for all subgroups H. We use this fact in proving the next theorem.

THEOREM 4.12   Let H be a normal subgroup of G. Then the cosets of H in G form a group with respect to the product of subsets as given in Definition 4.1.

Proof   Let H be a normal subgroup of G. We shall denote the set of all distinct cosets of H in G by G/H. Multiplication in G/H is associative, by part (a) of Theorem 4.2.

We need to show that the cosets of H in G are closed under the given product. Let aH and bH be arbitrary cosets of H in G. Using the associative property freely, we have

$$(aH)(bH) = a(Hb)H$$
$$= a(bH)H \quad \text{since } H \text{ is normal}$$
$$= (ab)H \cdot H$$
$$= abH \qquad \text{since } H^2 = H.$$

Thus G/H is closed, and $(aH)(bH) = abH$.

The coset $H = eH$ is an identity element since $(aH)(eH) = aeH = aH$ and $(eH)(aH) = eaH = aH$ for all aH in G/H.

The inverse of aH is $a^{-1}H$ since

$$(aH)(a^{-1}H) = aa^{-1}H = eH = H$$

and

$$(a^{-1}H)(aH) = a^{-1}aH = eH = H.$$

This completes the proof.   ∎

DEFINITION 4.13   If H is a normal subgroup of G, the group G/H that consists of the cosets of H in G is called the **quotient group** or **factor group** of G by H.

EXAMPLE 1   Let G be the octic group as given in Example 8 of Section 4.1:

$$G = \{e, \alpha, \alpha^2, \alpha^3, \beta, \gamma, \Delta, \theta\}.$$

It may be readily verified that $H = \{e, \gamma, \theta, \alpha^2\}$ is a normal subgroup of G. The distinct cosets of H in G are

$$H = eH = \gamma H = \theta H = \alpha^2 H = \{e, \gamma, \theta, \alpha^2\}$$

and

$$\alpha H = \alpha^3 H = \beta H = \Delta H = \{\alpha, \alpha^3, \beta, \Delta\}.$$

Thus $G/H = \{H, \alpha H\}$, and a multiplication table for G/H is given by

	H	αH
H	H	αH
αH	αH	H

$\square$

There is a very important and natural relation between the quotient groups of a group G and the homomorphisms from G to another group G'. A homomorphism is defined as follows.

DEFINITION 4.14 Let G be a group with respect to $\circledast$, and let G' be a group with respect to $\boxast$. A **homomorphism** from G to G' is a mapping $\phi : G \rightarrow G'$ such that

$$\phi(x \circledast y) = \phi(x) \boxast \phi(y)$$

for all x and y in G. If ϕ is a homomorphism from G to G' that is surjective, ϕ is called an **epimorphism**.

As we did with isomorphisms, we drop the special symbols $\circledast$ and $\boxast$ and simply write $\phi(xy) = \phi(x)\phi(y)$ for the given condition.

A homomorphism ϕ from G to G' need not be injective or surjective. If ϕ is both, that is, if ϕ is a bijection, then ϕ is an isomorphism as defined in Definition 3.21.

As with isomorphisms, we say that a homomorphism "preserves the group operation." Two simple consequences of this condition are that identities must correspond and inverses must be mapped onto inverses.

THEOREM 4.15 Let ϕ be a homomorphism from the group G to the group G'. If e denotes the identity in G and e' denotes the identity in G', then

(a) $\phi(e) = e'$, and
(b) $\phi(x^{-1}) = [\phi(x)]^{-1}$ for all x in G.

Proof   We have

$$e \cdot e = e \Rightarrow \phi(e \cdot e) = \phi(e)$$
$$\Rightarrow \phi(e) \cdot \phi(e) = \phi(e) \cdot e' \quad \text{since } \phi \text{ is a homomorphism}$$
$$\Rightarrow \phi(e) = e' \qquad\qquad \text{by the cancellation law.}$$

For any x in G,

$$x \cdot x^{-1} = e \Rightarrow \phi(x \cdot x^{-1}) = \phi(e) = e' \quad \text{by part (a)}$$
$$\Rightarrow \phi(x) \cdot \phi(x^{-1}) = e' \qquad \text{since } \phi \text{ is a homomorphism.}$$

Similarly, $x^{-1} \cdot x = e$ implies $\phi(x^{-1}) \cdot \phi(x) = e'$, and therefore $\phi(x^{-1}) = [\phi(x)]^{-1}$.   ∎

EXAMPLE 2   Consider the permutation group

$$G = S_3 = \{(1), (1,2,3), (1,3,2), (1,2), (1,3), (2,3)\}$$

and the multiplicative group

$$G' = \{[1], [2]\} \subseteq \mathbf{Z}_3.$$

The mapping $\phi : G \to G'$ defined by

$$\phi(1) = \phi(1,2,3) = \phi(1,3,2) = [1]$$
$$\phi(1,2) = \phi(1,3) = \phi(2,3) = [2]$$

can be shown by direct computation to be an epimorphism from G to G', but it is tedious to verify $\phi(xy) = \phi(x)\phi(y)$ for all 36 choices of the pair of factors x, y in S_3. As an alternative to this chore, we shall obtain another description of ϕ. We first note that if $\alpha = (1,2,3)$ and $\beta = (1,2)$, the elements of S_3 can be written as

$$(1) = \alpha^0 \beta^0 \qquad (1,2,3) = \alpha\beta^0 \qquad (1,3,2) = \alpha^2\beta^0$$
$$(1,2) = \alpha^0\beta \qquad\quad (1,3) = \alpha\beta \qquad\quad (2,3) = \alpha^2\beta.$$

We then make the following observations concerning S_3:

1.   Any element of S_3 can be written in the form $\alpha^i \beta^k$, with $i \in \{0,1,2\}$ and $k \in \{0,1\}$.
2. $\beta\alpha^k = \alpha^{-k}\beta$.
3.   Any $x \in S_3$ is either of the form $x = \alpha^i$ or of the form $x = \alpha^j\beta$.

Routine calculations will confirm that our mapping ϕ can be described by the rule

$$\phi(\alpha^i\beta^j) = [2]^j.$$

Having made these observations, we can now verify the equation $\phi(x)\phi(y) = \phi(xy)$ with a reasonable amount of work. For arbitrary x and y in S_3, we write either $x = \alpha^i$ or $x = \alpha^j\beta$, and $y = \alpha^k\beta^m$.

If $x = \alpha^i$, we have

$$\phi(xy) = \phi(\alpha^i \alpha^k \beta^m) = \phi(\alpha^{i+k} \beta^m) = [2]^m$$

and

$$\phi(x)\phi(y) = \phi(\alpha^i)\phi(\alpha^k \beta^m) = [2]^0 [2]^m = [2]^m.$$

If $x = \alpha^j \beta$, we have

$$
\begin{aligned}
\phi(xy) &= \phi(\alpha^j \beta \alpha^k \beta^m) \\
&= \phi(\alpha^j \alpha^{-k} \beta \beta^m) \\
&= \phi(\alpha^{j-k} \beta^{m+1}) \\
&= [2]^{m+1}
\end{aligned}
$$

and

$$
\begin{aligned}
\phi(x)\phi(y) &= \phi(\alpha^j \beta)\phi(\alpha^k \beta^m) \\
&= [2][2]^m \\
&= [2]^{m+1}.
\end{aligned}
$$

Thus $\phi(xy) = \phi(x)\phi(y)$ in all cases, and ϕ is a homomorphism (an epimorphism, actually) from G to G'. $\qquad\square$

If there exists an epimorphism from the group G to the group G', then G' is called a **homomorphic image** of G. Example 2 shows that $G' = \{[1], [2]\} \subseteq \mathbf{Z}_3$ is a homomorphic image of S_3.

Our next theorem shows that every quotient group G/H is a homomorphic image of G.

THEOREM 4.16   Let G be a group and let H be a normal subgroup of G. The mapping $\phi: G \to G/H$ defined by

$$\phi(a) = aH$$

is an epimorphism from G to G/H.

Proof   The rule $\phi(a) = aH$ clearly defines a mapping from G to G/H. For any a and b in G,

$$
\begin{aligned}
\phi(a) \cdot \phi(b) &= (aH)(bH) \\
&= abH \qquad \text{since } H \text{ is normal in } G \\
&= \phi(ab).
\end{aligned}
$$

Thus ϕ is a homomorphism. Every element of G/H is a coset of H in G that has the form aH for some a in G. For any such a, we have $\phi(a) = aH$. Therefore ϕ is an epimorphism. $\qquad\blacksquare$

EXAMPLE 3 Consider the octic group

$$G = \{e, \alpha, \alpha^2, \alpha^3, \beta, \gamma, \Delta, \theta\}$$

and its normal subgroup

$$H = \{e, \gamma, \theta, \alpha^2\}.$$

We saw in Example 1 that $G/H = \{H, \alpha H\}$. Theorem 4.16 assures us that the mapping $\phi: G \to G/H$ defined by

$$\phi(a) = aH$$

is an epimorphism. The values of ϕ are given in this case by

$$\phi(e) = \phi(\gamma) = \phi(\theta) = \phi(\alpha^2) = H$$
$$\phi(\alpha) = \phi(\alpha^3) = \phi(\beta) = \phi(\Delta) = \alpha H.$$

□

Theorem 4.16 says that every quotient group G/H is a homomorphic image of G. We shall see that, up to an isomorphism, these quotient groups give all of the homomorphic images of G. In order to prove this, we need the concept of the *kernel* of a homomorphism.

DEFINITION 4.17 Let ϕ be a homomorphism from the group G to the group G'. The **kernel** of ϕ is the set

$$\ker \phi = \{x \in G \mid \phi(x) = e'\}$$

where e' denotes the identity in G'.

THEOREM 4.18 For any homomorphism ϕ from the group G to the group G', $\ker \phi$ is a normal subgroup of G.

Proof The identity e is in $\ker \phi$ since $\phi(e) = e'$, so $\ker \phi$ is always nonempty. If $a \in \ker \phi$ and $b \in \ker \phi$, then $\phi(a) = e'$ and $\phi(b) = e'$ so that

$$\phi(ab) = \phi(a) \cdot \phi(b) = e' \cdot e' = e'$$

and therefore $ab \in \ker \phi$. By Theorem 4.15, $\phi(x) = e'$ implies $\phi(x^{-1}) = (e')^{-1} = e'$, so $\ker \phi$ contains inverses, and we have proved that $\ker \phi$ is a subgroup of G.

To show that $\ker \phi$ is normal, let $x \in G$ and $a \in \ker \phi$. Then

$$
\begin{aligned}
\phi(xax^{-1}) &= \phi(x)\phi(a)\phi(x^{-1}) && \text{since } \phi \text{ is a homomorphism} \\
&= \phi(x) \cdot e' \cdot \phi(x^{-1}) && \text{since } a \in \ker \phi \\
&= \phi(x) \cdot \phi(x^{-1}) && \\
&= e' && \text{by part (b) of Theorem 4.15.}
\end{aligned}
$$

Thus xax^{-1} is in ker ϕ, and ker ϕ is a normal subgroup by Theorem 4.9.   ∎

The mapping ϕ in Theorem 4.16 has H as its kernel, and thus shows that every normal subgroup of G is the kernel of a homomorphism. Combining this fact with Theorem 4.18, we see that the normal subgroups of G and the kernels of the homomorphisms from G to another group are the same subgroups of G.

We can now prove that every homomorphic image of G is isomorphic to a quotient group of G.

THEOREM 4.19   Let G and G' be groups with G' a homomorphic image of G. Then G' is isomorphic to a quotient group of G.

Proof   Let ϕ be an epimorphism from G to G', and let $K = $ ker ϕ. For each aK in G/K, define $\theta(aK)$ by

$$\theta(aK) = \phi(a).$$

First we need to prove that this rule defines a mapping. For any aK and bK in G/K,

$$
\begin{aligned}
aK = bK &\Leftrightarrow b^{-1}aK = K \\
&\Leftrightarrow b^{-1}a \in K \\
&\Leftrightarrow \phi(b^{-1}a) = e' \\
&\Leftrightarrow \phi(b^{-1})\phi(a) = e' \\
&\Leftrightarrow [\phi(b)]^{-1}\phi(a) = e' \\
&\Leftrightarrow \phi(a) = \phi(b) \\
&\Leftrightarrow \theta(aK) = \theta(bK).
\end{aligned}
$$

Thus θ is a well-defined mapping from G/K to G' and the $\Leftarrow$ part of the $\Leftrightarrow$ above shows that θ is injective as well.

We shall show that θ is an isomorphism from G/K to G'. Since

$$
\begin{aligned}
\theta(aK \cdot bK) &= \theta(abK) \\
&= \phi(ab) \\
&= \phi(a) \cdot \phi(b) \\
&= \theta(aK) \cdot \theta(bK),
\end{aligned}
$$

θ is a homomorphism. To show that θ is surjective, let a' be arbitrary in G'. Since ϕ is an epimorphism, there exists an element a in G such that $\phi(a) = a'$. Then aK is in G/K and

$$\theta(aK) = \phi(a) = a'.$$

Thus every element in G' is an image under θ, and this proves that θ is an isomorphism.   ∎

COROLLARY 4.20 If ϕ is an epimorphism from the group G to the group G', then G' is isomorphic to $G/\ker \phi$.

The corollary follows at once from the proof of the theorem. The statement in the corollary is often referred to as the **Fundamental Theorem of Homomorphisms**.

EXAMPLE 4 To illustrate Theorems 4.18 and 4.19, consider the groups from Example 2,

$$G = S_3 = \{(1), (1,2,3), (1,3,2), (1,2), (1,3), (2,3)\}$$

and

$$G' = \{[1], [2]\} \subseteq \mathbf{Z}_3.$$

The epimorphism $\phi: G \to G'$ is defined by

$$\phi(1) = \phi(1,2,3) = \phi(1,3,2) = [1]$$
$$\phi(1,2) = \phi(1,3) = \phi(2,3) = [2],$$

so the kernel of ϕ is the normal subgroup

$$K = \ker \phi$$
$$= \{(1), (1,2,3), (1,3,2)\}$$

of G. The quotient group G/K is given by

$$G/K = \{K, (1,2)K\},$$

where

$$(1,2)K = \{(1,2), (2,3), (1,3)\}.$$

The isomorphism $\theta: G/K \to G'$ has values

$$\theta(K) = \phi(1) = [1]$$
$$\theta((1,2)K) = \phi(1,2) = [2].$$ □

Using the results of this section, we can systematically find all of the homomorphic images of a group G. We now know that the homomorphic images of G are the same (in the sense of isomorphism) as the quotient groups of G.

EXAMPLE 5 Let $G = S_3$, the symmetric group on three elements. In order to find all the homomorphic images of G, we only need to find all of the normal subgroups H of G and form all possible quotient groups G/H. As we saw in Section 4.1, a complete list of the subgroups of G is provided by

$$H_1 = \{(1)\} \qquad H_4 = \{(1), (2, 3)\}$$
$$H_2 = \{(1), (1, 2)\} \qquad H_5 = \{(1), (1, 2, 3), (1, 3, 2)\}$$
$$H_3 = \{(1), (1, 3)\} \qquad H_6 = S_3.$$

Of these, H_1, H_5, and H_6 are the only normal subgroups. The possible homomorphic images of G, then, are

$$G/H_1 = \{H_1, (1, 2)H_1, (1, 3)H_1, (2, 3)H_1, (1, 2, 3)H_1, (1, 3, 2)H_1\}$$
$$G/H_5 = \{H_5, (1, 2)H_5\}$$
$$G/G = \{G\}.$$

Thus any homomorphic image of S_3 is either isomorphic to S_3, or to a cyclic group of order 2, or to a group with only the identity element.   □

EXERCISES 4.2

1   Let G be the octic group in Example 8 of Section 4.1, and let $H = \{e, \alpha^2\}$. Write out the distinct elements of G/H and make a multiplication table for G/H.

2   Let $G = A_4$ and assume that the four group

$$H = \{(1), (1, 2)(3, 4), (1, 3)(2, 4), (1, 4)(2, 3)\}$$

is a normal subgroup of A_4. Write out the distinct elements of G/H and make a multiplication table for G/H.

3   Find all homomorphic images of the octic group. (See Problem 10 of Exercises 4.1.)

4   Find all homomorphic images of A_4. (See Problem 11 of Exercises 4.1.)

5   Find all homomorphic images of the quaternion group. (See Problem 12 of Exercises 4.1.)

6   Find all homomorphic images of each group G in Problem 12 of Exercises 3.3.

7   Let $G = S_3$. For each H given below, show that the set of all left cosets of H in G does *not* form a group with respect to a product defined by $(aH)(bH) = abH$.
(a) $H = \{(1), (1, 2)\}$ (b) $H = \{(1), (1, 3)\}$ (c) $H = \{(1), (2, 3)\}$

8   Each of the following rules determines a mapping $\phi : G \to G$, where G is the group of all nonzero real numbers under multiplication. Decide in each case whether or not ϕ is a homomorphism, and state the kernel for those that are homomorphisms.
(a) $\phi(x) = |x|$ (b) $\phi(x) = 1/x$
(c) $\phi(x) = -x$ (d) $\phi(x) = x^2$

9   Assume that the four group

$$H = \{(1), (1, 2)(3, 4), (1, 3)(2, 4), (1, 4)(2, 3)$$

is a normal subgroup of S_4. Write out the distinct cosets of H in S_4 and construct a multiplication table for S_4/H.

10 Find an example of G, G', and ϕ such that G is a nonabelian group, G' is an abelian group, and ϕ is an epimorphism from G to G'.

11 If H is a subgroup of the group G such that $(aH)(bH) = abH$ for all left cosets aH and bH of H in G, prove that H is normal in G.

12 If H is a normal subgroup of the group G, prove that $(aH)^n = a^nH$ for every positive integer n.

13 Let G be a cyclic group. Prove that for every subgroup H of G, G/H is a cyclic group.

14 (a) Show that a cyclic group of order 8 has a cyclic group of order 4 as a homomorphic image.
(b) Show that a cyclic group of order 6 has a cyclic group of order 2 as a homomorphic image.

15 Suppose that ϕ is an epimorphism from the group G to the group G'. Prove that ϕ is an isomorphism if and only if $\ker \phi = \{e\}$, where e denotes the identity in G.

16 Suppose that G, G', and G'' are groups. If G' is a homomorphic image of G and G'' is a homomorphic image of G', prove that G'' is a homomorphic image of G.

17 If G is an abelian group and the group G' is a homomorphic image of G, prove that G' is abelian.

18 Assume that ϕ is an epimorphism from the group G to the group G'.
(a) Prove that if H is any subgroup of G, then $\phi(H)$ is a subgroup of G'.
(b) Prove that if K is any subgroup of G', then $\phi^{-1}(K)$ is a subgroup of G.
(c) Prove that the mapping $H \to \phi(H)$ is a bijection from the set of all subgroups of G that contain $\ker \phi$ to the set of all subgroups of G'.

19 Suppose ϕ is an epimorphism from the group G to the group G'. Let H be a normal subgroup of G containing $\ker \phi$ and let $H' = \phi(H)$.
(a) Prove that H' is a normal subgroup of G'.
(b) Prove that G/H is isomorphic to G'/H'.

20 Assume that the group G' is a homomorphic image of the group G.
(a) Prove that G' is cyclic if G is cyclic.
(b) Prove that $o(G')$ divides $o(G)$, whether G is cyclic or not.

21 (See Problem 24 of Exercises 4.1.) Let G be a group with center $\mathscr{C}(G) = C$. Prove that if G/C is cyclic, then G is abelian.

22 (See Problems 22 and 23 of Exercises 4.1.) If H and K are normal subgroups of the group G such that $G = HK$ and $H \cap K = \{e\}$, then G is said to be the **internal direct product** of H and K, and we write $G = H \times K$ to denote this. If $G = H \times K$, prove that $\phi : H \to G/K$ defined by $\phi(h) = hK$ is an isomorphism from H to G/K.

23 (See Problem 22.) If $G = H \times K$, prove that each element $g \in G$ can be written uniquely as $g = hk$ with $h \in H$ and $k \in K$.

24 (See Problems 17–20 of Exercises 4.1.) Let H be a subgroup of G and let K be a normal subgroup of G.
(a) Prove that the mapping $\phi : H \to HK/K$ defined by $\phi(h) = hK$ is an epimorphism from H to HK/K.
(b) Prove that $\ker \phi = H \cap K$.
(c) Prove that $H/H \cap K$ is isomorphic to HK/K.

25   Let H and K be arbitrary groups and let $H \otimes K$ denote the Cartesian product of H and K:

$$H \otimes K = \{(h, k) \mid h \in H \text{ and } k \in K\}.$$

Equality in $H \otimes K$ is defined by $(h, k) = (h', k')$ if and only if $h = h'$ and $k = k'$. Multiplication in $H \otimes K$ is defined by

$$(h_1, k_1)(h_2, k_2) = (h_1 h_2, k_1 k_2).$$

(a) Prove that $H \otimes K$ is a group. This group is called the **external direct product** of H and K.
(b) If e_1 and e_2 are the identity elements of H and K, respectively, show that $H' = \{(h, e_2) \mid h \in H\}$ is a normal subgroup of $H \otimes K$ that is isomorphic to H, and similarly that $K' = \{(e_1, k) \mid k \in K\}$ is a normal subgroup isomorphic to K.
(c) Prove that $H \otimes K / H'$ is isomorphic to K and $H \otimes K / K'$ is isomorphic to H.

4.3   DIRECT SUMS (OPTIONAL)

The over-all objective of this and the next section is to present some of the basic material on abelian groups. A tremendous amount of work has been done on the subject of abelian groups, and it is still an active area of research. One of the concepts fundamental to this subject is a *direct sum*, to be defined presently. Throughout this section, we write all abelian groups in additive notation.

We begin by defining the sum of a finite number of subgroups in an abelian group and showing that this sum is a subgroup.

DEFINITION 4.21   If $H_1, H_2, \ldots, H_n$ are subgroups of the abelian group G, then the **sum** $H_1 + H_2 + \cdots + H_n$ of these subgroups is defined by

$$H_1 + H_2 + \cdots + H_n = \{x \in G \mid x = h_1 + h_2 + \cdots + h_n \text{ with } h_i \in H_i\}.$$

THEOREM 4.22   If $H_1, H_2, \ldots, H_n$ are subgroups of the abelian group G, then $H_1 + H_2 + \cdots + H_n$ is a subgroup of G.

Proof   The sum $H_1 + H_2 + \cdots + H_n$ is clearly nonempty. For arbitrary

$$x = h_1 + h_2 + \cdots + h_n$$

with $h_i \in H_i$, the inverse

$$-x = (-h_1) + (-h_2) + \cdots + (-h_n)$$

is in the sum $H_1 + H_2 + \cdots + H_n$ since $-h_i \in H_i$ for each i. Also, if

$$y = h_1' + h_2' + \cdots + h_n'$$

with $h_i' \in H_i$, then

$$x + y = (h_1 + h_1') + (h_2 + h_2') + \cdots + (h_n + h_n')$$

is in the sum of the H_i since $h_i + h_i' \in H_i$ for each i. Thus $H_1 + H_2 + \cdots + H_n$ is a subgroup of G. ∎

The contents of Definition 4.10 and Theorem 4.11 may be restated as follows, with addition as the binary operation:

If A is a nonempty subset of the group G, then
the *subgroup of G generated by A* is the set

$$\langle A \rangle = \{ x \in G \,|\, x = a_1 + a_2 + \cdots + a_n \text{ with } a_i \in A \text{ or } -a_i \in A \}.$$

It is left as an exercise to prove that if $H_1, H_2, \ldots, H_n$ are subgroups of an abelian group G, then $G = H_1 + H_2 + \cdots + H_n$ if and only if G is generated by $\bigcup_{i=1}^{n} H_i$.

EXAMPLE 1   Let G be the group $G = \mathbf{Z}_{12}$ under addition, and consider the following sums of subgroups in G.

(a)   If

$$H_1 = \langle [3] \rangle = \{ [3], [6], [9], [0] \}$$

and

$$H_2 = \langle [2] \rangle = \{ [2], [4], [6], [8], [10], [0] \},$$

then

$$\begin{aligned} H_1 + H_2 &= \{ r[3] + s[2] \,|\, r, s \in \mathbf{Z} \} \\ &= \{ [3r + 2s] \,|\, r, s \in \mathbf{Z} \} \end{aligned}$$

is a subgroup. Since $[3(1) + 2(11)] = [25] = [1]$ in $\mathbf{Z}_{12}$ and $[1]$ generates $\mathbf{Z}_{12}$ under addition, we have

$$H_1 + H_2 = G.$$

(b)   Now let

$$K_1 = H_1 = \langle [3] \rangle,$$
$$K_2 = \langle [4] \rangle = \{ [4], [8], [0] \}.$$

The sum $K_1 + K_2$ is given by

$$\begin{aligned} K_1 + K_2 &= \{ u[3] + v[4] \,|\, u, v \in \mathbf{Z} \} \\ &= \{ [3u + 4v] \,|\, u, v \in \mathbf{Z} \}. \end{aligned}$$

Since $[3(-1) + 4(1)] = [1], [1] \in K_1 + K_2$ and hence

$K_1 + K_2 = G$.

(c)   With the same notation as in parts (a) and (b),

$H_2 + K_2 = H_2$

since $K_2 \subseteq H_2$. $\square$

We now consider the definition of a direct sum.

DEFINITION 4.23   If $H_1, H_2, \ldots, H_n$ are subgroups of the abelian group G, then $H_1 + H_2 + \cdots + H_n$ is a **direct sum** if and only if the expression for each x in the sum as

$x = h_1 + h_2 + \cdots + h_n$

with $h_i \in H_i$ is **unique**. We write

$H_1 \oplus H_2 \oplus \cdots \oplus H_n$

to indicate a direct sum.

The next theorem gives a simple fact about direct sums that can be very useful when we work with finite groups.

THEOREM 4.24   If $H_1, H_2, \ldots, H_n$ are finite subgroups of the abelian group G such that their sum is direct, then the order of $H_1 \oplus H_2 \oplus \cdots \oplus H_n$ is the product of the orders of the subgroups H_i:

$$o(H_1 \oplus H_2 \oplus \cdots \oplus H_n) = o(H_1)o(H_2)\cdots o(H_n).$$

Proof   With $h_i \in H_i$ in the expression

$x = h_1 + h_2 + \cdots + h_n,$

there are $o(H_i)$ choices for each h_i. Any change in one of the h_i produces a different element x, by the uniqueness property stated in Definition 4.23. Hence there are

$o(H_1)o(H_2)\cdots o(H_n)$

distinct elements x of the form $x = h_1 + h_2 + \cdots + h_n$, and the theorem follows.   ■

There are several equivalent ways to formulate the definition of direct sum. One of these is presented in the following theorem.

THEOREM 4.25 If each H_i is a subgroup of the abelian group G, the sum $H_1 + H_2 + \cdots + H_n$ is direct if and only if the following condition holds: Any equation of the form

$$h_1 + h_2 + \cdots + h_n = 0$$

with $h_i \in H_i$ implies that all $h_i = 0$.

Proof Assume first that the condition holds. If an element x in the sum of the H_i is written as

$$x = h_1 + h_2 + \cdots + h_n$$

and also as

$$x = h'_1 + h'_2 + \cdots + h'_n$$

with h_i and $h'_i \in H_i$ for each i, then

$$h_1 + h_2 + \cdots + h_n = h'_1 + h'_2 + \cdots + h'_n$$

and

$$(h_1 - h'_1) + (h_2 - h'_2) + \cdots + (h_n - h'_n) = 0.$$

The condition implies that $h_i - h'_i = 0$ and hence $h_i = h'_i$ for each i. Thus the sum $H_1 + H_2 + \cdots + H_n$ is direct.

        Conversely, suppose the sum $H_1 + H_2 + \cdots + H_n$ is direct. Then the identity element 0 in the sum can be written *uniquely* as

$$0 = 0 + 0 + \cdots + 0,$$

where the sum on the right indicates a choice of 0 as the term from each H_i. From the uniqueness property,

$$h_1 + h_2 + \cdots + h_n = 0$$

with $h_i \in H_i$ requires that all $h_i = 0$. ∎

Some intuitive feeling for the concept of a direct sum is provided by considering the special case where the sum has only two terms.

THEOREM 4.26 Let H_1 and H_2 be subgroups of the abelian group G. Then $G = H_1 \oplus H_2$ if and only if $G = H_1 + H_2$ and $H_1 \cap H_2 = \{0\}$.

Proof Assume first that $G = H_1 \oplus H_2$, and let $x \in H_1 \cap H_2$. Then $x = h_1$ for some $h_1 \in H_1$. Also, $x \in H_2$ and therefore $-x \in H_2$. Let

$h_2 = -x$. Then

$$h_1 + h_2 = x + (-x)$$
$$= 0$$

where $h_i \in H_i$, and this implies that $x = h_1 = h_2 = 0$, by Theorem 4.25.
Assume now that $G = H_1 + H_2$ and $H_1 \cap H_2 = \{0\}$. *If*

$$h_1 + h_2 = 0$$

with $h_i \in H_i$, then $h_1 = -h_2 \in H_1 \cap H_2$. Therefore $h_1 = 0$ and $h_2 = 0$.
By Theorem 4.25, $G = H_1 \oplus H_2$. ∎

EXAMPLE 2   In Example 1 we saw that the equations $H_1 + H_2 = G$
and $K_1 + K_2 = G$ were both valid. Since $H_1 \cap H_2 = \{[0], [6]\}$, the sum $H_1 + H_2$
is not direct. However, $K_1 \cap K_2 = \{[0]\}$, so $G = K_1 \oplus K_2$ in Example 1.   □

Theorem 4.26 can be generalized to the result stated in the next theorem.
A proof is requested in the exercises.

THEOREM 4.27   Let $H_1, H_2, \ldots, H_n$ be subgroups of the abelian
group G. The sum $H_1 + H_2 + \cdots + H_n$ is direct if and only if the
intersection of each H_j with the subgroup generated by $\bigcup_{i=1, i \neq j}^{n} H_i$ is
the identity subgroup $\{0\}$.

As a final result for this section, we prove the following theorem.

THEOREM 4.28   Let H_1 and H_2 be subgroups of the abelian group G
such that $G = H_1 \oplus H_2$. Then G/H_2 is isomorphic to H_1.

Proof   The rule $\phi(h_1) = h_1 + H_2$ defines a mapping ϕ from H_1 to
G/H_2.
Now

$$h_1 \in \ker \phi \Leftrightarrow \phi(h_1) = H_2$$
$$\Leftrightarrow h_1 + H_2 = H_2$$
$$\Leftrightarrow h_1 \in H_2$$
$$\Leftrightarrow h_1 = 0 \qquad \text{since } H_1 \cap H_2 = \{0\}.$$

Thus ϕ is injective. Let $g + H_2$ be arbitrary in G/H_2. Since
$G = H_1 \oplus H_2$, g can be written as $g = h_1 + h_2$ with $h_i \in H_i$.

Then

$$g + H_2 = (h_1 + h_2) + H_2$$
$$= h_1 + H_2 \qquad \text{since } h_2 + H_2 = H_2$$
$$= \phi(h_1),$$

and this shows that ϕ is surjective. Finally, ϕ is an isomorphism since

$$\phi(h_1 + h_1') = (h_1 + h_1')H_2$$
$$= (h_1 + H_2) + (h_1' + H_2)$$
$$= \phi(h_1) + \phi(h_1').$$ ∎

EXERCISES 4.3

1   If H_1 and H_2 are subgroups of the abelian group G such that $H_1 \subseteq H_2$, prove that $H_1 + H_2 = H_2$.

2   Suppose that H_1 and H_2 are subgroups of the abelian group G such that $G = H_1 \oplus H_2$. If K is a subgroup of G such that $K \supseteq H_1$, prove that $K = H_1 \oplus (K \cap H_2)$.

3   Assume that $H_1, H_2, \ldots, H_n$ are subgroups of the abelian group G such that the sum $H_1 + H_2 + \cdots + H_n$ is direct. If K_i is a subgroup of H_i for $i = 1, 2, \ldots, n$, prove that $K_1 + K_2 + \cdots + K_n$ is a direct sum.

4   Prove that if each H_i is a subgroup of the abelian group G, then $H_1 + H_2 + \cdots + H_n$ is the smallest subgroup of G that contains all the subgroups H_i.

5   If $H_1, H_2, \ldots, H_n$ are subgroups of the abelian group G, prove that $G = H_1 + H_2 + \cdots + H_n$ if and only if G is generated by $\bigcup_{i=1}^{n} H_i$.

6   Write $\mathbf{Z}_{20}$ as the direct sum of two of its nontrivial subgroups.

7   Let G be an abelian group of order mn, where m and n are relatively prime. If $H_1 = \{x \in G \mid mx = 0\}$ and $H_2 = \{x \in G \mid nx = 0\}$, prove that $G = H_1 \oplus H_2$.

8   Let H_1 and H_2 be cyclic subgroups of the abelian group G, where $H_1 \cap H_2 = \{0\}$. Prove that $H_1 \oplus H_2$ is cyclic if and only if $o(H_1)$ and $o(H_2)$ are relatively prime.

9   Show that $\mathbf{Z}_{15}$ is isomorphic to $\mathbf{Z}_3 \oplus \mathbf{Z}_5$, where the group operation in each of $\mathbf{Z}_{15}, \mathbf{Z}_3,$ and $\mathbf{Z}_5$ is addition.

10   Suppose that G and G' are abelian groups such that $G = H_1 \oplus H_2$ and $G' = H_1' \oplus H_2'$. If H_1 is isomorphic to H_1' and H_2 is isomorphic to H_2', prove that G is isomorphic to G'.

11   Suppose a is an element of order rs in an abelian group G. Prove that if r and s are relatively prime, then a can be written in the form $a = b_1 + b_2$ where b_1 has order r and b_2 has order s.

12   (See Problem 11.) Assume that a is an element of order $r_1 r_2 \cdots r_n$ in an abelian group where r_i and r_j are relatively prime if $i \neq j$. Prove that a can be written in the form

$$a = b_1 + b_2 + \cdots + b_n$$

where each b_i has order r_i.

13 Prove that if r and s are relatively prime positive integers, then any cyclic group of order rs is the direct sum of a cyclic group of order r and a cyclic group of order s.

14 Prove Theorem 4.27: If $H_1, H_2, \ldots, H_n$ are subgroups of the abelian group G, then the sum $H_1 + H_2 + \cdots + H_n$ is direct if and only if the intersection of each H_j with the subgroup generated by $\bigcup_{i=1, i \neq j}^{n} H_i$ is the identity subgroup $\{0\}$.

4.4 SOME RESULTS ON FINITE ABELIAN GROUPS (OPTIONAL)

The aim of this section is to sample the flavor of more advanced work in groups and, at the same time, to maintain an acceptable level of rigor in the presentation. We attempt to achieve this balance by restricting our attention to proofs of results for abelian groups. There are places where more general results hold, but their proofs are beyond the level of this text. In most instances of this sort, the more general results are stated informally and without proof.

The following definition of a p-group is fundamental to this entire section.

> **DEFINITION 4.29** If p is a prime, then a group G is called a **p-group** if each of its elements has an order that is a power of p.

A p-group can be finite or infinite. Although we do not prove it here, a finite group is a p-group if and only if its order is a power of p. Whether or not a group is abelian has nothing at all to do with being a p-group. This is brought out in the following example.

EXAMPLE 1 With $p = 2$ we can easily exhibit three p-groups of order 8.

(a) Consider first the cyclic group $C_8 = \langle a \rangle$ of order 8 generated by the permutation $a = (1, 2, 3, 4, 5, 6, 7, 8)$.

Each of a, a^3, a^5, and a^7 has order 8.
a^2 and a^6 have order 4.
a^4 has order 2.
The identity e has order 1.
Thus C_8 is a 2-group.

(b) Consider now the quaternion group $G = \{\pm 1, \pm i, \pm j, \pm k\}$ of Problem 21 in Exercises 3.1.

Each of the elements $\pm i, \pm j, \pm k$ has order 4.

-1 has order 2.

1 has order 1.

Hence G is another 2-group of order 8.

(c) Last, consider the octic group $G' = \{e, \alpha, \alpha^2, \alpha^3, \beta, \gamma, \Delta, \theta\}$ of Example 8 in Section 4.1.

Each of α and α^3 has order 4.

Each of $\alpha^2, \beta, \gamma, \Delta, \theta$ has order 2.

The identity e has order 1.

Thus G' is also a 2-group of order 8.

Of these three p-groups, C_8 is abelian while both G and G' are nonabelian. ☐

It may happen that G is not a p-group, yet some of its subgroups are p-groups. In connection with that possibility, we make the following definition.

DEFINITION 4.30 If G is a finite abelian group that has order that is divisible by the prime p, then $\boldsymbol{G_p}$ is the set of all elements of G that have orders that are powers of p.

As might be expected, the set G_p turns out to be a subgroup. For the remainder of this section, we write all abelian groups in additive notation.

THEOREM 4.31 The set G_p defined in Definition 4.30 is a subgroup of G.

Proof The identity 0 has order $1 = p^0$, so $0 \in G_p$. If $a \in G_p$, then a has order p^r for some nonnegative integer r. Since a and its inverse $-a$ have the same order, $-a$ is also in G_p. Let b be another element of G_p. Then b has order p^s for a nonnegative integer s. If t is the larger of r and s, then

$$p^t(a + b) = p^t a + p^t b$$
$$= 0 + 0$$
$$= 0.$$

This implies that the order of $a + b$ divides p^t and therefore is a power of p, since p is a prime. Thus $a + b \in G_p$, and G_p is a subgroup of G. ■

EXAMPLE 2 Consider the additive group $G = \mathbf{Z}_6$. The order of $\mathbf{Z}_6$ is 6, which is divisible by the primes 2 and 3. In this group,

Each of [1] and [5] has order 6.
Each of [2] and [4] has order 3.
[3] has order 2.
[0] has order 1.

For $p = 2$ or $p = 3$, the subgroups G_p are given by

$$G_2 = \{[3], [0]\}$$
$$G_3 = \{[2], [4], [0]\}.$$

The group G is not a p-group, but G_2 is a 2-subgroup of G and G_3 is a 3-subgroup of G. $\square$

If a group G has p-subgroups, certain of them are given special names, as described in the following definition.

DEFINITION 4.32 If p is a prime and m is a positive integer such that $p^m \mid o(G)$ and $p^{m+1} \nmid o(G)$, then a subgroup of G that has order p^m is called a **Sylow p-subgroup** of G.

EXAMPLE 3 In Example 2, G_2 is a Sylow 2-subgroup of G and G_3 is a Sylow 3-subgroup of G. As a less trivial example, consider the octic group from Example 8 of Section 4.1:

$$H = \{e, \alpha, \alpha^2, \alpha^3, \beta, \gamma, \Delta, \theta\}$$

where

$$e = (1) \qquad \alpha = (1, 2, 3, 4) \qquad \alpha^2 = (1, 3)(2, 4) \qquad \alpha^3 = (1, 4, 3, 2)$$
$$\beta = (1, 4)(2, 3) \qquad \gamma = (2, 4) \qquad \Delta = (1, 2)(3, 4) \qquad \theta = (1, 3).$$

The group H is a subgroup of order 2^3 in the symmetric group $G = S_4$, which has order $4! = 24$. Since $2^3 \mid o(S_4)$ and $2^4 \nmid o(S_4)$, the octic group is a Sylow 2-subgroup of S_4. $\square$

THEOREM 4.33 (**Cauchy's Theorem for Abelian Groups**) If G is an abelian group of order n and p is a prime such that $p \mid n$, then G has at least one element of order p.

Proof The proof is by induction on the order n of G, using the second principle of finite induction. For $n = 1$, the theorem holds by default.

Now let k be a positive integer, assume that the theorem is true for all positive integers $n < k$, and let G be an abelian group of order k. Also, suppose that the prime p is a divisor of k.

Consider first the case where G has only the trivial subgroups $\{0\}$ and G. Then any $a \neq 0$ in G must be a generator of G, $G = \langle a \rangle$. It follows from Problem 24 of Exercises 3.3 that the order k of G must be a prime. Since p divides this order, then p must equal k, and G actually has $p - 1$ elements of order p by Theorem 3.18.

Now consider the case where G has a nontrivial subgroup H; that is, $H \neq \{0\}$ and $H \neq G$ so that $1 < o(H) < k$. If $p \mid o(H)$, then H contains an element of order p by the induction hypothesis, and the theorem is true for G. Suppose then, that $p \nmid o(H)$. Since G is abelian, H is normal in G and the quotient group G/H has order

$$o(G/H) = \frac{o(G)}{o(H)}.$$

We have

$$o(G) = o(H)o(G/H),$$

so p divides the product $o(H)o(G/H)$. Since p is a prime and $p \nmid o(H)$, p must divide $o(G/H) < o(G) = k$. Applying the induction hypothesis, the abelian group G/H has an element $b + H$ of order p. Then

$$H = p(b + H) = pb + H,$$

and therefore $pb \in H$, where $b \notin H$. Let $r = o(H)$. The order of pb must be a divisor of r so that $r(pb) = 0$ and $p(rb) = 0$. Since p is a prime and $p \nmid r$, p and r are relatively prime. Hence there exist integers u and v such that $pu + rv = 1$.

The contention now is that the element $c = rb$ has order p. We have $pc = 0$, and we need to show that $c = rb \neq 0$. Assume the contrary, that $rb = 0$. Then

$$
\begin{aligned}
b &= 1b \\
&= (pu + rv)b \\
&= u(pb) + v(rb) \\
&= u(pb) + 0 \\
&= u(pb).
\end{aligned}
$$

Now $pb \in H$, and therefore $u(pb) \in H$. But $b \notin H$, so we have a contradiction. Thus $c = rb \neq 0$ is an element of order p in G, and the proof is complete. ∎

Cauchy's Theorem also holds for nonabelian groups, but we do not prove it here. The next theorem applies only to abelian groups.

THEOREM 4.34   If G is a finite abelian group and p is a prime such that $p \mid o(G)$, then G_p is a Sylow p-subgroup.

Proof   Assume that G is a finite abelian group such that p^m divides $o(G)$ but p^{m+1} does not divide $o(G)$. Then $o(G) = p^m k$, where p and k are relatively prime. We need to prove that G_p has order p^m.

We first argue that $o(G_p)$ is a power of p. If $o(G_p)$ had a prime factor q different from p, then G_p would have to contain an element of order q, according to Cauchy's Theorem. This would contradict the very definition of G_p, so we conclude that $o(G_p)$ is a power of p, say $o(G_p) = p^t$.

Suppose now that $o(G_p) < p^m$, that is, that $t < m$. Then the quotient group G/G_p has order $p^m k / p^t = p^{m-t} k$, which is divisible by p. Hence G/G_p contains an element $a + G_p$ of order p, by Theorem 4.33. Then

$$G_p = p(a + G_p) = pa + G_p$$

and this implies that $pa \in G_p$. Thus pa has order that is a power of p. This implies that a has order a power of p, and therefore $a \in G_p$, that is, $a + G_p = G_p$. This is a contradiction to the fact that $a + G_p$ has order p. Therefore $o(G_p) = p^m$ and G_p is a Sylow p-subgroup of G.   ∎

The next theorem shows the true significance of the Sylow p-subgroups in the structure of abelian groups.

THEOREM 4.35   Let G be an abelian group of order $n = p_1^{m_1} p_2^{m_2} \cdots p_r^{m_r}$, where the p_i are distinct primes and each m_i is a positive integer.
Then

$$G = G_{p_1} \oplus G_{p_2} \oplus \cdots \oplus G_{p_r}$$

where G_{p_i} is the Sylow p_i-subgroup of G that corresponds to the prime p_i.

Proof   Assume the hypothesis of the theorem. For each prime p_i, G_{p_i} is a Sylow p-subgroup of G by Theorem 4.34. Suppose an element $a_1 \in G_{p_1}$ is also in the subgroup generated by $G_{p_2}, G_{p_3}, \ldots, G_{p_r}$. Then

$$a_1 = a_2 + a_3 + \cdots + a_r,$$

where $a_i \in G_{p_i}$. Since G_{p_i} has order $p_i^{m_i}$, $p_i^{m_i} a_i = 0$ for $i = 2, \ldots, r$. Hence

$$p_2^{m_2} p_3^{m_3} \cdots p_r^{m_r} a_1 = 0.$$

Since the order of any $a_1 \in G_{p_1}$ is a power of p_1 and p_1 is relatively prime to $p_2^{m_2} p_3^{m_3} \cdots p_r^{m_r}$, this requires that $a_1 = 0$. A similar argument shows that the intersection of any G_{p_i} with the subgroup generated by the remaining subgroups

$$G_{p_1}, G_{p_2}, \ldots, G_{p_{i-1}}, G_{p_{i+1}}, \ldots, G_{p_r}$$

is the identity subgroup $\{0\}$. Hence the sum

$$G_{p_1} \oplus G_{p_2} \oplus \cdots \oplus G_{p_r}$$

is direct, and has order equal to the product of the orders $p_i^{m_i}$:

$$o(G_{p_1} \oplus G_{p_2} \oplus \cdots \oplus G_{p_r}) = p_1^{m_1} p_2^{m_2} \cdots p_r^{m_r} = o(G).$$

Therefore

$$G = G_{p_1} \oplus G_{p_2} \oplus \cdots \oplus G_{p_r}. \qquad \blacksquare$$

EXAMPLE 4   In Example 2, $G = G_2 \oplus G_3$. $\qquad \square$

Our next theorem is concerned with a class that is more general than finite abelian groups, the *finitely generated abelian groups*. An abelian group G is said to be finitely generated if there exists a set of elements $\{a_1, a_2, \ldots, a_n\}$ in G such that every $x \in G$ can be written in the form

$$x = z_1 a_1 + z_2 a_2 + \cdots + z_n a_n,$$

where the z_i are integers. The elements a_i are called **generators** of G and the set $\{a_1, a_2, \ldots, a_n\}$ is called a **generating set** for G. A finite abelian group G is surely a finitely-generated group since G itself is a generating set.

In a finitely-generated group, the Well-Ordering Principle assures us that there are generating sets that have the smallest possible number of elements. Such sets are called **minimal generating sets**. The number of elements in a minimal generating set for G is called the **rank** of G.

THEOREM 4.36   Any finitely-generated abelian group G (and therefore any finite abelian group) is a direct sum of cyclic groups.

Proof   The proof is by induction on the rank of G. If G has rank 1, then G is cyclic and the theorem is true.

Assume that the theorem is true for any group of rank $k - 1$, and let G be a group of rank k. We consider two cases.

Case 1   Suppose there exists a minimal generating set $\{a_1, a_2, \ldots, a_k\}$ for G such that any relation of the form

$$z_1 a_1 + z_2 a_2 + \cdots + z_k a_k = 0$$

with $z_i \in \mathbf{Z}$ implies that $z_1 a_1 = z_2 a_2 = \cdots = z_k a_k = 0$. Then

$$G = \langle a_1 \rangle + \langle a_2 \rangle + \cdots + \langle a_k \rangle,$$

and the theorem is true for this case.

Case 2   Suppose that Case 1 does not hold. That is, for any minimal generating set $\{a_1, a_2, \ldots, a_k\}$ of G, there exists a relation of the form

$$z_1 a_1 + z_2 a_2 + \cdots + z_k a_k = 0$$

with $z_i \in \mathbf{Z}$ such that some of the $z_i a_i \neq 0$. Among all the minimal generating sets and all the relations of this form, there exists a smallest positive integer $\bar{z}_i$ that occurs as a coefficient in one of these relations. Suppose this $\bar{z}_i$ occurs in a relation with the generating set $\{b_1, b_2, \ldots, b_k\}$. If necessary, the elements in $\{b_1, b_2, \ldots, b_k\}$ can be rearranged so that this smallest positive coefficient occurs with b_1, say as $\bar{z}_1$ in

$$\bar{z}_1 b_1 + \bar{z}_2 b_2 + \cdots + \bar{z}_k b_k = 0. \tag{1}$$

Now let $s_1, s_2, \ldots, s_k$ be any set of integers that occur as coefficients in a relation of the form

$$s_1 b_1 + s_2 b_2 + \cdots + s_k b_k = 0 \tag{2}$$

with these generators b_i. We shall show that $\bar{z}_1$ divides s_1. By the Division Algorithm, $s_1 = \bar{z}_1 q_1 + r_1$, where $0 \leq r_1 < \bar{z}_1$. Multiplying equation (1) by q_1 and subtracting the result from equation (2), we have

$$r_1 b_1 + (s_2 - \bar{z}_2 q_1) b_2 + \cdots + (s_k - \bar{z}_k q_1) b_k = 0.$$

The condition $0 \leq r_1 < \bar{z}_1$ forces $r_1 = 0$ by choice of $\bar{z}_1$ as the smallest positive integer in a relation of this form. Thus $\bar{z}_1$ is a factor of s_1.

We now show that $\bar{z}_1 \mid \bar{z}_i$ for $i = 2, \ldots, k$. Consider $\bar{z}_2$, for example. By the Division Algorithm, $\bar{z}_2 = \bar{z}_1 q_2 + r_2$, where $0 \leq r_2 < \bar{z}_1$. If we let $b_1' = b_1 + q_2 b_2$, then $\{b_1', b_2, \ldots, b_k\}$ is a minimal generating set for G, and

$$\bar{z}_1 b_1 + \bar{z}_2 b_2 + \cdots + \bar{z}_k b_k = 0$$
$$\Rightarrow \bar{z}_1 (b_1' - q_2 b_2) + \bar{z}_2 b_2 + \cdots + \bar{z}_k b_k = 0$$
$$\Rightarrow \bar{z}_1 b_1' + (\bar{z}_2 - \bar{z}_1 q_2) b_2 + \cdots + \bar{z}_k b_k = 0$$
$$\Rightarrow \qquad \bar{z}_1 b_1' + r_2 b_2 + \cdots + \bar{z}_k b_k = 0.$$

Now $r_2 \neq 0$ and $0 \leq r_2 < \bar{z}_1$ would contradict the choice of $\bar{z}_1$, so it must be that $r_2 = 0$ and $\bar{z}_1 \mid \bar{z}_2$. The same sort of argument can be applied to each of $\bar{z}_3, \ldots, \bar{z}_k$, so we have $\bar{z}_i = \bar{z}_1 q_i$ for $i = 2, \ldots, k$.

Substituting in equation (1), we obtain

$$\bar{z}_1 b_1 + \bar{z}_1 q_2 b_2 + \cdots + \bar{z}_1 q_k b_k = 0.$$

Let $c_1 = b_1 + q_2 b_2 + \cdots + q_k b_k$, and consider the set $\{c_1, b_2, \ldots, b_k\}$. This set generates G, and we have

$$\begin{aligned}
\bar{z}_1 c_1 &= \bar{z}_1 b_1 + \bar{z}_1 q_2 b_2 + \cdots + \bar{z}_1 q_k b_k \\
&= \bar{z}_1 b_1 + \bar{z}_2 b_2 + \cdots + \bar{z}_k b_k \\
&= 0.
\end{aligned}$$

If H denotes the subgroup of G that is generated by the set $\{b_2, \ldots, b_k\}$, then $G = \langle c_1 \rangle + H$ since the set $\{c_1, b_2, \ldots, b_k\}$ is a generating set for G. We shall show that the sum is direct.

If $s_1, s_2, \ldots, s_k$ are any integers such that

$$s_1 c_1 + s_2 b_2 + \cdots + s_k b_k = 0,$$

then substitution for c_1 yields

$$s_1 b_1 + (s_1 q_2 + s_2) b_2 + \cdots + (s_1 q_k + s_k) b_k = 0.$$

This implies that $\bar{z}_1$ divides s_1 and therefore $s_1 c_1 = 0$ since $\bar{z}_1 c_1 = 0$. Hence the sum is direct and

$$G = \langle c_1 \rangle \oplus H.$$

Since H has rank $k - 1$, the induction hypothesis applies to H, and H is a direct sum of cyclic groups. Therefore G is a direct sum of cyclic groups, and the theorem follows by induction. ∎

We can now give a complete description of the structure of any finite abelian group G. As in Theorem 4.35,

$$G = G_{p_1} \oplus G_{p_2} \oplus \cdots \oplus G_{p_r}$$

where G_{p_i} is the Sylow p_i-subgroup of order $p_i^{m_i}$ corresponding to the prime p_i. Each G_{p_i} can in turn be decomposed into a direct sum of cyclic subgroups $\langle a_{i,j} \rangle$, each of which has order a power of p_i:

$$G_{p_i} = \langle a_{i,1} \rangle \oplus \langle a_{i,2} \rangle \oplus \cdots \oplus \langle a_{i,t_i} \rangle,$$

where the product of the orders of the subgroups $\langle a_{i,j} \rangle$ is $p_i^{m_i}$. This description is frequently referred to as the **Fundamental Theorem on Finite Abelian Groups**. It can be used to describe systematically all the abelian groups of a given finite order, up to isomorphism.

EXAMPLE 5 For n a positive integer, let C_n denote a cyclic group of order n. If G is an abelian group of order $72 = 2^3 \cdot 3^2$, then G is the direct sum of its Sylow p-subgroups G_2 of order 2^3 and G_3 of order 3^2:

$$G = G_2 \oplus G_3.$$

Each of G_2 and G_3 is a sum of cyclic groups as described in the preceding paragraph. By considering all possibilities for the decompositions of G_2 and G_3, we deduce that any abelian group of order 72 is isomorphic to one of the following direct sums of cyclic groups:

$$C_{2^3} \oplus C_{3^2} \qquad\qquad C_{2^3} \oplus C_3 \oplus C_3$$
$$C_2 \oplus C_{2^2} \oplus C_{3^2} \qquad C_2 \oplus C_{2^2} \oplus C_3 \oplus C_3$$
$$C_2 \oplus C_2 \oplus C_2 \oplus C_{3^2} \quad C_2 \oplus C_2 \oplus C_2 \oplus C_3 \oplus C_3. \qquad \square$$

The main emphasis of this section has been on finite abelian groups, but the results presented here hardly scratch the surface. As an example of the interesting and important work that has been done on finite groups in general, we state the following theorem without proof.

THEOREM 4.37   (Sylow's Theorem) Let G be a finite group, and let p be a prime integer.

(a)   If m is a positive integer such that $p^m \mid o(G)$ and $p^{m+1} \nmid o(G)$, then G has a subgroup of order p^m.

(b)   For the same prime p, any two Sylow p-subgroups of G are conjugate subgroups.

(c)   If $p \mid o(G)$, the number n_p of distinct Sylow p-subgroups of G satisfies $n_p \equiv 1 \pmod{p}$.

The result in part (a) of Theorem 4.37 can be generalized to state that if $p^m \mid o(G)$ and $p^{m+1} \nmid o(G)$, then G has a subgroup of order p^k for any $k \in Z$ such that $0 \le k \le m$.

EXERCISES 4.4

1   Give an example of a p-group of order 9.

2   Find two p-groups of order 4 that are not isomorphic.

3   (a) Find all Sylow 3-subgroups of the alternating group A_4.
    (b) Find all Sylow 2-subgroups of A_4.

4   Find all Sylow 3-subgroups of the symmetric group S_4.

5   For each of the following Z_n, let G be the additive group $G = Z_n$ and write G as a direct sum of cyclic groups.
    (a) Z_{10} (b) Z_{15}
    (c) Z_{12} (d) Z_{18}

6   For each of the following values of n, describe all the abelian groups of order n, up to isomorphism.

(a) $n = 6$ (b) $n = 10$ (c) $n = 12$
(d) $n = 18$ (e) $n = 36$ (f) $n = 100$

7   Show that $\{a_1, a_2, \ldots, a_n\}$ is a generating set for the additive abelian group G if and only if $G = \langle a_1 \rangle + \langle a_2 \rangle + \cdots + \langle a_n \rangle$.

8   Give an example where G is a finite *nonabelian* group with order that is divisible by a prime p, and the set of all elements that have orders that are powers of p is *not* a subgroup of G.

9   If $p_1, p_2, \ldots, p_r$ are distinct primes, prove that any two abelian groups of order $n = p_1 p_2 \cdots p_r$ are isomorphic.

10   Suppose that the abelian group G can be written as the direct sum $G = C_{2^2} \oplus C_3 \oplus C_3$, where C_n is a cyclic group of order n.
(a) Prove that G has elements of order 12, but no element of order greater than 12.
(b) Find the number of distinct elements of G that have order 12.

11   Assume that G can be written as the direct sum $G = C_2 \oplus C_2 \oplus C_3 \oplus C_3$ where C_n is a cyclic group of order n.
(a) Prove that G has elements of order 6, but no element of order greater than 6.
(b) Find the number of distinct elements of G that have order 6.

12   Suppose that G is a *cyclic* group of order p^m, where p is a prime. If k is any integer such that $0 \leq k \leq m$, prove that G has a subgroup of order p^k.

13   Prove the result in Problem 12 for an arbitrary *abelian* group G of order p^m, where G is not necessarily cyclic.

14   Prove that G is an abelian group of order n and s is an integer that divides n, then G has a subgroup of order s.

Key Words and Phrases

Product of subsets
Left coset
Right coset
Index of a subgroup
Normal (invariant) subgroup
Subgroup generated by A
Quotient (factor) group
Homomorphism
Epimorphism
Homomorphic image
Kernel of a homomorphism

Fundamental Theorem of
  Homomorphisms
Sum of subgroups
Direct sum of subgroups
p-group
Sylow p-subgroup
Cauchy's Theorem for Abelian
  Groups
Generating set
Minimal generating set
Rank

References

Durbin, John R. *Modern Algebra*. New York: Wiley, 1979.

Fraleigh, John B. *A First Course in Abstract Algebra* (2nd ed.). Reading, Mass.: Addison-Wesley, 1976.

Hall, Marshall, Jr. *The Theory of Groups* (2nd ed.). New York: Chelsea, 1961.

Herstein, I. N. *Topics in Algebra* (2nd ed.). New York: Wiley, 1975.

Hillman, Abraham P. and Gerald L. Alexanderson. *A First Undergraduate Course in Abstract Algebra* (2nd ed.). Belmont, Calif: Wadsworth, 1978.

Jones, Burton W. *An Introduction to Modern Algebra.* New York: Macmillan, 1975.

McCoy, Neal H. *Introduction to Modern Algebra* (3rd ed.). Boston: Allyn and Bacon, 1975.

Maxfield, John E. and Margaret W. Maxfield. *Abstract Algebra and Solution by Radicals.* New York: Dover, 1983.

Schilling, Otto, F. G. and W. Stephen Piper. *Basic Abstract Algebra.* Boston: Allyn and Bacon, 1975.

Shapiro, Louis. *Introduction to Abstract Algebra.* New York: McGraw-Hill, 1975.

5 RINGS, INTEGRAL DOMAINS, AND FIELDS

5.1 DEFINITION OF A RING

A group is one of the simpler algebraic systems because it has only one binary operation. A step upward in the order of complexity is the *ring*. A ring has two binary operations, called *addition* and *multiplication*. Conditions are made on both binary operations, but fewer are made on multiplication. A full list of the conditions is in our formal definition.

DEFINITION 5.1(A) Suppose R is a set in which a relation of equality, denoted by $=$, and operations of addition and multiplication, denoted by $+$ and $\cdot$, respectively, are defined. Then R is a **ring** (with respect to these operations) if the conditions below are satisfied.

(1) R is **closed** under addition: $x \in R$ and $y \in R$ imply $x + y \in R$.

(2) Addition in R is **associative**: $x + (y + z) = (x + y) + z$ for all x, y, z in R.

(3) R has an **additive identity** 0: $x + 0 = 0 + x = x$ for all $x \in R$.

(4) R contains **additive inverses**: for x in R there exists $-x$ in R such that $x + (-x) = (-x) + x = 0$.

(5) Addition in R is **commutative**: $x + y = y + x$ for all x, y in R.

(6) R is **closed** under multiplication: $x \in R$ and $y \in R$ imply $x \cdot y \in R$.

(7) Multiplication in R is **associative**: $x \cdot (y \cdot z) = (x \cdot y) \cdot z$ for all x, y, z in R.

(8)   Two **distributive laws** hold in R: $x \cdot (y + z) = x \cdot y + x \cdot z$
and $(x + y) \cdot z = x \cdot z + y \cdot z$ for all x, y, z in R.
The notation xy will be used interchangeably with $x \cdot y$ to indicate multiplication.

As indicated in the definition, the additive identity of a ring is denoted by 0 and referred to as the **zero** of the ring. The additive inverse $-a$ is called the **negative** of a or the **opposite** of a. As in elementary algebra, we adhere to the convention that *multiplication takes precedence over addition*. That is, it is understood that in any expression involving both multiplication and addition, multiplications are performed first. Thus $xy + xz$ represents $(x \cdot y) + (x \cdot z)$, not $x(y + x)z$.

The statement of the definition can be shortened to a form that is easier to remember if we note that the first five conditions amount to the requirement that R be an abelian group under addition.

DEFINITION 5.1(B)   Suppose R is a set in which a relation of equality, denoted by $=$, and operations of addition and multiplication, denoted by $+$ and $\cdot$, respectively, are defined. Then R is a **ring** (with respect to these operations) if these conditions hold:
(1) R forms an **abelian group** with respect to **addition**.
(2) R is **closed** with respect to an **associative multiplication**.
(3)   Two **distributive laws** hold in R: $x \cdot (y + z) = x \cdot y + x \cdot z$
and $(x + y) \cdot z = x \cdot z + y \cdot z$ for all x, y, z in R.

EXAMPLE 1   Some simple examples of rings are provided by the familiar number systems with their usual operations of addition and multiplication. These are:

(a)   The set $\mathbf{Z}$ of all integers.
(b)   The set of all rational numbers.
(c)   The set of all real numbers.
(d)   The set of all complex numbers. $\square$

EXAMPLE 2   We shall verify that the set $\mathbf{E}$ of all even integers is a ring with respect to the usual addition and multiplication in $\mathbf{Z}$. The following conditions of Definition 5.1(A) are satisfied automatically since they hold throughout the ring $\mathbf{Z}$, which contains $\mathbf{E}$:
(2)   Addition in $\mathbf{E}$ is associative.
(5)   Addition in $\mathbf{E}$ is commutative.

(7)   Multiplication in **E** is associative.

(8)   The two distributive laws of Definition 5.1(A) hold in **E**.

The remaining conditions in Definition 5.1(A) may be checked as follows:

(1)   If $x \in \mathbf{E}$ and $y \in \mathbf{E}$, then $x = 2m$ and $y = 2n$ with m and n in **Z**. For the sum, we have $x + y = 2m + 2n = 2(m + n)$, which is in **E**. Thus **E** is closed under addition.

(3)   **E** contains the additive identity since $0 = (2)(0)$.

(4)   For any $x = 2k$ in **E**, the additive inverse of x is in **E** since $-x = 2(-k)$.

(6)   For $x = 2m$ and $y = 2n$ in **E**, the product $xy = 2(2mn)$ is in **E**, so **E** is closed under multiplication.   □

DEFINITION 5.2   Whenever a ring R_1 is a subset of a ring R_2 and has addition and multiplication as defined in R_2, we say that R_1 is a **subring** of R_2.

Thus the ring **E** of even integers is a subring of the ring **Z** of all integers. From Example 1 we see that the ring **Z** is a subring of the rational numbers, the rational numbers form a subring of the real numbers, and the real numbers form a subring of the complex numbers.

Generalizing from Example 2, we may observe that conditions (2), (5), (7), and (8) of Definition 5.1(A) are automatically satisfied in any subset of a ring, leaving only conditions (1), (3), (4), and (6) to be verified for the subset to form a subring. A slightly more efficient characterization of subrings is given in the following theorem, the proof of which is left as an exercise.

THEOREM 5.3   A subset S of the ring R is a subring of R if and only if these conditions are satisfied:

(a) S is nonempty.

(b) $x \in S$ and $y \in S$ imply $x + y$ and xy are in S.

(c) $x \in S$ implies $-x \in S$.

An even more efficient characterization of subrings is provided by the next theorem. The proof of this theorem is left as an exercise.

THEOREM 5.4   A subset S of the ring R is a subring of R if and only if these conditions are satisfied:

(a) S is nonempty.

(b) $x \in S$ and $y \in S$ imply $x - y$ and xy are in S.

EXAMPLE 3   Using Theorem 5.3 or Theorem 5.4, it is not difficult to verify the following examples of subrings.

(a)   The set of all real numbers of the form $m + n\sqrt{2}$, with $m \in \mathbf{Z}$ and $n \in \mathbf{Z}$, is a subring of the ring of all real numbers.

(b)   The set of all real numbers of the form $a + b\sqrt{2}$, with a and b rational numbers, is a subring of the real numbers.

(c)   The set of all real numbers of the form $a + b\sqrt[3]{2} + c\sqrt[3]{4}$, with a, b, and c rational numbers, is a subring of the real numbers.   □

The preceding examples of rings are all drawn from the number systems. The next example exhibits a class of rings with a different flavor: they are **finite rings** (that is, rings with a finite number of elements). The next example is also important because it presents the set $\mathbf{Z}_n$ of congruence classes modulo n for the first time in its proper context as a *ring*.

EXAMPLE 4   For $n > 1$, let $\mathbf{Z}_n$ denote the congruence classes of the integers modulo n:

$$\mathbf{Z}_n = \{[0], [1], [2], \ldots, [n-1]\}.$$

We have previously seen that the rules

$$[a] + [b] = [a + b] \quad \text{and} \quad [a] \cdot [b] = [ab]$$

define binary operations of addition and multiplication in $\mathbf{Z}_n$. We have seen that $\mathbf{Z}_n$ forms an abelian group under addition, with $[0]$ as the additive identity and $[-a]$ as the additive inverse of $[a]$. It has also been noted that this multiplication is associative. For arbitrary $[a], [b], [c]$ in $\mathbf{Z}_n$, we have

$$\begin{aligned}
[a] \cdot ([b] + [c]) &= [a] \cdot [b + c] \\
&= [a(b + c)] \\
&= [ab + ac] \\
&= [ab] + [ac] \\
&= [a] \cdot [b] + [a] \cdot [c],
\end{aligned}$$

so the left distributive law holds in $\mathbf{Z}_n$. The right distributive law can be verified in a similar way, and $\mathbf{Z}_n$ is a ring with respect to these operations.   □

Making use of some results from Chapter 1, we can obtain an example of a ring quite different from any of those discussed above.

EXAMPLE 5   Let U be a nonempty universal set, and let $\mathscr{P}(U)$ denote the collection of all subsets of U.

For arbitrary subsets A and B of U, let $A + B$ be defined as in Problem 26 of Exercises 1.1:

$$A + B = (A \cup B) - (A \cap B)$$
$$= \{x \in U \mid x \in A \text{ or } B, \text{ and } x \text{ is not in both } A \text{ and } B\}.$$

This rule defines an operation of addition on the subsets of U, and this operation is associative, by Problem 26(b) of Exercises 1.1. This addition is commutative since $A \cup B = B \cup A$ and $A \cap B = B \cap A$. The empty set $\emptyset$ is an additive identity since

$$\emptyset + A = A + \emptyset$$
$$= (A \cup \emptyset) - (A \cap \emptyset)$$
$$= A - \emptyset$$
$$= A.$$

An unusual feature here is that each subset A of U is its own additive inverse:

$$A + A = (A \cup A) - (A \cap A)$$
$$= A - A$$
$$= \emptyset.$$

We define multiplication in $\mathscr{P}(U)$ by

$$A \cdot B = A \cap B.$$

This multiplication is associative since

$$A \cdot (B \cdot C) = A \cap (B \cap C)$$
$$= (A \cap B) \cap C$$
$$= (A \cdot B) \cdot C.$$

The left distributive law $A \cap (B + C) = (A \cap B) + (A \cap C)$ is part (c) of Problem 26, Exercises 1.1, and the right distributive law follows from this one since forming intersections of sets is a commutative operation. Thus $\mathscr{P}(U)$ is a ring with respect to the operations $+$ and $\cdot$ as we have defined them. $\square$

DEFINITION 5.5 Let R be a ring. If there exists an element e in R such that $x \cdot e = e \cdot x = x$ for all x in R, then e is called a **unity** and R is a **ring with unity**. If multiplication in R is commutative, then R is called a **commutative ring**.

A ring may have either one of the properties in Definition 5.5 without the other, or it may have neither, or it may have both of the properties. These possibilities are illustrated in the following examples.

EXAMPLE 6   The ring **Z** of all integers has both properties, so **Z** is a commutative ring with a unity. As other examples of this type, $\mathbf{Z}_n$ is a commutative ring with unity [1], and $\mathscr{P}(U)$ is a commutative ring with the subset U as unity. $\qquad\square$

EXAMPLE 7   The ring **E** of all even integers is a commutative ring, but **E** does not have a unity. $\qquad\square$

EXAMPLE 8   A 2×2 **matrix** (plural: **matrices**) over the set S is a rectangular array of four elements from S, arranged in two rows and two columns. Let M denote the set of all 2×2 matrices over the real numbers. That is, M is the set of all 2×2 matrices of the form $\begin{bmatrix} a & b \\ c & d \end{bmatrix}$ where a, b, c, and d are real numbers. Equality in M is defined by

$$\begin{bmatrix} a & b \\ c & d \end{bmatrix} = \begin{bmatrix} x & y \\ z & w \end{bmatrix} \quad \text{if and only if } a = x, b = y, c = z, \text{ and } d = w.$$

Addition and multiplication are defined in M by

$$\begin{bmatrix} a & b \\ c & d \end{bmatrix} + \begin{bmatrix} x & y \\ z & w \end{bmatrix} = \begin{bmatrix} a + x & b + y \\ c + z & d + w \end{bmatrix}$$

and

$$\begin{bmatrix} a & b \\ c & d \end{bmatrix} \cdot \begin{bmatrix} x & y \\ z & w \end{bmatrix} = \begin{bmatrix} ax + bz & ay + bw \\ cx + dz & cy + dw \end{bmatrix}.$$

We do not verify the details here, but M is a ring, the *ring of all 2×2 matrices over the real numbers*. The ring M has $\begin{bmatrix} 1 & 0 \\ 0 & 1 \end{bmatrix}$ as a unity. It is a noncommutative ring, as can be seen from the computations

$$\begin{bmatrix} 1 & 1 \\ -1 & 1 \end{bmatrix}\begin{bmatrix} 1 & 2 \\ 2 & 4 \end{bmatrix} = \begin{bmatrix} 3 & 6 \\ 1 & 2 \end{bmatrix}$$

and

$$\begin{bmatrix} 1 & 2 \\ 2 & 4 \end{bmatrix}\begin{bmatrix} 1 & 1 \\ -1 & 1 \end{bmatrix} = \begin{bmatrix} -1 & 3 \\ -2 & 6 \end{bmatrix}. \qquad\square$$

EXAMPLE 9   Let S be the set of all 2×2 matrices over the ring **E** of all even integers.

$$S = \left\{ \begin{bmatrix} a & b \\ c & d \end{bmatrix} \,\middle|\, a, b, c, \text{ and } d \text{ are in } \mathbf{E} \right\}.$$

Equality, addition, and multiplication are defined in S by the rules given in Example 8. It is left as an exercise to verify that S is a noncommutative ring that does not have a unity. □

The definition of a unity allows the possibility of more than one unity in a ring. However, this possibility cannot happen.

THEOREM 5.6   If R is a ring that has a unity, the unity is unique.

Proof   Suppose that both e and e' are unity elements in a ring R. Consider the product $e \cdot e'$ in R. On the one hand, we have $e \cdot e' = e$ since e' is a unity. On the other hand, $e \cdot e' = e'$ since e is a unity. Thus

$$e = e \cdot e' = e'$$

and the unity is unique. ∎

In general discussions we shall denote a unity by e. When a ring R has a unity, it is in order to consider the existence of multiplicative inverses.

DEFINITION 5.7   Let R be a ring with unity e, and let $a \in R$. If there is an element x in R such that $ax = xa = e$, then x is a **multiplicative inverse of a**.

As with the unity, a multiplicative inverse of an element is unique whenever it exists. The proof of this is left as an exercise.

THEOREM 5.8   Suppose R is a ring with unity e. If an element $a \in R$ has a multiplicative inverse, the multiplicative inverse of a is unique.

We shall use the standard notation a^{-1} to denote the multiplicative inverse of a, if the inverse exists.

EXAMPLE 10   Some elements in a ring R may have multiplicative inverses while others do not. In the ring $\mathbf{Z}_{10}$, [1] and [9] are their own multiplicative inverses, and [3] and [7] are inverses of each other. All other elements of $\mathbf{Z}_{10}$ do not have multiplicative inverses. □

Since every ring R forms an abelian group with respect to addition, many of our results for groups have immediate applications concerning addition in a ring. For example, Theorem 3.3 gives these results:

1. The zero element in R is unique.
2. For each x in R, $-x$ is unique.
3. For each x in R, $-(-x) = x$.
4. For any x and y in R, $-(x + y) = -y - x$.
5. If a, x, and y are in R and $a + x = a + y$, then $x = y$.

Whenever both addition and multiplication are involved, the results are not so direct, but they turn out much as we might expect. One basic result of this type is that a product is 0 if one of the factors is 0.

THEOREM 5.9   If R is a ring, then

$$a \cdot 0 = 0 \cdot a = 0$$

for all $a \in R$.

Proof   Let a be arbitrary in R. We reduce $a \cdot 0$ to 0 by using various conditions in Definition 5.1(A), as indicated:

$$
\begin{aligned}
a \cdot 0 &= a \cdot 0 + 0 & &\text{by condition (3)} \\
&= a \cdot 0 + \{a \cdot 0 + [-(a \cdot 0)]\} & &\text{by condition (4)} \\
&= (a \cdot 0 + a \cdot 0) + [-(a \cdot 0)] & &\text{by condition (2)} \\
&= [a \cdot (0 + 0)] + [-(a \cdot 0)] & &\text{by condition (8)} \\
&= a \cdot 0 + [-(a \cdot 0)] & &\text{by condition (3)} \\
&= 0 & &\text{by condition (4).}
\end{aligned}
$$

Similar steps can be used to reduce $0 \cdot a$ to 0.   ∎

Theorem 5.9 says that a product is 0 if one of the factors is 0. Note that the converse is not true: a product may be 0 when neither factor is 0. An illustration is provided by $[2] \cdot [5] = [0]$ in $\mathbf{Z}_{10}$.

DEFINITION 5.10   Let R be a ring and let $a \in R$. If $a \neq 0$ and if there exists an element $b \neq 0$ in R such that either $ab = 0$ or $ba = 0$, then a is called a **proper divisor of zero**, or a **zero divisor**.

If we compare the steps used in the proof of Theorem 5.9 to the last part of the proof of Theorem 2.2, we see that they are essentially the same. In the same fashion, the proof of the first part of the next theorem is parallel to another part of the proof of Theorem 2.2. The same sort of similarity exists between Problems 1–10 of Exercises 2.1 and the remaining parts of the theorem, and their proofs are left as exercises.

THEOREM 5.11   For arbitrary x, y, and z in a ring R, the following equalities hold:

(a) $(-x)y = -(xy)$
(b) $x(-y) = -(xy)$
(c) $(-x)(-y) = xy$
(d) $x(y-z) = xy - xz$
(e) $(x-y)z = xz - yz$

Proof of (a)   Since the additive inverse $-(xy)$ of the element xy is unique, we only need to show that $xy + (-x)y = 0$. We have

$$
\begin{aligned}
xy + (-x)y &= [x + (-x)]y \quad &&\text{by the right distributive law} \\
&= 0 \cdot y &&\text{by the definition of } -x \\
&= 0 &&\text{by Theorem 5.9.} \qquad\blacksquare
\end{aligned}
$$

Even though a ring does not form a group with respect to multiplication, both associative laws in a ring R can be generalized by the procedure followed in Definition 3.5 and Theorem 3.6. For a positive integer $n \geq 2$, the expressions $a_1 + a_2 + \cdots + a_n$ and $a_1 a_2 \cdots a_n$ are defined recursively by

$$ a_1 + a_2 + \cdots + a_k + a_{k+1} = (a_1 + a_2 + \cdots + a_k) + a_{k+1} $$

and

$$ a_1 a_2 \cdots a_k a_{k+1} = (a_1 a_2 \cdots a_k) a_{k+1}. $$

The details are too repetitive to present here, so we accept the following theorem without proof.

THEOREM 5.12   (**Generalized Associative Laws**) Let $n \geq 2$ be a positive integer, and let $a_1, a_2, \ldots, a_n$ denote elements of a ring R. For any positive integer m such that $1 \leq m < n$,

$$ (a_1 + a_2 + \cdots + a_m) + (a_{m+1} + \cdots + a_n) = a_1 + a_2 + \cdots + a_n $$

and

$$ (a_1 a_2 \cdots a_m)(a_{m+1} \cdots a_n) = a_1 a_2 \cdots a_n. $$

Generalized distributive laws also hold in an arbitrary ring. This fact is stated in the following theorem, with the proofs left as exercises.

THEOREM 5.13   Let $n \geq 2$ be a positive integer, and let $b, a_1, a_2, \ldots, a_n$ denote elements of a ring R. Then we have

(a) $b(a_1 + a_2 + \cdots + a_n) = ba_1 + ba_2 + \cdots + ba_n$, and
(b) $(a_1 + a_2 + \cdots + a_n)b = a_1 b + a_2 b + \cdots + a_n b$.

EXERCISES 5.1

1 Confirm the statements made in Example 3 by proving that the following sets are subrings of the ring of all real numbers.
   (a) The set of all real numbers of the form $m + n\sqrt{2}$, with $m \in \mathbf{Z}$ and $n \in \mathbf{Z}$
   (b) The set of all real numbers of the form $a + b\sqrt{2}$, with a and b rational numbers
   (c) The set of all real numbers of the form $a + b\sqrt[3]{2} + c\sqrt[3]{4}$, with a, b, and c rational numbers

2 Decide if each of the following sets is a ring with respect to the usual operations of addition and multiplication. If it is not a ring, state at least one condition in Definition 5.1(A) that fails to hold.
   (a) The set of all integers that are multiples of 5
   (b) The set of all real numbers of the form $m + n\sqrt{3}$, with $m \in \mathbf{Z}$ and $n \in \mathbf{Z}$
   (c) The set of all real numbers of the form $a + \sqrt[3]{5}$, where a and b are rational numbers
   (d) The set of all real numbers of the form $a + b\sqrt[3]{5} + c\sqrt[3]{25}$, where a, b, and c are rational numbers
   (e) The set of all positive real numbers
   (f) The set of all complex numbers of the form $m + ni$, where $m \in \mathbf{Z}$ and $n \in \mathbf{Z}$
   (g) The set of all real numbers of the form $m + n\sqrt{2}$, where $m \in \mathbf{E}$ and $n \in \mathbf{Z}$
   (h) The set of all real numbers of the form $m + n\sqrt{2}$, where $m \in \mathbf{Z}$ and $n \in \mathbf{E}$

3 Let $U = \{a, b\}$. Using addition and multiplication as they are defined in Example 5, construct addition and multiplication tables for the ring $\mathscr{P}(U)$ that consists of the elements $\varnothing, A = \{a\}, B = \{b\}, U$.

4 Follow the instructions in Problem 3 and use the universal set $U = \{a, b, c\}$.

5 Let $U = \{a, b\}$. Define addition and multiplication in $\mathscr{P}(U)$ by $C + D = C \cup D$ and $CD = C \cap D$. Decide if $\mathscr{P}(U)$ is a ring with respect to these operations. If it is not, state a condition in Definition 5.1(A) that fails to hold.

6 Work Problem 5 using $U = \{a\}$.

7 Supply the details necessary to prove that M in Example 8 is a noncommutative ring with unity.

8 Verify that S in Example 9 is a noncommutative ring that does not have a unity.

9 Find all zero divisors in $\mathbf{Z}_n$ for the following values of n.
   (a) $n = 6$ (b) $n = 8$
   (c) $n = 10$ (d) $n = 12$
   (e) $n = 14$ (f) n a prime integer

10 For the given value of n, find the elements of $\mathbf{Z}_n$ that have multiplicative inverses.
   (a) $n = 6$ (b) $n = 8$
   (c) $n = 10$ (d) $n = 12$
   (e) $n = 14$ (f) n a prime integer

11 Prove Theorem 5.3: A subset S of the ring R is a subring of R if and only if these conditions are satisfied:
   (a) S is nonempty.
   (b) $x \in S$ and $y \in S$ imply $x + y$ and xy are in S.
   (c) $x \in S$ implies $-x \in S$.

12 Prove Theorem 5.4: A subset S of the ring R is a subring of R if and only if these conditions are satisfied:
(a) S is nonempty.
(b) $x \in S$ and $y \in S$ imply $x - y$ and xy are in S.

13 Assume R is a ring with unity e. Prove Theorem 5.8: If $a \in R$ has a multiplicative inverse, the multiplicative inverse of a is unique.

14 (See Example 4.) Prove the right distributive law in $\mathbf{Z}_n$:
$([a] + [b]) \cdot [c] = [a] \cdot [c] + [b] \cdot [c].$

15 Complete the proof of Theorem 5.9 by showing that $0 \cdot a = 0$ for any a in a ring R.

16 Let R be a ring, and let x, y, and z be arbitrary elements of R. Complete the proof of Theorem 5.11 by proving the following statements.
(a) $x(-y) = -(xy)$ (b) $(-x)(-y) = xy$
(c) $x(y - z) = xy - xz$ (d) $(x - y)z = xz - yz$

17 Suppose that G is a group with respect to addition, with identity element 0. Define a multiplication in G by $ab = 0$ for all $a, b \in G$. Show that G forms a ring with respect to these operations.

18 If R_1 and R_2 are subrings of the ring R, prove that $R_1 \cap R_2$ is a subring of R.

19 Find subrings R_1 and R_2 of $\mathbf{Z}$ such that $R_1 \cup R_2$ is not a subring of $\mathbf{Z}$.

20 If a and b are elements of the ring R that have multiplicative inverses in R, prove that the product ab also has a multiplicative inverse in R.

21 Prove that if the element a of a ring R has a multiplicative inverse in R, then a is not a zero divisor in R.

22 Suppose that $a, b,$ and c are elements of a ring R such that $ab = ac$. Prove that if a has a multiplicative inverse, then $b = c$.

23 For a fixed element a of a ring R, prove that the set $\{x \in R \mid ax = 0\}$ is a subring of R.

24 Consider the set $R = \{[0], [2], [4], [6], [8]\} \subseteq \mathbf{Z}_{10}$.
(a) Construct addition and multiplication tables for R, using the operations as defined in $\mathbf{Z}_{10}$.
(b) Observe that R is a commutative ring with unity $[6]$.
(c) Is R a subring of $\mathbf{Z}_{10}$? If not, give a reason.
(d) Does R have zero divisors?
(e) Which elements of R have multiplicative inverses?

25 Consider the set $S = \{[0], [2], [4], [6], [8], [10], [12], [14], [16]\} \subseteq \mathbf{Z}_{18}$. Using addition and multiplication as defined in $\mathbf{Z}_{18}$, answer the following questions.
(a) Is S a ring? If not, give a reason.
(b) Is S a commutative ring with unity? If not, give a reason.
(c) Is S a subring of $\mathbf{Z}_{18}$? If not, give a reason.
(d) Does S have zero divisors?
(e) Which elements of S have multiplicative inverses?

26 The addition table and part of the multiplication table for the ring $R = \{a, b, c\}$ are given in Figure 5.1. Use the distributive laws to complete the multiplication table.

+	a	b	c		·	a	b	c
a	a	b	c		a	a	a	a
b	b	c	a		b	a	c	
c	c	a	b		c	a		

FIGURE 5.1

27 The addition table and part of the multiplication table for the ring $R = \{a, b, c, d\}$ are given in Figure 5.2. Use the distributive laws to complete the multiplication table.

+	a	b	c	d		·	a	b	c	d
a	a	b	c	d		a	a	a	a	a
b	b	c	d	a		b	a	c		
c	c	d	a	b		c	a		a	
d	d	a	b	c		d	a		a	c

FIGURE 5.2

28 (See Examples 8 and 9). Let R be the set of all 2×2 matrices over $\mathbf{Z}$. Assume that R is a ring with respect to addition and multiplication as defined in Example 8, and consider the subset S of R that consists of all 2×2 matrices of the form $\begin{bmatrix} a & b \\ 0 & c \end{bmatrix}$, where $a, b, c \in \mathbf{Z}$.

(a) Show that S is a noncommutative ring with unity.

(b) Which elements of S have multiplicative inverses?

29 Consider the set T of all 2×2 matrices of the form $\begin{bmatrix} a & a \\ b & b \end{bmatrix}$, where a and b are real numbers. Addition and multiplication in T are as defined in Example 8.

(a) Show that T is a ring that does not have a unity.

(b) Show that T is not a commutative ring.

30 Prove the following equalities in an arbitrary ring R.

(a) $(x + y)(z + w) = (xz + xw) + (yz + yw)$

(b) $(x + y)(z - w) = (xz + yz) - (xw + yw)$

(c) $(x - y)(z - w) = (xz + yw) - (xw + yz)$

(d) $(x + y)(x - y) = (x^2 - y^2) + (yx - xy)$

31 (a) Prove Theorem 5.13(a).

(b) Prove Theorem 5.13(b).

32 Let R and S be arbitrary rings. In the Cartesian product $R \times S$ of R and S, define

$(r, s) = (r', s')$ if and only if $r = r'$ and $s = s'$,

$(r_1, s_1) + (r_2, s_2) = (r_1 + r_2, s_1 + s_2)$,

$(r_1, s_1)(r_2, s_2) = (r_1 r_2, s_1 s_2)$.

Prove that the Cartesian product is a ring with respect to these operations. It is called the **direct sum** of R and S and is denoted by $R \oplus S$.

33 (See Problem 32.) Write out the elements of $\mathbf{Z}_2 \oplus \mathbf{Z}_2$ and construct addition and multiplication tables for this ring. [Suggestion: Write 0 for [0], 1 for [1] in $\mathbf{Z}_2$.]

34   Suppose R is a ring in which all elements x satisfy $x^2 = x$. (Such a ring is called a **Boolean ring**.)
(a) Prove that $x = -x$ for each $x \in R$. [Hint: Consider $(x + x)^2$.]
(b) Prove that R is commutative. [Hint: Consider $(x + y)^2$.]

5.2   INTEGRAL DOMAINS AND FIELDS

In the preceding section we defined the terms *ring with unity*, *commutative ring*, and *zero divisors*. All three of these terms are used in defining an integral domain.

> **DEFINITION 5.14**   Let D be a ring. Then D is an **integral domain** provided these conditions hold:
> (1) D is a commutative ring.
> (2) D has a unity e, and $e \neq 0$.
> (3) D has no zero divisors.

Note that the requirement $e \neq 0$ means that an integral domain must have at least two elements.

EXAMPLE 1   The ring $\mathbf{Z}$ of all integers is an integral domain, but the ring $\mathbf{E}$ of all even integers is not an integral domain since it does not contain a unity. As familiar examples of integral domains we can list the set of all rational numbers, the set of all real numbers, and the set of all complex numbers, all of these with their usual operations.   □

EXAMPLE 2   The ring $\mathbf{Z}_{10}$ is a commutative ring with a unity, but the presence of zero divisors such as [2] and [5] prevents $\mathbf{Z}_{10}$ from being an integral domain. Considered as a possible integral domain, the ring M of all 2×2 matrices with real numbers as elements fails on two counts: multiplication is not commutative and it has zero divisors.   □

In Example 4 of Section 5.1 we saw that $\mathbf{Z}_n$ is a ring for every value of $n > 1$. Moreover, $\mathbf{Z}_n$ is a commutative ring since

$$[a] \cdot [b] = [ab] = [ba] = [b] \cdot [a]$$

for all $[a]$, $[b]$ in $\mathbf{Z}_n$. Since $\mathbf{Z}_n$ has [1] as the unity, $\mathbf{Z}_n$ is an integral domain if and only if it has no zero divisors. The following theorem characterizes these $\mathbf{Z}_n$, and it provides us with a large class of *finite integral domains* (that is, integral domains that have a finite number of elements).

THEOREM 5.15   For $n > 1$, $\mathbf{Z}_n$ is an integral domain if and only if n is a prime.

Proof   From the discussion above it is clear that we only need to prove that $\mathbf{Z}_n$ has no zero divisors if and only if n is a prime.

Suppose first that n is a prime. Let $[a] \neq [0]$ in $\mathbf{Z}_n$, and suppose $[a][b] = [0]$ for some $[b]$ in $\mathbf{Z}_n$. Now $[a][b] = [0]$ implies $[ab] = [0]$, and therefore $n \mid ab$. However, $[a] \neq [0]$ means that $n \nmid a$. Thus $n \mid ab$ and $n \nmid a$. Since n is a prime, this implies that $n \mid b$, by Theorem 2.14; that is, $[b] = [0]$. We have shown that if $[a] \neq [0]$, the only way that $[a][b]$ can be $[0]$ is for $[b]$ to be $[0]$. Therefore $\mathbf{Z}_n$ has no zero divisors, and is an integral domain.

Suppose now that n is not a prime. Then n has divisors other than ± 1 and $\pm n$, so there are integers a and b such that

$$n = ab, \quad \text{where } 1 < a < n \text{ and } 1 < b < n.$$

This means that $[a] \neq [0]$, $[b] \neq 0$, but

$$[a][b] = [ab] = [n] = [0].$$

Therefore $[a]$ is a zero divisor in $\mathbf{Z}_n$, and $\mathbf{Z}_n$ is not an integral domain.

Combining the two cases, we see that n is a prime if and only if $\mathbf{Z}_n$ is an integral domain.   ∎

One direct consequence of the absence of zero divisors in an integral domain is that the cancellation law for multiplication must hold.

THEOREM 5.16   If a, b, and c are elements of an integral domain D such that $a \neq 0$ and $ab = ac$, then $b = c$.

Proof   Suppose a, b, and c are elements of an integral domain D such that $a \neq 0$ and $ab = ac$. Now

$$ab = ac \Rightarrow ab - ac = 0$$
$$\Rightarrow a(b - c) = 0.$$

Since $a \neq 0$ and D has no zero divisors, it must be that $b - c = 0$, and hence $b = c$.   ∎

It can be shown that if the cancellation law holds in a commutative ring, then the ring cannot have zero divisors. The proof of this is left as an exercise.

To require that a ring have no zero divisors is equivalent to requiring that a product of nonzero elements must always be different from 0. Or, stated another way, a product that is 0 must have at least one factor equal to 0.

A *field* is another special type of ring, and we shall examine the relationship between a field and an integral domain. We begin with a definition.

DEFINITION 5.17 Let F be a ring. Then F is a **field** provided these conditions hold:
(1) F is a commutative ring.
(2) F has a unity e, and $e \neq 0$.
(3) Every nonzero element of F has a multiplicative inverse.

The rational numbers, the real numbers, and the complex numbers are familiar examples of fields.

Part of the relation between fields and integral domains is stated in the following theorem.

THEOREM 5.18 Every field is an integral domain.

Proof Let F be a field. To prove that F is an integral domain, we only need to show that F has no zero divisors. Suppose a and b are elements of F such that $ab = 0$. If $a \neq 0$, then $a^{-1} \in F$ and

$$ab = 0 \Rightarrow a^{-1}(ab) = a^{-1} \cdot 0$$
$$\Rightarrow (a^{-1}a)b = 0$$
$$\Rightarrow eb = 0$$
$$\Rightarrow b = 0.$$

Similarly, if $b \neq 0$, then $a = 0$. Therefore F has no zero divisors and is an integral domain. ∎

It is certainly not true that every integral domain is a field. For example, the set $\mathbf{Z}$ of all integers forms an integral domain, and the integers 1 and -1 are the only elements of $\mathbf{Z}$ that have multiplicative inverses. It is perhaps surprising, but an integral domain with a finite number of elements is always a field. This is the other part of the relationship between a field and an integral domain.

THEOREM 5.19 Every finite integral domain is a field.

Proof Assume D is a finite integral domain. Let n be the number of distinct elements in D, say

$$D = \{d_1, d_2, \ldots, d_n\}$$

where the d_i are the distinct elements of D. Now let a be any nonzero element of D, and consider the set of products

$$\{ad_1, ad_2, \ldots, ad_n\}.$$

These products are all distinct, for $a \neq 0$ and $ad_r = ad_s$ would imply $d_r = d_s$, by Theorem 5.16, and the d_i are all distinct. These n products are all contained in D and no two of them are equal. Hence they are the same as the elements of D, except possibly for order. This means that every element of D appears somewhere in the list

$$ad_1, ad_2, \ldots, ad_n.$$

In particular, the unity e is one of these products. That is, $ad_k = e$ for some d_k. Since multiplication is commutative in D, we have $d_k a = ad_k = e$, and d_k is a multiplicative inverse of a. Thus D is a field. ∎

COROLLARY 5.20 $\mathbf{Z}_n$ is a field if and only if n is a prime.

Proof This follows at once from Theorems 5.15 and 5.19. ∎

We have seen that the elements of a ring form an abelian group with respect to addition. A similar comparison can be made for the nonzero elements of a field. It is readily seen that the nonzero elements form an abelian group with respect to multiplication. The definition of a field can thus be reformulated as follows: A **field** is a set of elements in which equality, addition, and multiplication are defined such that the following conditions hold:

(1) F forms an abelian group with respect to addition.
(2) The nonzero elements of F form an abelian group with respect to multiplication.
(3) The distributive law $x(y + z) = xy + xz$ holds for all x, y, z in F.

EXERCISES 5.2

1 Decide which of the following are integral domains and which are the fields with respect to the usual operations of addition and multiplication. State a reason for each that fails to be an integral domain or a field.
  (a) The set of all real numbers of the form $m + n\sqrt{2}$, where m and n are integers
  (b) The set of all real numbers of the form $a + b\sqrt{2}$, where a and b are rational numbers
  (c) The set of all real numbers of the form $a + b\sqrt[3]{2}$, where a and b are rational numbers
  (d) The set of all real numbers of the form $a + b\sqrt[3]{2} + c\sqrt[3]{4}$, where $a, b,$ and c are rational numbers
  (e) The set of all complex numbers of the form $m + ni$, where $m \in \mathbf{Z}$ and $n \in \mathbf{Z}$
  (f) The set of all complex numbers of the form $m + ni$, where $m \in \mathbf{E}$ and $n \in \mathbf{E}$ (**E** is the ring of all even integers.)
  (g) The set of all complex numbers of the form $a + bi$, where a and b are rational numbers
  (h) The set of all real numbers of the form $m + n\sqrt{2}$, where $m \in \mathbf{Z}$ and $n \in \mathbf{E}$

2   Consider the set $R = \{[0], [2], [4], [6], [8]\} \subseteq \mathbf{Z}_{10}$, with addition and multiplication as defined in $\mathbf{Z}_{10}$.
(a) Is R an integral domain? If not, give a reason.
(b) Is R a field? If not, give a reason.

3   Consider the set $S = \{[0], [2], [4], [6], [8], [10], [12], [14], [16]\} \subseteq \mathbf{Z}_{18}$, with addition and multiplication as defined in $\mathbf{Z}_{18}$.
(a) Is S an integral domain? If not, give a reason.
(b) Is S a field? If not, give a reason.

4   Let $S = \{(0,0), (1,1), (0,1), (1,0)\}$, where $0 = [0]$ and $1 = [1]$ are the elements of $\mathbf{Z}_2$. Equality, addition, and multiplication are defined in S as follows:

$(a, b) = (c, d)$ if and only if $a = c$ and $b = d$ in $\mathbf{Z}_2$,
$(a, b) + (c, d) = (a + c, b + d)$,
$(a, b) \cdot (c, d) = (ad + bc + bd, ad + bc + ac)$.

(a) Prove that multiplication in S is associative.
Assume that S is a ring and consider these questions, giving a reason for any negative answers.
(b) Is S a commutative ring?
(c) Does S have a unity?
(d) Is S an integral domain?
(e) Is S a field?

5   Let W be the set of all ordered pairs (x, y) of integers x and y. Equality, addition, and multiplication are defined as follows:

$(x, y) = (z, w)$ if and only if $x = z$ and $y = w$ in $\mathbf{Z}$,
$(x, y) + (z, w) = (x + z, y + w)$,
$(x, y) \cdot (z, w) = (xz - yw, xw + yz)$.

Given that W is a ring, determine whether or not W is commutative and whether or not W has a unity. Justify your decisions.

6   Let S be the set of all 2×2 matrices of the form $\begin{bmatrix} x & 0 \\ x & 0 \end{bmatrix}$, where x is a real number. Assume that S is a ring with respect to addition and multiplication as defined in Example 8 of Section 5.1. Answer the following questions, and give a reason for any negative answers.
(a) Is S a commutative ring?
(b) Does S have a unity?
(c) Is S an integral domain?
(d) Is S a field?

7   Let R be the set of all matrices of the form $\begin{bmatrix} a & -b \\ b & a \end{bmatrix}$, where a and b are integers.
Assume that R is a ring with respect to addition and multiplication as defined in Example 8 of Section 5.1. Determine whether or not R is commutative, and whether or not R has a unity.

8   Let R be a commutative ring in which the cancellation law for multiplication holds. That is, if a, b, and c are elements of R, then $a \neq 0$ and $ab = ac$ always imply $b = c$. Prove that R has no zero divisors.

9   Prove that if a subring R of an integral domain D contains the unity element of D, then R is an integral domain.

10   If e is the unity in an integral domain D, prove that $(-e)a = -a$ for all $a \in D$.

11   An element x of a ring is called **idempotent** if $x^2 = x$. Prove that the only idempotent elements in an integral domain are 0 and e.

12   (a) Give an example where a and b are not zero divisors in a ring R, but the sum $a + b$ is a zero divisor.
   (b) Prove that the set of all elements in a ring R that are not zero divisors is closed under multiplication.

13   Find the multiplicative inverse of the given element.
   (a) $[11]$ in $\mathbf{Z}_{317}$ (b) $[11]$ in $\mathbf{Z}_{138}$ (c) $[8]$ in $\mathbf{Z}_{21}$

14   (See Problem 34 of Exercises 5.1.) Let R be a Boolean ring with unity e. Prove that every element of R except 0 is a zero divisor.

15   If $a \neq 0$ in a field F, prove that for every $b \in F$ the equation $ax = b$ has a unique solution for x in F.

16   Suppose S is a subset of a field F that contains at least two elements and satisfies the following conditions:
   (a) $x \in S$ and $y \in S$ imply $x - y \in S$;
   (b) $x \in S$ and $y \neq 0 \in S$ imply $xy^{-1} \in S$.
   Prove that S is a field.

5.3   THE FIELD OF QUOTIENTS OF AN INTEGRAL DOMAIN

The example of an integral domain that is the most familiar to us is the set $\mathbf{Z}$ of all integers, and the most familiar example of a field is the set of all rational numbers. There is a very natural and intimate relationship between these two systems. In fact, a rational number is by definition a quotient $\dfrac{a}{b}$ of integers a and b, with $b \neq 0$; that is, the set of rational numbers is the set of all quotients of integers with nonzero denominators. For this reason, the set of rational numbers is frequently referred to as "the quotient field of the integers." In this section we shall see that an analogous field of quotients can be constructed for an arbitrary integral domain.

Before we proceed with the presentation of this construction, let us review the basic definitions of equality, addition, and multiplication in the rational numbers. We recall that, for rational numbers $\dfrac{a}{b}$ and $\dfrac{c}{d}$,

$$\frac{a}{b} = \frac{c}{d} \quad \text{if and only if} \quad ad = bc,$$

$$\frac{a}{b} + \frac{c}{d} = \frac{ad + bc}{bd},$$

$$\frac{a}{b} \cdot \frac{c}{d} = \frac{ac}{bd}.$$

Note that the definitions of equality, addition, and multiplication for rational numbers are based on the corresponding definitions for the integers. These definitions guide our construction of the quotient field for an arbitrary integral domain D.

DEFINITION 5.21 Let D be an integral domain. The **set of quotients** for D is the set Q of all ordered pairs (a, b) of elements $a \in D$, $b \in D$, with $b \neq 0$. **Equality** in Q is defined by

$$(a, b) = (c, d) \quad \text{if and only if} \quad ad = bc.$$

Addition in Q is defined by

$$(a, b) + (c, d) = (ad + bc, bd)$$

and **multiplication** in Q is defined by

$$(a, b) \cdot (c, d) = (ac, bd).$$

The verification that the equality defined in Q has the reflexive, symmetric, and transitive properties is left as an exercise.

We shall at times need to use the fact that, for any $x \neq 0$ in D and any (a, b) in Q,

$$(a, b) = (ax, bx).$$

This follows at once from the equality $a(bx) = b(ax)$ in the integral domain D.

We shall verify that the addition we have defined is in fact a binary operation on Q. To do this, we need to show that the sum of two elements is unique (or well-defined). Suppose that $(a, b) = (x, y)$ and $(c, d) = (z, w)$ in Q. We need to show that $(a, b) + (c, d) = (x, y) + (z, w)$. We have

$$(a, b) + (c, d) = (ad + bc, bd)$$

and

$$(x, y) + (z, w) = (xw + yz, yw).$$

To prove these elements equal, we need

$$(ad + bc)yw = bd(xw + yz)$$

or

$$adyw + bcyw = bdxw + bdyz.$$

We have

$$(a, b) = (x, y) \Rightarrow ay = bx$$
$$\Rightarrow (ay)(dw) = (bx)(dw)$$
$$\Rightarrow adyw = bdxw$$

and

$$(c, d) = (z, w) \Rightarrow cw = dz$$
$$\Rightarrow (cw)(by) = (dz)(by)$$
$$\Rightarrow bcyw = bdyz.$$

By adding corresponding sides of equations, we obtain

$$adyw + bcyw = bdxw + bdyz.$$

Thus $(a, b) + (c, d) = (x, y) + (z, w)$.

It can be similarly shown that multiplication as given in Definition 5.21 is a binary operation on Q. With these preliminaries out of the way, we can now state our theorem.

THEOREM 5.22 Let D be an integral domain. The set Q as given in Definition 5.21 is a field.

Proof We first consider the postulates for addition. For arbitrary (a, b) and (c, d) in Q, we have $b \neq 0$ and $d \neq 0$ in D. Since D is an integral domain, $b \neq 0$ and $d \neq 0$ imply $bd \neq 0$, so $(a, b) + (c, d) = (ad + bc, bd)$ is an element of Q. Thus Q is closed under addition. It is left as an exercise to prove that addition is associative. Let $(0, b)$ denote any element of Q with the first element 0. Any pair of the form $(0, d)$ with $d \neq 0$ is equal to $(0, b)$, and this is the zero element of Q since

$$(x, y) + (0, b) = (x \cdot b + y \cdot 0, \, y \cdot b) = (xb, yb) = (x, y).$$

The last equality follows from the fact that $b \neq 0$, as was pointed out just after Definition 5.21. Routine calculations show that $(-a, b)$ is the additive inverse of (a, b) in Q, and that addition in Q is commutative.

The fact that $b \neq 0$ and $d \neq 0$ imply $bd \neq 0$ assures us that Q is closed under the product $(a, b) \cdot (c, d) = (ac, bd)$. The verification of the associative property for multiplication is left as an exercise.

We shall verify the left distributive property, and leave the other as an exercise. Let (x, y), (z, w), and (u, v) denote arbitrary elements of Q. We have

$$(x, y) \cdot [(z, w) + (u, v)] = (x, y) \cdot (zv + wu, wv)$$
$$= (xzv + xwu, ywv)$$

and

$$(x, y) \cdot (z, w) + (x, y) \cdot (u, v) = (xz, yw) + (xu, yv)$$
$$= (xyzv + xywu, y^2wv)$$
$$= (y[xzv + xwu], y[ywv]).$$

Comparing the results of these two calculations, we see that the last one differs from the first only in that both elements in the pair have been multiplied by y. Since (x, y) in Q requires $y \neq 0$, these results are equal.

Since multiplication in D is commutative, we have

$$(a, b) \cdot (c, d) = (ac, bd)$$
$$= (ca, db) = (c, d) \cdot (a, b).$$

Thus Q is a commutative ring.

Let $b \neq 0$ in D, and consider the element (b, b) in Q. For any (x, y) in Q we have

$$(x, y) \cdot (b, b) = (xb, yb)$$
$$= (x, y)$$

so (b, b) is a right identity for multiplication. Since multiplication is commutative, (b, b) is a nonzero unity for Q.

We have seen that the zero element of Q is any pair of the form $(0, b)$. Thus any nonzero element has the form (c, d), with both c and d nonzero. But then (d, c) is also in Q, and

$$(c, d) \cdot (d, c) = (cd, dc)$$
$$= (d, d),$$

so (d, c) is the multiplicative inverse of (c, d) in Q. This completes the proof that Q is a field. ∎

Note that in the proof of Theorem 5.22 the unity e of D did not appear explicitly anywhere. In fact, the construction yields a field if we start with a commutative ring that has no zero divisors instead of with an integral domain. However, we make use of the unity of D in Theorem 5.24.

The concept of an isomorphism can be applied to rings as well as to groups. The definition is a very natural extension of the concept of a group isomorphism. Since there are two binary operations involved in the definition of a ring, we simply require that both operations be preserved.

DEFINITION 5.23 Let R and R' denote two rings. A mapping $\phi : R \rightarrow R'$ is a **ring isomorphism** from R to R' provided the following conditions hold:
(1) ϕ is a bijection from R to R'.
(2) $\phi(x + y) = \phi(x) + \phi(y)$ for all x and y in R.
(3) $\phi(x \cdot y) = \phi(x) \cdot \phi(y)$ for all x and y in R.
If an isomorphism from R to R' exists, we say that R is **isomorphic** to R'.

Of course, the term *ring isomorphism* may be applied to systems that are more than a ring; that is, there may be a ring isomorphism that involves integral domains or fields. The relation of being isomorphic is reflexive, symmetric, and transitive on rings, just as it was with groups.

The field of quotients Q of an integral domain D has a significant feature that has not yet been brought to light. In the sense of isomorphism, it contains the integral domain D. More precisely, Q contains a subring D' that is isomorphic to D.

THEOREM 5.24 Let D and Q be as given in Definition 5.21, and let e denote the unity of D. The set D' that consists of all elements of Q that have the form (x, e) is a subring of Q, and D is isomorphic to D'.

Proof Referring to Definition 5.1(A), we see that conditions (2), (5), (7), and (8) are automatically satisfied in D', and we only need to check conditions (1), (3), (4), and (6).

For arbitrary (x, e) and (y, e) in D', we have

$$(x, e) + (y, e) = (x \cdot e + y \cdot e, e \cdot e)$$
$$= (x + y, e)$$

and D' is closed under addition. The element $(0, e)$ is in D', so D' contains the zero element of Q. For (x, e) in D', the additive inverse is $(-x, e)$, an element of D'. Lastly, the calculation

$$(x, e) \cdot (y, e) = (xy, e)$$

shows that D' is closed under multiplication. Thus D' is a subring of Q.

To prove that D is isomorphic to D', we use the natural mapping $\phi : D \to D'$ defined by

$$\phi(x) = (x, e).$$

The mapping ϕ is obviously a bijection. Since

$$\phi(x + y) = (x + y, e)$$
$$= (x, e) + (y, e)$$
$$= \phi(x) + \phi(y)$$

and

$$\phi(x \cdot y) = (xy, e)$$
$$= (x, e) \cdot (y, e)$$
$$= \phi(x) \cdot \phi(y),$$

ϕ is a ring isomorphism from D to D'. ■

Thus the quotient field Q contains D in the sense of isomorphism. We say that D is **embedded** in Q, or that Q is an **extension** of D. More generally, if S is a ring that contains a subring R' that is isomorphic to a given ring R, we say that R is **embedded** in S, or that S is an **extension** of R.

There is one more observation about Q that should be made. For any nonzero (b, e) in D', the multiplicative inverse of (b, e) in Q is $(b, e)^{-1} = (e, b)$, and every element of Q can be written in the form

$$(a, b) = (a, e) \cdot (e, b) = (a, e) \cdot (b, e)^{-1}.$$

If the isomorphism ϕ in the proof of Theorem 5.24 is used to identify x in D with (x, e) in D', then every element of Q can be identified as a quotient ab^{-1} of elements a and b of D, with $b \neq 0$.

From this, it follows that any field F that contains the integral domain D must also contain Q, because F must contain b^{-1} for each $b \neq 0$ in D and must contain the product ab^{-1} for all $a \in D$. Thus Q is the smallest field that contains D.

If the construction presented in this section is carried out beginning with $D = \mathbf{Z}$, the field $\mathbf{Q}$ of rational numbers is obtained with the elements written as (a, b) instead of a/b. The isomorphism ϕ in the proof of Theorem 5.24 maps an integer x onto $(x, 1)$, which is playing the role of $x/1$ in the notation, and we end up with the integers embedded in the rational numbers. The construction of the rational numbers from the integers is in this way a special case of the procedure described here.

EXERCISES 5.3

1   Prove that the equality defined in Q by Definition 5.21 is reflexive, symmetric, and transitive.

2   Prove that the multiplication given in Definition 5.21 is a binary operation on Q. That is, prove that multiplication is well-defined in Q.

3   Prove that addition is associative in Q.

4   Show that $(-a, b)$ is the additive inverse of (a, b) in Q.

5   Prove that addition is commutative in Q.

6   Prove that multiplication in Q is associative.

7   Prove the right distributive property in Q:

$$[(x, y) + (z, w)] \cdot (u, v) = (x, y) \cdot (u, v) + (z, w) \cdot (u, v).$$

8   Prove that the relation of being isomorphic is an equivalence relation on the set of all rings.

9   Assume that the ring R is isomorphic to the ring R'. Prove that if R is commutative, then R' is commutative.

10   Let W be the ring in Problem 5 of Exercises 5.2, and let R be the ring in Problem 7 of the same exercise set. Given that W and R are isomorphic rings, define an isomorphism from W to R and prove that your mapping is an isomorphism.

11 Since this section presents a method for constructing a field of quotients for an arbitrary integral domain D, we might ask what happens if D is already a field. As an example, consider the situation when $D = \mathbf{Z}_3$.
   (a) With $D = \mathbf{Z}_3$, write all of the ordered pairs in Q, sort these elements according to equality, and then list all of the distinct elements of Q.
   (b) Exhibit an isomorphism from Q to D.

12 Work Problem 11 with $D = \mathbf{Z}_5$.

13 Prove that if D is a field to begin with, the field of quotients Q is isomorphic to D.

14 Let F denote the field of quotients of the integral domain $D = \mathbf{E}$, the set of all even integers. Prove that F is isomorphic to the field of rational numbers.

15 Let D be the set of all complex numbers of the form $m + ni$, where $m \in \mathbf{Z}$ and $n \in \mathbf{Z}$. Carry out the construction of the quotient field Q for this integral domain, and show that this quotient field is isomorphic to the set of all complex numbers of the form $a + bi$, where a and b are rational numbers.

16 Prove that any field that contains an integral domain D must contain a subfield isomorphic to the quotient field Q of D.

17 Assume R is a ring, and let S be the set of all ordered pairs (m, x) where $m \in \mathbf{Z}$ and $x \in R$. Equality in S is defined by

$$(m, x) = (n, y) \quad \text{if and only if} \quad m = n \text{ and } x = y.$$

Addition and multiplication in S are defined by

$$(m, x) + (n, y) = (m + n, x + y)$$

and

$$(m, x) \cdot (n, y) = (mn, my + nx + xy),$$

where my and nx are *multiples* of y and x in the ring R.
   (a) Prove that S is a ring with unity.
   (b) Prove that $\phi: R \to S$ defined by $\phi(x) = (0, x)$ is an isomorphism from R to a subring R' of S. This result shows that any ring can be embedded in a ring that has a unity.

5.4   ORDERED INTEGRAL DOMAINS

In Section 2.1, we assumed that the set $\mathbf{Z}$ of all integers satisfied a list of five postulates. The last two of these postulates led to the introduction of the order relation "greater than" in $\mathbf{Z}$, and to the proof of the Well-Ordering Theorem (Theorem 2.7). In this section, we follow a development along similar lines in a more general setting.

DEFINITION 5.25 An integral domain D is an **ordered integral domain** if D contains a subset D^+ that has the following properties.
(1) D^+ is closed under addition.

(2) D^+ is closed under multiplication.

(3)   For each $x \in D$, one and only one of the following statements is true:

$$x \in D^+; \qquad x = 0; \qquad -x \in D^+.$$

Such a subset D^+ is called a **set of positive elements** for D.

Analogous to the situation in **Z**, condition (3) in Definition 5.25 is referred to as the **law of trichotomy**, and an element $x \in D$ such that $-x \in D^+$ is called a **negative element** of D.

EXAMPLE 1   The integral domain **Z** is, of course, an example of an ordered integral domain. With their usual sets of positive elements, the set of all rational numbers and the set of all real numbers furnish two other examples of ordered integral domains.   □

Later, we shall see that not all integral domains are ordered integral domains.

Following the same sort of procedure that we followed with the integers, we can use the set of positive elements in an ordered integral domain D to define the order relation "greater than" in D.

DEFINITION 5.26   Let D be an ordered integral domain with D^+ as the set of positive elements. The relation "greater than," denoted by $>$, is defined on elements x and y of D by

$$x > y \quad \text{if and only if} \quad x - y \in D^+.$$

The symbol $>$ is read "is greater than." Similarly, $<$ is read "is less than." We define $x < y$ if and only if $y > x$. As direct consequences of the definition, we have

$$x > 0 \quad \text{if and only if} \quad x \in D^+$$

and

$$x < 0 \quad \text{if and only if} \quad -x \in D^+.$$

The three properties of D^+ in Definition 5.25 translate at once into the following properties of $>$ in D.

(1)   If $x > 0$ and $y > 0$, then $x + y > 0$.

(2)   If $x > 0$ and $y > 0$, then $xy > 0$.

(3)   For each $x \in D$, one and only one of the following statements is true:

$$x > 0; \qquad x = 0; \qquad x < 0.$$

The other basic properties of $>$ are stated in the next theorem. We prove the first two and leave the proofs of the others as exercises.

THEOREM 5.27   Suppose that D is an ordered integral domain. The relation $>$ has the following properties, where x, y, and z are arbitrary elements of D.

(a)   If $x > y$, then $x + z > y + z$.
(b)   If $x > y$ and $z > 0$, then $xz > yz$.
(c)   If $x > y$ and $y > z$, then $x > z$.
(d)   One and only one of the following statements is true:

$x > y$; $x = y$; $x < y$.

Proof of (a)   If $x > y$, then $x - y \in D^+$, by Definition 5.26. Since

$$(x + z) - (y + z) = x + z - y - z$$
$$= x - y,$$

this means that $(x + z) - (y + z) \in D^+$, and therefore $x + z > y + z$.

Proof of (b)   Suppose $x > y$ and $z > 0$. Then $x - y \in D^+$ and $z \in D^+$. Part (2) of Definition 5.25 requires that D^+ be closed under multiplication, so the product $(x - y)z$ must be in D^+. Since $(x - y)z = xz - yz$, we have $xz - yz \in D^+$, and therefore $xz > yz$.   ∎

Our main goal in this section is to characterize the integers as an ordered integral domain that has a certain type of set of positive elements. As a first step in this direction, we prove the following simple theorem, which may be compared to Theorem 2.5.

THEOREM 5.28   For any $x \neq 0$ in an ordered integral domain D, $x^2 \in D^+$.

Proof   Suppose $x \neq 0$ in D. By part (3) of Definition 5.25, either $x \in D^+$ or $-x \in D^+$. If $x \in D^+$, then $x^2 = x \cdot x$ is in D^+ since D^+ is closed under multiplication. If $-x \in D^+$, then $x^2 = x \cdot x = (-x)(-x)$ is in D^+, again by closure of D^+ under multiplication. In either case, we have $x^2 \in D^+$.   ∎

COROLLARY 5.29   In any ordered integral domain, $e \in D^+$.

Proof   This follows from the fact that $e = e^2$.   ∎

The preceding theorem and its corollary can be used to show that the set $\mathscr{C}$ of all complex numbers does not form an ordered integral domain. Suppose, to the contrary, that $\mathscr{C}$ does contain a set $\mathscr{C}^+$ of positive elements. By Corollary 5.29, $1 \in \mathscr{C}^+$, and therefore $-1 \notin \mathscr{C}^+$ by the law of trichotomy. Theorem 5.28 requires, however, that $i^2 = -1$ be in $\mathscr{C}^+$, and we have a contradiction. Therefore $\mathscr{C}$ does not contain a set of positive elements. In other words, *it is impossible to impose an order relation on the set of all complex numbers.*

We shall need the following definition.

DEFINITION 5.30 A nonempty subset S of an ordered integral domain D is **well-ordered** if for every nonempty subset T of S, there is an element $m \in T$ such that $m \leq x$ for all $x \in T$. Such an element m is called a **least element** of T.

Thus $S \neq \varnothing$ in D is well-ordered if every nonempty subset of S contains a least element. We proved in Theorem 2.7 that the set of all positive integers is well-ordered.

The next step toward our characterization of the integers is the following theorem.

THEOREM 5.31 If D is an ordered integral domain in which the set D^+ of positive elements is well-ordered, then
(a) e is the least element of D^+, and
(b) $D^+ = \{ne \mid n \in \mathbf{Z}^+\}$.

Proof We have $e \in D^+$ by Corollary 5.29. To prove that e is the least element of D^+, let T be the set of all $x \in D^+$ such that $e > x > 0$, and assume that T is nonempty. Since D^+ is well-ordered, T has a least element m, and

$$e > m > 0.$$

Using Theorem 5.27(b) and multiplying by m, we have

$$m \cdot e > m^2 > m \cdot 0.$$

That is,

$$m > m^2 > 0,$$

and this contradicts the choice of m as the least element of T. Therefore T is empty, and e is the least element of D^+.

Now let S be the set of all $m \in \mathbf{Z}^+$ such that $me \in D^+$. We have $1 \in S$ since $1e = e \in D^+$. Assume that $k \in S$. Then $ke \in D^+$, and this implies that

$$(k + 1)e = ke + e$$

is in S since D^+ is closed under addition. Thus $k \in S$ implies $k + 1 \in S$, and $S = \mathbf{Z}^+$ by the induction postulate for the positive integers. This proves that

$$D^+ \supseteq \{ne \mid n \in \mathbf{Z}^+\}.$$

In order to prove that $D^+ \subseteq \{ne \mid n \in \mathbf{Z}^+\}$, let L be the set of all elements of D^+ that are not of the form ne with $n \in \mathbf{Z}^+$, and suppose that L is nonempty. Since D^+ is well-ordered, L has a least element ℓ. It must be that

$$\ell > e$$

since e is the least element of D^+, and therefore $\ell - e > 0$. Now

$$\begin{aligned} e > 0 &\Rightarrow e + (-e) > 0 + (-e) \quad \text{by Theorem 5.27(a)} \\ &\Rightarrow 0 > -e \\ &\Rightarrow \ell > \ell - e \quad\quad\quad\quad\quad \text{by Theorem 5.27(a).} \end{aligned}$$

Thus we have $\ell > \ell - e > 0$. By choice of ℓ as least element of L, $\ell - e \notin L$, so

$$\ell - e = pe \quad \text{for some } p \in \mathbf{Z}^+.$$

This implies that

$$\begin{aligned} \ell &= pe + e \\ &= (p + 1)e, \quad \text{where } p + 1 \in \mathbf{Z}^+, \end{aligned}$$

and we have a contradiction to the fact that ℓ is an element that cannot be written in the form ne with $n \in \mathbf{Z}^+$. Therefore $L = \varnothing$, and $D^+ = \{ne \mid n \in \mathbf{Z}^+\}$. ∎

We can now give the characterization of the integers toward which we have been working.

THEOREM 5.32   If D is an ordered integral domain in which the set D^+ of positive elements is well-ordered, then D is isomorphic to the ring $\mathbf{Z}$ of all integers.

Proof   We first show that

$$D = \{ne \mid n \in \mathbf{Z}\}.$$

For an arbitrary $x \in D$, the law of trichotomy requires that exactly one of the following holds:

$$x \in D^+; \qquad x = 0; \qquad -x \in D^+.$$

If $x \in D^+$, then $x = ne$ for some $n \in \mathbf{Z}^+$, by Theorem 5.31(b). If $x = 0$, then $x = 0e$. Finally, if $-x \in D^+$, then $-x = me$ for $m \in \mathbf{Z}^+$, and therefore[†] $x = -(me) = (-m)e$, where $-m \in \mathbf{Z}$. Hence $D = \{ne \mid n \in \mathbf{Z}\}$.

Consider now the rule defined by

$$\phi(ne) = n,$$

for any ne in D. To demonstrate that this rule is well-defined, it is sufficient to show that each element of D can be written as ne in only one way. To do this, suppose $me = ne$. Without loss of generality, we may assume that $m \geq n$. Now

$$me = ne \Rightarrow me - ne = 0$$
$$\Rightarrow (m - n)e = 0.$$

If $m - n > 0$, then $(m - n)e \in D^+$ by Theorem 5.31(b). Therefore it must be that $m - n = 0$ and $m = n$. This shows that the rule $\phi(ne) = n$ defines a mapping ϕ from D to $\mathbf{Z}$.

If $\phi(me) = \phi(ne)$, then $m = n$, so $me = ne$. Hence ϕ is injective. An arbitrary $n \in \mathbf{Z}$ is the image of $ne \in D$ under ϕ, so ϕ is a bijection.

That ϕ is a ring isomorphism follows from the equalities

$$\phi(me + ne) = \phi[(m + n)e]$$
$$= m + n$$
$$= \phi(me) + \phi(ne)$$

and

$$\phi(me \cdot ne) = \phi[(mn)e]$$
$$= mn$$
$$= \phi(me) \cdot \phi(ne). \qquad \blacksquare$$

EXERCISES 5.4

1   Complete the proof of Theorem 5.27 by proving the following statements, where x, y, and z are arbitrary elements of an ordered integral domain D.
(a) If $x > y$ and $y > z$, then $x > z$.
(b) One and only one of the following statements is true:

$$x > y; \qquad x = y; \qquad x < y.$$

[†] The equality $-(me) = (-m)e$ is the additive form of the familiar property of exponents $(a^m)^{-1} = a^{-m}$ in a group.

2   Prove the following statements for arbitrary elements x, y, z of an ordered integral domain D.

(a) If $x > y$ and $z < 0$, then $xz < yz$.

(b) If $x > y$ and $z > w$, then $x + z > y + w$.

(c) If $x > y > 0$, then $x^2 > y^2$.

(d) If $x \neq 0$ in D, then $x^{2n} > 0$ for every positive integer n.

(e) If $x > 0$ and $xy > xz$, then $y > z$.

3   Prove the following statements for arbitrary elements in an ordered integral domain.

(a) $a > b$ implies $-b > -a$.

(b) $a > e$ implies $a^2 > a$.

(c) If $a > b$ and $c > d$, where a, b, c, and d are all positive elements, then $ac > bd$.

4   If a and b have multiplicative inverses in an ordered integral domain and $a > b > 0$, prove that $b^{-1} > a^{-1} > 0$.

5   Prove that the equation $x^2 + e = 0$ has no solution in an ordered integral domain.

6   Prove that if a is any element of an ordered integral domain D, then there exists an element $b \in D$ such that $b > a$. (Thus D has no greatest element, and *no finite integral domain can be an ordered integral domain*.)

7   For an element x of an ordered integral domain D, the **absolute value** $|x|$ is defined by

$$|x| = \begin{cases} x & \text{if } x \geq 0 \\ -x & \text{if } 0 > x. \end{cases}$$

(a) Prove that $-|x| \leq x \leq |x|$ for all $x \in D$.

(b) Prove that $|xy| = |x| \cdot |y|$ for all $x, y \in D$.

(c) Prove that $|x + y| \leq |x| + |y|$ for all $x, y \in D$.

8   If x and y are elements of an ordered integral domain D, prove the following inequalities.

(a) $x^2 - 2xy + y^2 \geq 0$

(b) $x^2 + y^2 \geq xy$

(c) $x^2 + y^2 \geq -xy$

9   An **ordered field** is an ordered integral domain that is also a field. In the quotient field Q of an ordered integral domain D, define Q^+ by $Q^+ = \{(a, b) \mid ab \in D^+\}$. Prove that Q^+ is a set of positive elements for Q, and hence that Q is an ordered field.

10   (See Problem 9.) According to Definition 5.26, $>$ is defined in Q by $(a, b) > (c, d)$ if and only if $(a, b) - (c, d) \in Q^+$. Show that $(a, b) > (c, d)$ if and only if $abd^2 - cdb^2 \in D^+$.

11   (See Problems 9 and 10.) If each $x \in D$ is identified with (x, e) in Q, prove that $D^+ \subseteq Q^+$. (This means that the order relation defined in Problem 10 coincides in D with the original order relation in D. We say that the ordering in Q is an **extension** of the ordering in D.)

12   Prove that if x and y are rational numbers such that $x > y$, then there exists a rational number z such that $x > z > y$. (This means that between any two distinct rational numbers there is another rational number.)

13 (a) If D is an ordered integral domain, prove that each element in the quotient field Q of D can be written in the form (a, b) with $b > 0$ in D.

(b) If $(a, b) \in Q$ with $b > 0$ in D, prove that $(a, b) \in Q^+$ if and only if $a > 0$ in D.

14 (See Problem 13.) If (a, b) and $(c, d) \in Q$ with $b > 0$ and $d > 0$ in D, prove that $(a, b) > (c, d)$ if and only if $ad > bc$ in D.

15 (See Problem 13.) If x and y are positive rational numbers, prove that there exists a positive integer n such that $nx > y$. This property is called the **Archimedean Property** of the rational numbers. [Hint: Write $x = a/b$ and $y = c/d$ with each of $a, b, c, d \in \mathbf{Z}^+$.]

Key Words and Phrases

Ring	Field
Zero of a ring	Quotient field
Negative of an element	Ring isomorphism
Subring	Isomorphic rings
Finite ring	Embedded
Ring with unity	Extension
Commutative ring	Ordered integral domain
Multiplicative inverse	Set of positive elements
Proper divisor of zero	Law of Trichotomy
Zero divisor	Negative element
Generalized associative laws	Greater than, less than
Generalized distributive laws	Well-ordered subset
Integral domain	Least element

References

Bundrick, Charles M. and John J. Leeson. *Essentials of Abstract Algebra.* Monterey, Calif.: Brooks-Cole, 1972.

Durbin, John R. *Modern Algebra.* New York: Wiley, 1979.

Fraleigh, John B. *A First Course in Abstract Algebra* (2nd ed.). Reading, Mass.: Addison-Wesley, 1976.

Herstein, I. N. *Topics in Algebra* (2nd ed.). New York: Wiley, 1975.

Hillman, Abraham P. and Gerald L. Alexanderson. *A First Undergraduate Course in Abstract Algebra* (2nd ed.) Belmont, Calif.: Wadsworth, 1978.

Larsen, Max D. *Introduction to Modern Algebraic Concepts.* Reading, Mass.: Addison-Wesley, 1969.

McCoy, Neal H. *Introduction to Modern Algebra* (3rd ed.). Boston: Allyn and Bacon, 1975.

McCoy, Neal H. *Rings and Ideals* (Carus Mathematical Monograph No. 8). Washington, D.C.: The Mathematical Association of America, 1968.

McCoy, Neal H. *The Theory of Rings.* New York: Chelsea, 1972.

Mitchell, A. Richard and Roger W. Mitchell. *An Introduction to Abstract Algebra.* Monterey, Calif.: Brooks-Cole, 1970.

Schilling, Otto F. G. and W. Stephen Piper. *Basic Abstract Algebra.* Boston: Allyn and Bacon, 1975.

6 MORE ON RINGS

6.1 IDEALS AND QUOTIENT RINGS

In this chapter, we develop some theory of rings that parallels the theory of groups presented in Chapter 4. We shall see that the concept of an *ideal* in a ring is analogous to that of a *normal subgroup* in a group.

> **DEFINITION 6.1(A)**   A subset I of a ring R is an **ideal** of R if the following conditions hold:
> (a) I is a subring of R.
> (b) $x \in I$ and $r \in R$ imply xr and rx are in I.

Note that the second condition in this definition requires more than closure of I under multiplication. It requires that I "absorb" multiplication by arbitrary elements of R, both on the right and on the left.

In more advanced study of rings, the type of subring described in Definition 6.1(A) is referred to as a "two-sided" ideal, and terms that are more specialized are introduced: a **right ideal** of R is a subring S of R such that $xr \in S$ for all $x \in S, r \in R$, and a **left ideal** of R is a subring S of R such that $rx \in S$ for all $x \in S, r \in R$. We only mention these terms in passing here, and observe that these distinctions cannot be made in a commutative ring.

The subrings $I = \{0\}$ and $I = R$ are always ideals of a ring R. These ideals are labeled **trivial**.

If R is a ring with unity e and I is an ideal of R that contains e, then

$$e \in I \quad \text{and} \quad r \in R \Rightarrow er = r = re \quad \text{is in } I,$$

so it must be that $I = R$. That is, the only ideal of R that contains e is the ring R itself.

EXAMPLE 1 In Section 5.1 we saw that the set $\mathbf{E}$ of all even integers is a subring of the ring $\mathbf{Z}$ of all integers. To show that condition (b) of Definition 6.1(A) holds, let $x \in \mathbf{E}$ and $m \in \mathbf{Z}$. Since $x \in \mathbf{E}$, $x = 2k$ for some integer k. We have

$$xm = mx = m(2k) = 2(mk),$$

so $xm = mx$ is in $\mathbf{E}$. Thus $\mathbf{E}$ is an ideal of $\mathbf{Z}$. □

In combination with Theorem 5.3, Definition 6.1(A) provides the following check list of conditions that must be satisfied in order that a subset I of a ring R be an ideal:

(1) I is nonempty.
(2) $x \in I$ and $y \in I$ imply $x + y$ and xy are in I.
(3) $x \in I$ implies $-x \in I$.
(4) $x \in I$ and $r \in R$ imply xr and rx are in I.

The multiplicative closure in the second condition is implied by the fourth condition, so it may be deleted to obtain an alternate form of the definition of an ideal.

DEFINITION 6.1(B) A subset I of a ring R is an **ideal** of R provided the following conditions are satisfied:

(1) I is nonempty.
(2) $x \in I$ and $y \in I$ imply $x + y \in I$.
(3) $x \in I$ implies $-x \in I$.
(4) $x \in I$ and $r \in R$ imply xr and rx are in I.

A more efficient check list is given in Problem 1 at the end of this section.

EXAMPLE 2 In Problem 28 of Exercises 5.1, we saw that the set

$$S = \left\{ \begin{bmatrix} a & b \\ 0 & c \end{bmatrix} \middle| a, b, c \in \mathbf{Z} \right\}$$

forms a noncommutative ring with respect to the operations of matrix addition and multiplication. In this ring S, consider the subset

$$I = \left\{ \begin{bmatrix} 0 & b \\ 0 & 0 \end{bmatrix} \middle| b \in \mathbf{Z} \right\},$$

which is clearly nonempty. Since

$$\begin{bmatrix} 0 & x \\ 0 & 0 \end{bmatrix} + \begin{bmatrix} 0 & y \\ 0 & 0 \end{bmatrix} = \begin{bmatrix} 0 & x + y \\ 0 & 0 \end{bmatrix},$$

I is closed under addition. And since

$$-\begin{bmatrix} 0 & b \\ 0 & 0 \end{bmatrix} = \begin{bmatrix} 0 & -b \\ 0 & 0 \end{bmatrix},$$

I contains the additive inverse of each of its elements. For arbitrary $\begin{bmatrix} x & y \\ 0 & z \end{bmatrix}$ in S,

we have

$$\begin{bmatrix} 0 & b \\ 0 & 0 \end{bmatrix}\begin{bmatrix} x & y \\ 0 & z \end{bmatrix} = \begin{bmatrix} 0 & bz \\ 0 & 0 \end{bmatrix} \quad \text{and} \quad \begin{bmatrix} x & y \\ 0 & z \end{bmatrix}\begin{bmatrix} 0 & b \\ 0 & 0 \end{bmatrix} = \begin{bmatrix} 0 & xb \\ 0 & 0 \end{bmatrix},$$

and both of these products are in I. Thus I is an ideal of S. □

EXAMPLE 3   Example 8 in Section 5.1 introduced the ring M of all 2×2 matrices over the real numbers $\mathcal{R}$, and Problem 29 of Exercises 5.1 introduced the subring T of M, given by

$$T = \left\{ \begin{bmatrix} a & a \\ b & b \end{bmatrix} \middle| a, b \in \mathcal{R} \right\}.$$

For arbitrary $\begin{bmatrix} a & a \\ b & b \end{bmatrix} \in T, \begin{bmatrix} x & y \\ z & w \end{bmatrix} \in M$, the product

$$\begin{bmatrix} x & y \\ z & w \end{bmatrix}\begin{bmatrix} a & a \\ b & b \end{bmatrix} = \begin{bmatrix} xa + yb & xa + yb \\ za + wb & za + wb \end{bmatrix}$$

is in T, so T absorbs multiplication on the left by elements of M. However, the product

$$\begin{bmatrix} a & a \\ b & b \end{bmatrix}\begin{bmatrix} x & y \\ z & w \end{bmatrix} = \begin{bmatrix} ax + az & ay + aw \\ bx + bz & by + bw \end{bmatrix}$$

is *not* always in T, and T does not absorb multiplication on the right by elements of M. This failure keeps T from being an ideal[†] of M. □

[†] T could be said to be a *left ideal* of M.

Example 1 may be generalized to the set of all multiples of any fixed integer n. That is, the set $\{nk \mid k \in \mathbf{Z}\}$ of all multiples of n is an ideal of $\mathbf{Z}$. Instead of proving this fact, we establish the following more general result.

EXAMPLE 4 Let R be a commutative ring with unity e. For any fixed $a \in R$, we shall show that the set

$$(a) = \{ar \mid r \in R\}$$

is an ideal of R.

This set is nonempty since $a = ae$ is in (a). Let $x = ar$ and $y = as$ be arbitrary elements of (a), where $r \in R$, $s \in R$. Then

$$x + y = ar + as = a(r + s)$$

where $r + s \in R$, so (a) is closed under addition. We also have

$$-x = -(ar) = a(-r)$$

where $-r \in R$, so (a) contains additive inverses. For arbitrary $t \in R$,

$$tx = xt = (ar)t = a(rt)$$

where $rt \in R$. Thus $tx = xt$ is in (a) for arbitrary $x \in (a)$, $t \in R$, and (a) is an ideal of R. □

This example leads to the following definition.

DEFINITION 6.2 If a is a fixed element of the commutative ring R with unity, the ideal

$$(a) = \{ar \mid r \in R\},$$

which consists of all multiples of a by elements r of R, is called the **principal ideal** generated by a in R.

The next theorem gives an indication of the importance of principal ideals.

THEOREM 6.3 In the ring $\mathbf{Z}$ of integers, every ideal is a principal ideal.

Proof The trivial ideal $\{0\}$ is certainly a principal ideal, $\{0\} = (0)$. Consider then an ideal I of $\mathbf{Z}$ such that $I \neq \{0\}$. Since $I \neq \{0\}$, I contains an integer $m \neq 0$. And since I contains both m and $-m$, it must contain some positive integers. Let n be the least positive integer in I. (Such an n exists, by the Well-Ordering Theorem.) For an arbitrary $k \in I$, the Division Algorithm asserts that there are integers q and r such that

$$k = nq + r, \quad \text{with } 0 \le r < n.$$

Solving for r, we have

$$r = k - nq,$$

and this equation shows that $r \in I$ since k and n are in I and I is an ideal. That is, r is an element of I such that $0 \le r < n$, where r is the *least positive element* of I. This forces the equality $r = 0$, and therefore $k = nq$. It follows that every element of I is a multiple of n, and therefore $I = (n)$. ∎

Part of the analogy between ideals of a ring and normal subgroups of a group lies in the fact that ideals form the basis for a quotient structure much like the quotient group formed from the cosets of a normal subgroup.

To begin with, a ring R is an abelian group under addition, and any ideal I of R is a normal subgroup of this additive group. Thus we may consider the additive quotient group R/I that consists of all the cosets

$$r + I = I + r = \{r + x \mid x \in I\}$$

of I in R. From our work in Chapter 4, we know that

$$a + I = b + I \quad \text{if and only if} \quad a - b \in I,$$
$$(a + I) + (b + I) = (a + b) + I,$$

and that R/I is an abelian group with respect to this operation of addition.

In order to make a ring from these cosets, we consider a multiplication defined by

$$(a + I)(b + I) = ab + I.$$

We must show that this multiplication is well-defined. That is, we need to show that if

$$a + I = a' + I \quad \text{and} \quad b + I = b' + I$$

then

$$ab + I = a'b' + I.$$

Now

$$a + I = a' + I \Rightarrow a = a' + x, \quad \text{where } x \in I$$
$$b + I = b' + I \Rightarrow b = b' + y, \quad \text{where } y \in I.$$

Thus

$$ab = (a' + x)(b' + y) = a'b' + a'y + xb' + xy.$$

Since $x \in I$, $y \in I$, and I is an ideal, each of $a'y$, xb', and xy are in I. Therefore their sum

$$z = a'y + xb' + xy$$

is in I, and $z + I = I$. This gives

$$ab + I = a'b' + z + I = a'b' + I$$

and our product is well-defined.

THEOREM 6.4   Let I be an ideal of the ring R. Then the set R/I of additive cosets $r + I$ of I in R forms a ring with respect to coset addition

$$(a + I) + (b + I) = (a + b) + I$$

and our product is well-defined.

$$(a + I)(b + I) = ab + I.$$

Proof   Assume I is an ideal of R. We noted earlier that the additive quotient group R/I is an abelian group with respect to addition.
        We have already proved that the product

$$(a + I)(b + I) = ab + I$$

is well-defined in R/I, and closure under multiplication is automatic from the definition of this product. That the product is associative follows from

$$
\begin{aligned}
(a + I)[(b + I)(c + I)] &= (a + I)(bc + I) \\
&= a(bc) + I \\
&= (ab)c + I \quad \text{since multiplication is associative} \\
&\qquad\qquad\quad \text{in } R \\
&= (ab + I)(c + I) \\
&= [(a + I)(b + I)](c + I).
\end{aligned}
$$

Verifying the left distributive law, we have

$$
\begin{aligned}
(a + I)[(b + I) + (c + I)] &= (a + I)[(b + c) + I] \\
&= a(b + c) + I \\
&= (ab + ac) + I \quad \text{from the left distributive} \\
&\qquad\qquad\qquad\quad \text{law in } R \\
&= (ab + I) + (ac + I) \\
&= (a + I)(b + I) + (a + I)(c + I).
\end{aligned}
$$

The proof of the right distributive law is similar. Leaving that as an exercise, we conclude that R/I is a ring.   ∎

DEFINITION 6.5   If I is an ideal of the ring R, the ring R/I described in Theorem 6.4 is called[†] the **quotient ring** of R by I.

EXAMPLE 5   In the ring $\mathbf{Z}$ of integers, consider the principal ideal

$$(4) = \{4k \mid k \in \mathbf{Z}\}.$$

The distinct elements of the ring $\mathbf{Z}/(4)$ are

$$(4) = \{\ldots, -8, -4, 0, 4, 8, \ldots\}$$
$$1 + (4) = \{\ldots, -7, -3, 1, 5, 9, \ldots\}$$
$$2 + (4) = \{\ldots, -6, -2, 2, 6, 10, \ldots\}$$
$$3 + (4) = \{\ldots, -5, -1, 3, 7, 11, \ldots\}.$$

We see, then, that these cosets are the same as the elements of $\mathbf{Z}_4$:

$$(4) = [0], \qquad 1 + (4) = [1], \qquad 2 + (4) = [2], \qquad 3 + (4) = [3].$$

Moreover, the addition

$$\{a + (4)\} + \{b + (4)\} = \{a + b\} + (4)$$

agrees exactly with

$$[a] + [b] = [a + b]$$

in $\mathbf{Z}_4$, and the multiplication

$$\{a + (4)\}\{b + (4)\} = ab + (4)$$

agrees exactly with

$$[a][b] = [ab]$$

in $\mathbf{Z}_4$. Thus $\mathbf{Z}/(4)$ is our old friend $\mathbf{Z}_4$. Put another way, $\mathbf{Z}_4$ is the quotient ring of the integers $\mathbf{Z}$ by the ideal (4).   □

The specific case in Example 5 generalizes at once to an arbitrary integer $n > 1$, and we see that $\mathbf{Z}_n$ is the quotient ring of $\mathbf{Z}$ by the ideal (n). This is our final and best description of $\mathbf{Z}_n$.

As a final remark to this section, we note that

$$(a + I)(b + I) = ab + I$$
$$\neq \{xy \mid x \in a + I \text{ and } y \in b + I\}.$$

[†] R/I is also known as "the ring of residue classes modulo the ideal I."

As a particular instance, consider $I = (4)$ as in Example 5. We have

$$(0 + I)(0 + I) = 0 + I = I,$$

but

$$\{xy \mid x \in 0 + I \text{ and } y \in 0 + I\} = \{16r \mid r \in \mathbf{Z}\}$$

since $x = 4p$ and $y = 4q$ for $p, q \in \mathbf{Z}$ imply $xy = 16pq$.

EXERCISES 6.1

1   Let I be a subset of the ring R. Prove that I is an ideal of R if and only if I is nonempty and $x - y$, xr, and rx are in I for all $x, y \in I, r \in R$.

2   Complete the proof of Theorem 6.4 by proving the right distributive law in R/I.

3   If I_1 and I_2 are two ideals of the ring R, prove that $I_1 \cap I_2$ is an ideal of R.

4   If $\{I_\lambda\}$, $\lambda \in \mathcal{L}$, is an arbitrary collection of ideals I_λ of the ring R, prove that $\bigcap_{\lambda \in \mathcal{L}} I_\lambda$ is an ideal of R.

5   Find two ideals I_1 and I_2 of the ring $\mathbf{Z}$ such that:
(a) $I_1 \cup I_2$ is *not* an ideal of $\mathbf{Z}$.
(b) $I_1 \cup I_2$ is an ideal of $\mathbf{Z}$.

6   If I_1 and I_2 are two ideals of the ring R, prove that the set

$$I_1 + I_2 = \{x + y \mid x \in I_1, y \in I_2\}$$

is an ideal of R that contains each of I_1 and I_2.

7   Prove that if R is a field, then R has only the trivial ideals $\{0\}$ and R.

8   Let I be an ideal in a commutative ring R with unity. Prove that if I contains an element a that has a multiplicative inverse, then $I = R$.

9   In the ring $\mathbf{Z}$ of integers, prove that every subring is an ideal.

10   Let m and n be nonzero integers. Prove that $(m) \subseteq (n)$ if and only if n divides m.

11   If a and b are nonzero integers and m is the least common multiple of a and b, prove that $(a) \cap (b) = (m)$.

12   Prove that every ideal of $\mathbf{Z}_n$ is a principal ideal. [Hint: See Corollary 3.19.]

13   Find all distinct principal ideals of $\mathbf{Z}_n$ for the given value of n.
(a) $n = 7$ (b) $n = 11$ (c) $n = 12$
(d) $n = 18$ (e) $n = 20$ (f) $n = 24$

14   If R is a commutative ring and a is a fixed element of R, prove that the set $I_a = \{x \in R \mid ax = 0\}$ is an ideal of R.

15   (See Problem 28 of Exercises 5.1.) Given that the set

$$S = \left\{ \begin{bmatrix} x & y \\ 0 & z \end{bmatrix} \,\middle|\, x, y, z \in \mathbf{Z} \right\}$$

is a ring with respect to matrix addition and multiplication, show that

$$I = \left\{ \begin{bmatrix} a & b \\ 0 & 0 \end{bmatrix} \,\middle|\, a, b \in \mathbf{Z} \right\}$$

is an ideal of S.

16   (a) Show that the set

$$R = \left\{ \begin{bmatrix} x & 0 \\ y & 0 \end{bmatrix} \middle| x, y \in \mathbf{Z} \right\}$$

is a ring with respect to matrix addition and multiplication.
(b) Is R commutative?
(c) Does R have a unity?
(d) Decide whether or not the set

$$U = \left\{ \begin{bmatrix} 0 & 0 \\ a & 0 \end{bmatrix} \middle| a \in \mathbf{Z} \right\}$$

is an ideal of R and justify your answer.

17   For a fixed element a of a commutative ring R, prove that the set $I = \{ar \mid r \in R\}$ is an ideal of R. [Hint: Compare this with Example 4, and note that the element a itself may not be in this set I.]

18   Let R be a commutative ring that does not have a unity. For a fixed $a \in R$, prove that the set

$$(a) = \{na + ra \mid n \in \mathbf{Z}, r \in R\}$$

is an ideal of R that contains the element a. [This ideal is called the **principal ideal** of R that is *generated by a*.]

19   An element a of a ring R is called **nilpotent** if $a^n = 0$ for some positive integer n. Show that the set of all nilpotent elements in a commutative ring R forms an ideal of R.

20   If I is an ideal of R, prove that the set

$$K_I = \{x \in R \mid xa = 0 \text{ for all } a \in I\}$$

is an ideal of R.

21   Let R be a commutative ring with unity whose only ideals are $\{0\}$ and R itself. Prove that R is a field. [Hint: See Problem 17.]

22   Suppose that R is a commutative ring with unity and that I is an ideal of R. Prove that the set of all $x \in R$ such that $x^n \in I$ for some positive integer n is an ideal of R.

23   An ideal I of a commutative ring R is a **prime ideal** if $ab \in I$ implies either $a \in I$ or $b \in I$. Let R be a commutative ring with unity, and suppose $I \neq R$ is an ideal of R. Prove that R/I is an integral domain if and only if I is a prime ideal.

24   Prove that an ideal $(n) \neq \{0\}$ of $\mathbf{Z}$ is a prime ideal if and only if n is a prime integer.

6.2   RING HOMOMORPHISMS

We turn our attention now to ring homomorphisms and their relations to ideals and quotient rings.

DEFINITION 6.6   If R and R' are rings, a **ring homomorphism** from R to R' is a mapping $\theta: R \to R'$ such that

$$\theta(x + y) = \theta(x) + \theta(y) \quad \text{and} \quad \theta(xy) = \theta(x)\theta(y)$$

for all x and y in G.

That is, a ring homomorphism is a mapping from one ring to another that preserves both ring operations. This situation is analogous to the one where a homomorphism from one group to another preserves the group operation, and it explains the use of the term "homomorphism" in both situations. It is sometimes desirable to use one of the terms "group homomorphism" or "ring homomorphism" for clarity, but in many cases the context makes the meaning clear for the single word "homomorphism." If only groups are under consideration, then "homomorphism" means "group homomorphism," and if rings are under consideration, "homomorphism" means "ring homomorphism."

Some terminology for special types of homomorphisms is given in the following definition.

DEFINITION 6.7   Let θ be a homomorphism from the ring R to the ring R'.

(1)   If θ is surjective, then θ is called an **epimorphism** and R' is called a **homomorphic image** of R.

(2)   If θ is a bijection (both surjective and injective), then θ is an **isomorphism**.

EXAMPLE 1   Consider the mapping $\theta: \mathbf{Z} \to \mathbf{Z}_n$ defined by

$$\theta(a) = [a].$$

Since

$$\theta(a + b) = [a + b] = [a] + [b] = \theta(a) + \theta(b)$$

and

$$\theta(ab) = [ab] = [a][b] = \theta(a)\theta(b)$$

for all a and b in $\mathbf{Z}$, θ is a homomorphism from $\mathbf{Z}$ to $\mathbf{Z}_n$. In fact, θ is an *epimorphism* and $\mathbf{Z}_n$ is a *homomorphic image* of $\mathbf{Z}$. □

EXAMPLE 2   Consider $\theta: \mathbf{Z}_6 \to \mathbf{Z}_6$ defined by

$$\theta([a]) = 4[a].$$

It follows from

$$\theta([a] + [b]) = 4([a] + [b])$$
$$= 4[a] + 4[b]$$
$$= \theta([a]) + \theta([b])$$

that θ preserves addition. For multiplication, we have

$$\theta([a][b]) = \theta([ab]) = 4[ab] = [4ab]$$

and

$$\theta([a])\theta([b]) = (4[a])(4[b]) = 16[ab] = [16ab] = [4ab]$$

since $[16] = [4]$ in $\mathbf{Z}_6$. Thus θ is a homomorphism. It can be verified that $\theta(\mathbf{Z}_6) = \{[0], [2], [4]\}$, and we see that θ is neither surjective nor injective.   □

THEOREM 6.8   If θ is a homomorphism from the ring R to the ring R', then
(a) $\theta(0) = 0$, and
(b) $\theta(-r) = -\theta(r)$ for all $r \in R$.

Proof   The statement in part (a) follows from

$$\theta(0) = \theta(0) + 0$$
$$= \theta(0) + \theta(0) - \theta(0)$$
$$= \theta(0 + 0) - \theta(0)$$
$$= \theta(0) - \theta(0)$$
$$= 0.$$

To prove part (b), we observe that

$$\theta(r) + \theta(-r) = \theta[r + (-r)]$$
$$= \theta(0)$$
$$= 0.$$

Since the additive inverse is unique in the additive group of R',
$$-\theta(r) = \theta(-r).$$ ∎

Under a ring homomorphism, images of subrings are subrings and inverse images of subrings are also subrings. This is the content of the next theorem.

THEOREM 6.9   Suppose θ is a homomorphism from the ring R to the ring R'.
(a)   If S is a subring of R, then $\theta(S)$ is a subring of R'.
(b)   If S' is a subring of R', then $\theta^{-1}(S')$ is a subring of R.

Proof To prove part (a), suppose S is a subring of R. We shall verify that the conditions of Theorem 5.3 are satisfied by $\theta(S)$. The element $\theta(0) = 0$ is in $\theta(S)$, so $\theta(S)$ is nonempty. Let x' and y' be arbitrary elements of $\theta(S)$. Then there exist elements $x, y \in S$ such that $\theta(x) = x'$ and $\theta(y) = y'$. Since S is a subring, $x + y$ and xy are in S. Therefore

$$\theta(x + y) = \theta(x) + \theta(y)$$
$$= x' + y'$$

and

$$\theta(xy) = \theta(x)\theta(y) = x'y'$$

are in $\theta(S)$, and $\theta(S)$ is closed under addition and multiplication. Since $-x$ is in S and

$$\theta(-x) = -\theta(x) = -x',$$

we have $-x' \in \theta(S)$, and it follows that $\theta(S)$ is a subring of R'.

To prove part (b), assume that S' is a subring of R'. We have 0 in $\theta^{-1}(S')$ since $\theta(0) = 0$, so $\theta^{-1}(S')$ is nonempty. Let $x \in \theta^{-1}(S')$ and $y \in \theta^{-1}(S')$. This implies that $\theta(x) \in S'$ and $\theta(y) \in S'$. Hence $\theta(x) + \theta(y) = \theta(x + y)$ and $\theta(x)\theta(y) = \theta(xy)$ are in S' since S' is a subring. Now

$$\theta(x + y) \in S' \Rightarrow x + y \in \theta^{-1}(S')$$

and

$$\theta(xy) \in S' \Rightarrow xy \in \theta^{-1}(S').$$

We also have

$$\theta(x) \in S' \Rightarrow -\theta(x) = \theta(-x) \in S'$$
$$\Rightarrow -x \in \theta^{-1}(S'),$$

and $\theta^{-1}(S')$ is a subring of R by Theorem 5.3. ∎

DEFINITION 6.10 If θ is a homomorphism from the ring R to the ring R', the **kernel** of θ is the set

$$\ker \theta = \{x \in R \mid \theta(x) = 0\}.$$

EXAMPLE 3 In Example 1, the epimorphism $\theta : \mathbf{Z} \to \mathbf{Z}_n$ is defined by $\theta(a) = [a]$. Now $\theta(a) = [0]$ if and only if a is a multiple of n, so

$$\ker \theta = \{\ldots, -2n, -n, 0, n, 2n, \ldots\}$$

for this θ.

In Example 2, the homomorphism $\theta: \mathbf{Z}_6 \to \mathbf{Z}_6$ defined by $\theta([a]) = 4[a]$ has kernel given by

$$\ker \theta = \{[0], [3]\}. \qquad \Box$$

In these two examples, $\ker \theta$ is an ideal of the domain of θ. This is true in general for homomorphisms, according to the following theorem.

THEOREM 6.11 If θ is any homomorphism from the ring R to the ring R', then $\ker \theta$ is an ideal of R, and $\ker \theta = \{0\}$ if and only if θ is injective.

Proof Under the hypothesis, we know that $\ker \theta$ is a subring of R' from Theorem 6.9. For any $x \in \ker \theta$ and $r \in R$, we have

$$\theta(xr) = \theta(x)\theta(r)$$
$$= 0 \cdot \theta(r) = 0,$$

and similarly $\theta(rx) = 0$. Thus xr and rx are in $\ker \theta$, and $\ker \theta$ is an ideal of R.

If θ is injective, then $\theta(x) \in \ker \theta$ implies $\theta(x) = 0 = \theta(0)$ and therefore $x = 0$. Hence $\ker \theta = \{0\}$ if θ is injective. Conversely, if $\ker \theta = \{0\}$, then

$$\theta(x) = \theta(y) \Rightarrow \theta(x) - \theta(y) = 0$$
$$\Rightarrow \theta(x - y) = 0$$
$$\Rightarrow x - y = 0$$
$$\Rightarrow x = y.$$

This means that θ is injective if $\ker \theta = \{0\}$, and the proof is complete. ∎

EXAMPLE 4 This example illustrates the last part of Theorem 6.11 and provides a nice example of a ring isomorphism.

For the set $U = \{a, b\}$, the power set of U is $\mathscr{P}(U) = \{\varnothing, A, B, U\}$, where $A = \{a\}$ and $B = \{b\}$. With addition defined by

$$X + Y = (X \cup Y) - (X \cap Y)$$

and multiplication by

$$X \cdot Y = X \cap Y,$$

$\mathscr{P}(U)$ forms a ring, as we saw in Example 5 of Section 5.1. Addition and multiplication tables for $\mathscr{P}(U)$ are given in Figure 6.1.

+	∅	A	B	U
∅	∅	A	B	U
A	A	∅	U	B
B	B	U	∅	A
U	U	B	A	∅

·	∅	A	B	U
∅	∅	∅	∅	∅
A	∅	A	∅	A
B	∅	∅	B	B
U	∅	A	B	U

FIGURE 6.1

The ring $R = \mathbf{Z}_2 \oplus \mathbf{Z}_2$ was introduced in Problems 32 and 33 of Exercises 5.1. If we write 0 for [0] and 1 for [1] in $\mathbf{Z}_2$, the elements of R are given by $R = \{(0,0), (1,0), (0,1), (1,1)\}$. Tables for R are displayed in Figure 6.2.

+	(0,0)	(1,0)	(0,1)	(1,1)
(0,0)	(0,0)	(1,0)	(0,1)	(1,1)
(1,0)	(1,0)	(0,0)	(1,1)	(0,1)
(0,1)	(0,1)	(1,1)	(0,0)	(1,0)
(1,1)	(1,1)	(0,1)	(1,0)	(0,0)

·	(0,0)	(1,0)	(0,1)	(1,1)
(0,0)	(0,0)	(0,0)	(0,0)	(0,0)
(1,0)	(0,0)	(1,0)	(0,0)	(1,0)
(0,1)	(0,0)	(0,0)	(0,1)	(0,1)
(1,1)	(0,0)	(1,0)	(0,1)	(1,1)

FIGURE 6.2

Consider the mapping $\theta: \mathscr{P}(U) \to R$ defined by

$$\theta(\varnothing) = (0,0), \qquad \theta(A) = (1,0), \qquad \theta(B) = (0,1), \qquad \theta(U) = (1,1).$$

If each element x in the tables for $\mathscr{P}(U)$ is replaced by $\theta(x)$, the resulting tables agree completely with those in Figure 6.2. Thus θ is an isomorphism. We note that the kernel of θ consists of the zero element in $\mathscr{P}(U)$. □

We know now that every kernel of a homomorphism from a ring R is an ideal of R. The next theorem shows that every ideal of R is a kernel of a homomorphism from R. This means that the ideals of R and the kernels of the homomorphisms from R to another ring are the same subrings of R.

THEOREM 6.12 If I is an ideal of the ring R, the mapping $\theta: R \to R/I$ defined by

$$\theta(r) = r + I$$

is an epimorphism from R to R/I.

Proof The rule $\theta(r) = r + I$ clearly defines a surjective mapping θ from R to R/I. Since

$$\theta(x + y) = (x + y) + I$$
$$= (x + I) + (y + I)$$
$$= \theta(x) + \theta(y)$$

and

$$\theta(xy) = xy + I$$
$$= (x + I)(y + I)$$
$$= \theta(x)\theta(y),$$

θ is indeed an epimorphism from R to R/I. ∎

The last theorem shows that every quotient ring of a ring R is a homomorphic image of R. A result in the opposite direction is given in the next theorem.

THEOREM 6.13 If a ring R' is a homomorphic image of the ring R, then R' is isomorphic to a quotient ring of R.

Proof Suppose θ is an epimorphism from R to R', and let $K = \ker \theta$. For each $a + K$ in R/K, define $\phi(a + K)$ by

$$\phi(a + K) = \theta(a).$$

To prove that this rule defines a mapping, let $a + K$ and $b + K$ be arbitrary elements of R/K. Then

$$a + K = b + K \Leftrightarrow a - b \in K$$
$$\Leftrightarrow \theta(a - b) = 0$$
$$\Leftrightarrow \theta(a) = \theta(b)$$
$$\Leftrightarrow \phi(a + K) = \phi(b + K).$$

This shows that ϕ is well-defined and injective as well. From the definition of ϕ, it follows that $\phi(R/K) = \theta(R)$. But $\theta(R) = R'$ since θ is an epimorphism. Thus ϕ is surjective, and consequently is a bijection from R/K to R'.

For arbitrary $a + K$ and $b + K$ in R/K,

$$\phi[(a + K) + (b + K)] = \phi[(a + b) + K]$$
$$= \theta(a + b)$$
$$= \theta(a) + \theta(b) \quad \text{since } \theta \text{ is an epimorphism}$$
$$= \phi(a + K) + \phi(b + K)$$

and

$$\phi[(a + K)(b + K)] = \phi(ab + K)$$
$$= \theta(ab)$$
$$= \theta(a)\theta(b) \quad \text{since } \theta \text{ is an epimorphism}$$
$$= \phi(a + K)\phi(b + K).$$

Thus ϕ is an isomorphism from R/K to R'. ∎

As an immediate consequence of the proof of this theorem, we have the following corollary. The statement of the corollary is frequently called the **Fundamental Theorem of Ring Homomorphisms**.

COROLLARY 6.14   If θ is an epimorphism from the ring R to the ring R', then R' is isomorphic to $R/\ker \theta$.

We see now that, in the sense of isomorphism, the homomorphic images of a ring R are the same as the quotient rings of R. This gives a systematic way to search for all the homomorphic images of a given ring. To illustrate the usefulness of this method, we shall find all the homomorphic images of the ring $\mathbf{Z}$ of integers.

EXAMPLE 5   In order to find all homomorphic images of $\mathbf{Z}$, we shall find all possible ideals of $\mathbf{Z}$ and form all possible quotient rings. According to Theorem 6.3, every ideal of $\mathbf{Z}$ is a principal ideal.

For the trivial ideal $(0) = \{0\}$, we obtain the quotient ring $\mathbf{Z}/(0)$, which is isomorphic to $\mathbf{Z}$ since $a + (0) = b + (0)$ if and only if $a = b$. For the other trivial ideal $(1) = \mathbf{Z}$, we obtain the quotient ring $\mathbf{Z}/\mathbf{Z}$, which has only one element and is isomorphic to $\{0\}$. As shown in the proof of Theorem 6.3, any nontrivial ideal I of $\mathbf{Z}$ has the form $I = (n)$ for some positive integer $n > 1$. For these ideals, we obtain the quotient rings[†] $\mathbf{Z}/(n) = \mathbf{Z}_n$. Thus the homomorphic images of $\mathbf{Z}$ are $\mathbf{Z}$ itself, $\{0\}$, and the rings $\mathbf{Z}_n$. $\qquad\qquad\square$

EXERCISES 6.2

Unless otherwise stated, R and R' denote arbitrary rings throughout this set of exercises.

1   Suppose θ is an epimorphism from R to R'. Prove that R' is commutative if R is commutative.

2   Prove that if θ is an epimorphism from R to R' and if R has a unity e, then $\theta(e)$ is a unity in R'.

3   (See Problem 2.) Suppose that θ is an epimorphism from R to R' and that R has a unity. Prove that if a^{-1} exists for $a \in R$, then $[\theta(a)]^{-1}$ exists, and $[\theta(a)]^{-1} = \theta(a^{-1})$.

4   (See Problem 15 of Exercises 6.1.) Assume that the set

$$S = \left\{ \begin{bmatrix} x & y \\ 0 & z \end{bmatrix} \middle| x, y, z \in \mathbf{Z} \right\}$$

is a ring with respect to matrix addition and multiplication.

[†] See the paragraph immediately following Example 5 in Section 6.1.

(a) Verify that the mapping $\theta : S \to \mathbf{Z}$ defined by $\theta \left(\begin{bmatrix} x & y \\ 0 & z \end{bmatrix} \right) = z$ is an epimorphism from S to $\mathbf{Z}$.

(b) Describe ker θ and exhibit an isomorphism from $S/\text{ker }\theta$ to $\mathbf{Z}$.

5 (See Problem 16 of Exercises 6.1.) Assume that the set

$$R = \left\{ \begin{bmatrix} x & 0 \\ y & 0 \end{bmatrix} \middle| x, y \in \mathbf{Z} \right\}$$

is a ring with respect to matrix addition and multiplication.

(a) Verify that the mapping $\theta : S \to \mathbf{Z}$ defined by $\theta \left(\begin{bmatrix} x & 0 \\ y & 0 \end{bmatrix} \right) = x$ is an epimorphism from R to $\mathbf{Z}$.

(b) Describe ker θ and exhibit an isomorphism from $R/\text{ker }\theta$ to $\mathbf{Z}$.

6 For any $a \in \mathbf{Z}$, let $[a]_6$ denote $[a]$ in $\mathbf{Z}_6$ and $[a]_2$ denote $[a]$ in $\mathbf{Z}_2$.

(a) Prove that the mapping $\theta : \mathbf{Z}_6 \to \mathbf{Z}_2$ defined by $\theta([a]_6) = [a]_2$ is a homomorphism.

(b) Find ker θ.

7 In the field $\mathscr{C}$ of complex numbers, show that the mapping θ that maps each complex number onto its conjugate, $\theta(a + bi) = a - bi$, is an isomorphism from $\mathscr{C}$ to $\mathscr{C}$.

8 (See Example 3 of Section 5.1.) Let S denote the subring of the real numbers that consists of all real numbers of the form $m + n\sqrt{2}$, with $m \in \mathbf{Z}$ and $n \in \mathbf{Z}$. Prove that $\theta(m + n\sqrt{2}) = m - n\sqrt{2}$ defines an isomorphism from S to S.

9 Consider the mapping $\theta : \mathbf{Z}_{12} \to \mathbf{Z}_{12}$ defined by $\theta([a]) = 4[a]$. Decide whether or not θ is a homomorphism, and justify your answer.

10 Figure 6.3 gives addition and multiplication tables for the ring $R = \{a, b, c\}$ in Problem 26 of Exercises 5.1. Use these tables together with addition and multiplication tables for $\mathbf{Z}_3$ to find an isomorphism from R to $\mathbf{Z}_3$.

+	a	b	c
a	a	b	c
b	b	c	a
c	c	a	b

·	a	b	c
a	a	a	a
b	a	c	b
c	a	b	c

FIGURE 6.3

11 Figure 6.4 gives addition and multiplication tables for the ring $R = \{a, b, c, d\}$ in Problem 27 of Exercises 5.1. Construct addition and multiplication tables for the subring $R' = \{[0], [2], [4], [6]\}$ of $\mathbf{Z}_8$, and find an isomorphism from R to R'.

+	a	b	c	d
a	a	b	c	d
b	b	c	d	a
c	c	d	a	b
d	d	a	b	c

·	a	b	c	d
a	a	a	a	a
b	a	c	a	c
c	a	a	a	a
d	a	c	a	c

FIGURE 6.4

12 (See Problem 32 of Exercises 5.1.) Let R_1 be the subring of $R \oplus R'$ that consists of all elements of the form $(r, 0)$, where $r \in R$. Prove that R_1 is isomorphic to R.

13 Each of the following rules determines a mapping $\theta : \mathcal{R} \to \mathcal{R}$, where $\mathcal{R}$ is the field of real numbers. Decide in each case whether or not θ preserves addition, whether or not θ preserves multiplication, and whether or not θ is a homomorphism.

(a) $\theta(x) = |x|$

(b) $\theta(x) = 2x$

(c) $\theta(x) = -x$

(d) $\theta(x) = x^2$

(e) $\theta(x) = \begin{cases} 0 & \text{if } x = 0 \\ \dfrac{1}{x} & \text{if } x \neq 0 \end{cases}$

(f) $\theta(x) = x + 1$

14 For each given value of n, find all homomorphic images of $\mathbf{Z}_n$. [Hint: See Problem 12 of Exercises 6.1.]

(a) $n = 6$

(b) $n = 8$

(c) $n = 10$

(d) $n = 12$

(e) $n = 15$

(f) $n = 18$

15 Suppose F is a field and θ is an epimorphism from F to a ring S such that $\ker \theta \neq F$. Prove that θ is an isomorphism and that S is a field.

16 Assume that θ is an epimorphism from R to R'. Prove the following statements.

(a) If I is an ideal of R, then $\theta(I)$ is an ideal of R'.

(b) If I' is an ideal of R', then $\theta^{-1}(I')$ is an ideal of R.

(c) The mapping $I \to \theta(I)$ is a bijection from the set of ideals I of R that contain $\ker \theta$ to the set of all ideals of R'.

17 In the ring $\mathbf{Z}$ of integers, let new operations of addition and multiplication be defined by

$$x \oplus y = x + y + 1 \quad \text{and} \quad x \odot y = xy + x + y,$$

where x and y are arbitrary integers and $x + y$ and xy denote the usual addition and multiplication in $\mathbf{Z}$.

(a) Prove that the integers form a ring R' with respect to $\oplus$ and $\odot$.

(b) Identify the zero element and unity of R'.

(c) Prove that $\mathbf{Z}$ is isomorphic to R'.

18 Let K and I be ideals of the ring R. Prove that $K/K \cap I$ is isomorphic to $(K + I)/I$. [Hint: See Problem 24 of Exercises 4.2.]

6.3 THE CHARACTERISTIC OF A RING

In this section we focus on the fact that the elements of a ring R form an abelian group under addition.

When the binary operation in a group G is multiplication, each element a of G generates a cyclic group $\langle a \rangle$ that consists of all integral powers of a. If there are positive integers n such that $a^n = e$ and m is the smallest such positive integer, then m is the (multiplicative) *order* of a.

When the binary operation in a group is addition, the cyclic subgroup $\langle a \rangle$ consists of all integral multiples ka of a. If there are positive integers n such that $na = 0$ and m is the smallest such positive integer, then m is the (additive) *order* of a. In a sense, the characteristic of a ring is a generalization from this idea.

DEFINITION 6.15 If there are positive integers n such that $nx = 0$ for *all* x in the ring R, then the smallest positive integer m such that $mx = 0$ for all $x \in R$ is called the **characteristic** of R. If no such positive integer exists, then R is said to be of **characteristic zero**.

It is logical in the last case to call zero the characteristic of R since $n = 0$ is the only integer such that $nx = 0$ for all $x \in R$.

EXAMPLE 1 The ring $\mathbf{Z}$ of integers has characteristic zero since $nx = 0$ for all $x \in \mathbf{Z}$ requires that $n = 0$. For the same reason, the field $\mathcal{R}$ of real numbers and the field $\mathcal{C}$ of complex numbers each have characteristic zero. $\square$

EXAMPLE 2 Consider the ring $\mathbf{Z}_6$. For the various elements of $\mathbf{Z}_6$, we have

$$1[0] = [0] \qquad 6[1] = [0] \qquad 3[2] = [0]$$
$$2[3] = [0] \qquad 3[4] = [0] \qquad 6[5] = [0].$$

Although smaller positive integers work for some individual elements of $\mathbf{Z}_6$, the smallest positive integer m such that $m[a] = [0]$ for all $[a] \in \mathbf{Z}_6$ is $m = 6$. Thus $\mathbf{Z}_6$ has characteristic 6. This example generalizes readily, and we see that $\mathbf{Z}_n$ has characteristic n. $\square$

THEOREM 6.16 Let R be a ring with unity e. If e has finite additive order m, then m is the characteristic of R.

Proof Suppose R is a ring with unity e and that e has finite additive order m. Then m is the least positive integer such that $me = 0$. For arbitrary $x \in R$,

$$mx = m(ex) = (me)x = 0 \cdot x = 0.$$

Thus $mx = 0$ for all $x \in R$, and m is the smallest positive integer for which this is true. By Definition 6.15, R has characteristic m. $\blacksquare$

In connection with the last theorem, we note that if R has a unity e and e does not have finite additive order, then R has characteristic zero. In either case, the characteristic can be determined simply by investigating the additive order of e.

THEOREM 6.17   The characteristic of an integral domain is either zero or a prime integer.

Proof   Let D be an integral domain. As mentioned before, D has characteristic zero if the additive order of the unity e is not finite. Suppose then, that e has finite additive order m. By Theorem 6.16, D has characteristic m, and we only need to show that m is a prime integer. Assume, to the contrary, that m is not a prime and $m = rs$ for positive integers r and s such that $1 < r < m$ and $1 < s < m$. Then we have $re \neq 0$ and $se \neq 0$, but

$$(re)(se) = (rs)e^2 = (rs)e = me = 0.$$

This is a contradiction to the fact that D is an integral domain. Therefore m is a prime integer and the proof is complete.   ∎

If the characteristic of a ring R is zero, it follows that R has an infinite number of elements. However, the converse is not true. R may have an infinite number of elements and not have characteristic zero. This is illustrated in the next example.

EXAMPLE 3   Consider the ring $\mathscr{P}(\mathbf{Z})$ of all subsets of the integers $\mathbf{Z}$, with operations

$$X + Y = (X \cup Y) - (X \cap Y)$$
$$X \cdot Y = X \cap Y$$

for all X, Y in $\mathscr{P}(\mathbf{Z})$. The ring $\mathscr{P}(\mathbf{Z})$ has an infinite number of elements, yet

$$X + X = (X \cup X) - (X \cap X)$$
$$= X - X$$
$$= \varnothing$$

where $\varnothing$ is the zero element for $\mathscr{P}(\mathbf{Z})$. Thus $\mathscr{P}(\mathbf{Z})$ has characteristic 2.   □

THEOREM 6.18   An integral domain with characteristic zero contains a subring that is isomorphic to $\mathbf{Z}$, and an integral domain with positive characteristic p contains a subring that is isomorphic to $\mathbf{Z}_p$.

Proof   Let D be an integral domain with unity e. Define the mapping $\theta : \mathbf{Z} \to D$ by

$$\theta(n) = ne$$

for each $n \in \mathbf{Z}$. Since

$$\theta(m + n) = (m + n)e = me + ne = \theta(m) + \theta(n)$$

and

$$\theta(mn) = (mn)e = mne^2 = (me)(ne) = \theta(m)\theta(n),$$

θ is a homomorphism from $\mathbf{Z}$ to D. By Theorem 6.9(a), $\theta(\mathbf{Z})$ is a subring of D.

Suppose D has characteristic zero. Then $ne = 0$ if and only if $n = 0$, and it follows that ker $\theta = \{0\}$. According to Theorem 6.11, this means that θ is injective and therefore an isomorphism from $\mathbf{Z}$ to the subring $\theta(\mathbf{Z})$ of D.

Suppose now that D has characteristic p. Then p is the additive order of e, and $ne = 0$ if and only if $p \mid n$, by Theorem 3.14(b). In this case, we have ker $\theta = (p)$, the set of all multiples of p in $\mathbf{Z}$. By Corollary 6.14, the subring $\theta(\mathbf{Z})$ of D is isomorphic to $\mathbf{Z}/(p) = \mathbf{Z}_p$. ∎

The terms *embedded* and *extension* were introduced in connection with quotient fields in Section 5.3. Stated in these terms, Theorem 6.18 says that any integral domain with characteristic zero has $\mathbf{Z}$ embedded in it, and any integral domain with characteristic p has $\mathbf{Z}_p$ embedded in it.

In Problem 17 of Exercises 5.3, a construction was given by which an arbitrary ring can be embedded in a ring with unity. The next theorem is an improvement on this statement.

THEOREM 6.19   Any ring R can be embedded in a ring S with unity that has the same characteristic as R.

Proof   If R has characteristic zero, Problem 17 of Exercises 5.3 gives a construction whereby R can be embedded in a ring S with unity. To see that the ring S has characteristic zero, we observe that

$$n(1, 0) = (n, 0) = (0, 0)$$

if and only if $n = 0$.

Suppose now that R has characteristic n. We follow the same type of construction as before with $\mathbf{Z}$ replaced by $\mathbf{Z}_n$. Let S be the set of all ordered pairs $([m], x)$ where $[m] \in \mathbf{Z}_n$ and $x \in R$. Equality in S is defined by

$$([m], x) = ([k], y) \quad \text{if and only if } [m] = [k] \text{ and } x = y.$$

Addition and multiplication are defined by

$$([m], x) + ([k], y) = ([m + k], x + y)$$

and

$$([m], x) \cdot ([k], y) = ([mk], my + kx + xy).$$

It is straightforward to show that S forms an abelian group with respect to addition, the zero element being $([0], 0)$. This is left as an exercise in Problem 16.

The rule for multiplication yields an element of S, but we need to show that this element is unique. To do this, let $([m_1], x_1) = ([m_2], x_2)$ and $([k_1], y_1) = ([k_2], y_2)$. Then $[m_1] = [m_2], x_1 = x_2$, $[k_1] = [k_2]$, and $y_1 = y_2$ from the definition of equality. Using the definition of multiplication and these equalities, we get

$$([m_1], x_1) \cdot ([k_1], y_1) = ([m_1 k_1], m_1 y_1 + k_1 x_1 + x_1 y_1)$$

and

$$([m_2], x_2) \cdot ([k_2], y_2) = ([m_2 k_2], m_2 y_2 + k_2 x_2 + x_2 y_2)$$
$$= ([m_1 k_1], m_2 y_1 + k_2 x_1 + x_1 y_1).$$

Comparing the results of these two computations, we see that we need

$$m_2 y_1 + k_2 x_1 = m_1 y_1 + k_1 x_1$$

to conclude that the results are equal. Now

$$[m_1] = [m_2] \Rightarrow m_2 - m_1 = pn \quad \text{for some } p \in \mathbf{Z}$$
$$\Rightarrow m_2 = m_1 + pn.$$

Therefore

$$m_2 y_1 = (m_1 + pn) y_1$$
$$= m_1 y_1 + npy_1$$
$$= m_1 y_1$$

since py_1 is in R and R has characteristic n. Similarly, $k_2 x_1 = k_1 x_1$, and we conclude that the product is well-defined.

Verifying that multiplication is associative, we have

$$([m], x)\{([k], y)([r], z)\} = ([m], x)([kr], kz + ry + yz)$$
$$= ([mkr], mkz + mry + myz + krx + kxz$$
$$+ rxy + xyz)$$
$$= ([mk], my + kx + xy) \cdot ([r], z)$$
$$= \{([m], x)([k], y)\}([r], z).$$

The left distributive law follows from

$$([m], x)\{([k], y) + ([r], z)\} = ([m], x)([k + r], y + z)$$
$$= ([mk + mr], my + mz + kx + rx + xy$$
$$+ xz)$$
$$= ([mk], my + kx + xy) + ([mr], mz + rx$$
$$+ xz)$$
$$= ([m], x)([k], y) + ([m], x)([r], z).$$

The verification of the right distributive law is similar to this, and is left as an exercise.

The argument up to this point shows that S is a ring. Since each of $\mathbf{Z}_n$ and R has characteristic n,

$$n([m], x) = (n[m], nx) = ([0], 0)$$

for all $([m], x)$ in S, and n is the least positive integer for which this is true. Thus S has characteristic n.

Consider now the mapping $\theta : R \to S$ defined by $\theta(x) = ([0], x)$ for all $x \in R$. Since

$$\theta(x) = \theta(y) \Leftrightarrow ([0], x) = ([0], y) \Leftrightarrow x = y,$$

θ is a bijection from R to $\theta(R)$. Now

$$\theta(x + y) = ([0], x + y) = ([0], x) + ([0], y) = \theta(x) + \theta(y)$$

and

$$\theta(xy) = ([0], xy) = ([0], x)([0], y) = \theta(x)\theta(y),$$

so θ is an isomorphism from R to $\theta(R)$, and $\theta(R)$ is a subring of S by Theorem 6.9(a). This shows that R is embedded in S.   ∎

EXERCISES 6.3

1   Find the characteristic of the following rings. ($R \oplus S$ is defined in Problem 32 of Exercises 5.1.)
   (a) $\mathbf{Z}_2 \oplus \mathbf{Z}_2$ (b) $\mathbf{Z}_3 \oplus \mathbf{Z}_3$ (c) $\mathbf{Z}_2 \oplus \mathbf{Z}_3$
   (d) $\mathbf{Z}_2 \oplus \mathbf{Z}_4$ (e) $\mathbf{Z}_4 \oplus \mathbf{Z}_6$

2   Let D be an integral domain with positive characteristic. Prove that all nonzero elements of D have the same additive order.

3   Show by example that the statement in Problem 2 is no longer true if "an integral domain" is replaced by "a ring."

4   Suppose that R and S are rings with positive characteristics m and n, respectively. If k is the least common multiple of m and n, prove that $R \oplus S$ has characteristic k.

5 Prove that if both of R and S in Problem 4 are integral domains, then $R \oplus S$ has characteristic mn if $m \neq n$.

6 Prove that the characteristic of a field is either 0 or a prime.

7 Let D be an integral domain with four elements, $D = \{0, e, a, b\}$, where e is the unity.
   (a) Prove that D has characteristic 2.
   (b) Construct an addition table for D.

8 Prove that $\mathbf{Z}_n$ has a nonzero element whose additive order is less than n if and only if n is not a prime integer.

9 Let R be a ring with more than one element that has no zero divisors. Prove that the characteristic of R is either zero or a prime integer.

10 A **Boolean ring** is a ring in which all elements x satisfy $x^2 = x$. Prove that every Boolean ring has characteristic 2.

11 Suppose R is a ring with positive characteristic n. Prove that if I is any ideal of R, then n is a multiple of the characteristic of I.

12 If F is a field with positive characteristic p, prove that the set
$$\{e, 2e, 3e, \ldots, (p-1)e\}$$
of multiples of the unity e forms a subfield of F.

13 If p is a positive prime integer, prove that any field with p elements is isomorphic to $\mathbf{Z}_p$.

14 Let I be the set of all elements of a ring R that have finite additive order. Prove that I is an ideal of R.

15 Prove that if a ring R has a finite number of elements, then the characteristic of R is a positive integer.

16 As in the proof of Theorem 6.19, let $S = \{([m], x) \,|\, [m] \in \mathbf{Z}_n \text{ and } x \in R\}$. Prove that S forms an abelian group with respect to addition.

17 With S as in Problem 16, prove that the right distributive law holds in S.

18 With S as in Problem 16, prove that the set $R' = \{([0], x) \,|\, x \in R\}$ is an ideal of S.

19 Prove that every ordered integral domain has characteristic zero.

Key Words and Phrases

Ideal	Isomorphism
Trivial ideals	Homomorphic image
Principal ideal	Kernel
Quotient ring	Fundamental Theorem of Ring
Ring homomorphism	Homomorphisms
Epimorphism	Characteristic of a ring

References

Bundrick, Charles M. and John J. Leeson. *Essentials of Abstract Algebra.* Monterey, Calif.: Brooks-Cole, 1972.

Durbin, John R. *Modern Algebra.* New York: Wiley, 1979.

Fraleigh, John B. *A First Course in Abstract Algebra* (2nd ed.). Reading, Mass.: Addison-Wesley, 1976.

Herstein, I. N. *Topics in Algebra* (2nd ed.). New York: Wiley, 1975.

Hillman, Abraham P. and Gerald L. Alexanderson. *A First Undergraduate Course in Abstract Algebra* (2nd ed.). Belmont, Calif.: Wadsworth, 1978.

McCoy, Neal H. *Introduction to Modern Algebra* (3rd ed.). Boston: Allyn and Bacon, 1975.

McCoy, Neal H. *Rings and Ideals* (Carus Mathematical Monograph No. 8). Washington, D.C.: The Mathematical Association of America, 1968.

McCoy, Neal H. *The Theory of Rings*. New York: Chelsea, 1972.

7 REAL AND
COMPLEX NUMBERS

7.1   THE FIELD OF REAL NUMBERS

At this point, it is possible to fit some of the familiar number systems into the structures developed in the preceding chapters.

In Theorem 5.32, the ring $\mathbf{Z}$ of all integers was characterized as an ordered integral domain in which the set of positive elements is well-ordered. By "characterized," we mean that any ordered integral domain in which the set of positive elements is well-ordered must be isomorphic to the ring $\mathbf{Z}$ of all integers.

At the end of Section 5.3 we noted that the construction of the rational numbers from $\mathbf{Z}$ is a special case of the procedure described in that section. That is, the set $\mathbf{Q}$ of all rational numbers is the quotient field of $\mathbf{Z}$ and therefore is the smallest field that contains $\mathbf{Z}$. From a more abstract point of view, the field of rational numbers can be characterized as the smallest ordered field. That is, any ordered field must contain a subfield that is isomorphic to $\mathbf{Q}$. (See Problems 22–24 of the exercises.)

The main goal of this section is to present a similar characterization for the field of real numbers. The following definition is essential.

DEFINITION 7.1   Let S be a nonempty subset of an ordered field F. An element u of F is an **upper bound** of S if $u \geq x$ for all $x \in S$. An element u of F is a **least upper bound** of S if these conditions are satisfied:

(1) u is an upper bound of S.

(2)   If $b \in F$ is an upper bound of S, then $b \geq u$.

The phrase "least upper bound" is abbreviated "l.u.b."

EXAMPLE 1   Let $F = \mathbf{Q}$ be the field of rational numbers, and let S be the set of all negative rational numbers.

If a is any negative rational number, then there exists $b \in \mathbf{Q}$ such that $0 > b > a$, by Problem 12 of Exercises 5.4. Thus no negative rational number is an upper bound of S. However, any positive rational number u is an upper bound of S since

$$u > 0 > x \quad \text{for all } x \in S.$$

The rational number 0 is also an upper bound of S since $0 > x$ for all $x \in S$. In fact, 0 is a least upper bound of S in $\mathbf{Q}$.   □

If $u \in F$ and $v \in F$ are both least upper bounds of the nonempty subset S of an ordered field F, then the second condition in Definition 7.1 requires both $v \geq u$ and $u \geq v$. Therefore $u = v$, and the least upper bound of S in F is unique whenever it exists.

Later we shall exhibit a nonempty subset of $\mathbf{Q}$ that has an upper bound in $\mathbf{Q}$ but does not have a least upper bound in $\mathbf{Q}$. The following theorem will be needed.

THEOREM 7.2   There is no rational number x such that $x^2 = 2$.

Proof   Assume that the theorem is false. That is, assume a rational number x exists such that $x^2 = 2$. We may assume, without loss of generality, that $x = p/q$ is expressed in *lowest terms* as a quotient of integers p and q. That is,

$$\left(\frac{p}{q}\right)^2 = 2,$$

with 1 as the greatest common divisor of p and q. This implies that

$$p^2 = 2q^2.$$

Hence 2 divides p^2, and since 2 is a prime, this implies that 2 divides p, by Theorem 2.14. Let $p = 2r$, where $r \in \mathbf{Z}$. Then we have

$$(2r)^2 = 2q^2$$
$$4r^2 = 2q^2$$

and therefore

$$2r^2 = q^2.$$

This implies, however, that 2 divides q, by another application of Theorem 2.14. Thus 2 is a common divisor of p and q, and we have a contradiction to the fact that 1 is the greatest common divisor of p and q. This contradiction establishes the theorem. ∎

EXAMPLE 2 Let

$$S = \{x \in \mathbf{Q} \mid x > 0 \text{ and } x^2 \leq 2\}.$$

We shall show that S is a nonempty subset of $\mathbf{Q}$ that has an upper bound in $\mathbf{Q}$ but does not have a l.u.b. in $\mathbf{Q}$.

The set S is nonempty since 1 is in S. The rational number 3 is an upper bound of S in $\mathbf{Q}$ since $x \geq 3$ requires $x^2 \geq 9$ by Problem 2(c) of Exercises 5.4.

It is not so easy to show that S does not have a l.u.b. in $\mathbf{Q}$. As a start, we shall prove the following two statements for *positive* $u \in \mathbf{Q}$:

1. If u is not an upper bound of S, then $u^2 < 2$.
2. If $u^2 < 2$, then u is not an upper bound of S.

Consider statement 1. If $u \in \mathbf{Q}$ is not an upper bound of S, then there exists $x \in S$ such that $0 < u < x$. By Problem 2(c) of Exercises 5.4, this implies that $u^2 < x^2$. Since $x^2 \leq 2$ for all $x \in S$, we have $u^2 < 2$.

To prove statement 2, suppose that $u \in \mathbf{Q}$ is positive and $u^2 < 2$. Then $\dfrac{2 - u^2}{2u + 1}$ is a positive rational number. By Problem 12 of Exercises 5.4, there exists a rational number d such that

$$0 < d < \min\left\{1, \frac{2 - u^2}{2u + 1}\right\}$$

where $\min\left\{1, \dfrac{2 - u^2}{2u + 1}\right\}$ denotes the smaller of the two numbers in braces. If we now put $v = u + d$, then v is a positive rational number, $v > u$, and

$$
\begin{aligned}
v^2 &= u^2 + 2ud + d^2 \\
&< u^2 + 2ud + d &\quad \text{since } 0 < d < 1 \text{ implies } 0 < d^2 < d \\
&= u^2 + (2u + 1)d \\
&< u^2 + (2u + 1) \cdot \frac{2 - u^2}{2u + 1} &\quad \text{since } d < \frac{2 - u^2}{2u + 1} \\
&= 2.
\end{aligned}
$$

Thus v is an element of S such that $v > u$, and u is not an upper bound of S.

Having established statements 1 and 2, we may combine them with Theorem 7.2 and obtain the following statement:

3. A positive $u \in \mathbf{Q}$ is an upper bound of S if and only if $u^2 > 2$.

With this fact at hand, we can now show that S does not have a l.u.b. in $\mathbf{Q}$.

Suppose $u \in \mathbf{Q}$ is an upper bound of S. Then u is positive since all elements of S are positive, and $u^2 > 2$ by statement 3. Let

$$w = u - \frac{u^2 - 2}{2u}$$

$$= \frac{u^2 + 2}{2u}$$

$$= \frac{u}{2} + \frac{1}{u}.$$

Then w is a positive rational number and $w < u$ since $\dfrac{u^2 - 2}{2u}$ is positive. Now

$$w^2 = \left(u - \frac{u^2 - 2}{2u} \right)^2$$

$$= u^2 - (u^2 - 2) + \left(\frac{u^2 - 2}{2u} \right)^2$$

$$= 2 + \left(\frac{u^2 - 2}{2u} \right)^2$$

$$> 2,$$

so w is an upper bound of S by statement 3. Since $w < u$, we have that u is not a least upper bound of S. Since u was an arbitrary upper bound of S in $\mathbf{Q}$, this proves that S does not have a l.u.b. in $\mathbf{Q}$. $\qquad\square$

Example 2 establishes a very significant deficiency in the field $\mathbf{Q}$ of rational numbers, namely that some nonempty sets of rational numbers have an upper bound in $\mathbf{Q}$ but fail to have a least upper bound in $\mathbf{Q}$. The next definition gives a designation for those ordered fields that do not have this deficiency.

DEFINITION 7.3 Let F be an ordered field. Then F is **complete** if every nonempty subset of F that has an upper bound in F has a least upper bound in F.

The basic difference between the field of rational numbers and the field of real numbers is that the real number field is complete. It is possible to construct the field of real numbers from the field of rational numbers, but this construction is too lengthy and difficult to be included here, and it is more properly a part of that area of mathematics known as *analysis*. The method of construction most commonly used is one that is credited to Richard Dedekind (1831–1916) and utilizes what are called *Dedekind cuts*. In our treatment, we shall assume the validity of the following theorem.

THEOREM 7.4   There exists a field $\mathscr{R}$, called the **field of real numbers**, that is a complete ordered field. Any complete ordered field F has the following properties:

(a) F is isomorphic to $\mathscr{R}$.

(b) F contains a subfield that is isomorphic to the field **Q** of rational numbers, and the ordering in F is an extension of the ordering in this subfield.

The set of all real numbers may be represented geometrically by setting up a one-to-one correspondence between real numbers and the points on a straight line. To begin, we select a point on a horizontal line, designate it as the *origin,* and let this point correspond to the number 0. A second point is now chosen to the right of the origin and we let this point correspond to the number 1. The distance between the two points corresponding to 0 and 1 is now taken as one unit of measure. Points on the line located successively 1 unit farther to the right are made to correspond to the positive integers 2, 3, 4, ... in succession. With the same unit of measure and beginning at the origin, points on the line located successively 1 unit farther to the left are made to correspond to the negative integers $-1, -2, -3, \ldots$ (see Figure 7.1). This sets up a one-to-one correspondence between the set **Z** of all integers and some of the points on the line.

Points on the line that correspond to nonintegral rational numbers are now located by using distances proportional to their expressions as quotients a/b of integers a and b, and by using directions to the right for positive numbers, and to the left for negative numbers. For example, the point corresponding to 3/2 is located midway between the points that correspond to 1 and 2, whereas the point corresponding to $-3/2$ is located midway between those that correspond to -1 and -2. In this manner, a one-to-one correspondence is established between the set **Q** of rational numbers and a subset of the points on the line.

It is not very difficult to demonstrate that there are points on the line that do not correspond to any rational number. This can be done by considering a

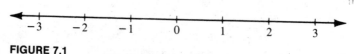

FIGURE 7.1

right triangle with each leg one unit in length (see Figure 7.2). By the Pythagorean Theorem, the length h of the hypotenuse of the triangle in Figure 7.2 satisfies the equation $h^2 = 2$. There is a point on the line located at a distance h units to the right of the origin, but this point cannot correspond to a rational number, by Theorem 7.2.

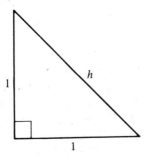

FIGURE 7.2

The foregoing demonstration shows that there are gaps in the rational numbers, even though any two distinct rational numbers have another rational number located between them (see Problem 12 of Exercises 5.4). We *assume* now that the one-to-one correspondence that we have set up between the rational numbers and points on the line can be extended to the set of all real numbers and the set of all points on the line. The points that do not correspond to rational numbers are assumed to correspond to real numbers that are not rational, that is, to *irrational* numbers. For example, the discussion in the preceding paragraph located the point that corresponds to the irrational number $h = \sqrt{2}$.

One more aspect of the real numbers is worthy of mention: the *decimal representation* of real numbers. Here we assume that each real number can be represented by a decimal expression that either terminates, such as

$$\frac{9}{8} = 1.125,$$

or continues without end, as do the repeating decimal[†]

$$\frac{14}{11} = 1.272727\cdots$$
$$= 1.\overline{27}$$

and the nonrepeating decimal

$$\sqrt{2} = 1.4142136\cdots.$$

[†] The bar above 27 indicates that the digits 27 repeat endlessly.

The decimal expression for a rational number a/b may be found by long division. For example, for the rational number 14/11, long division yields

$$
\begin{array}{r}
1.27 \\
11\overline{\smash{\big)}14.00} \\
\underline{11} \\
30 \\
\underline{22} \\
80 \\
\underline{77} \\
3
\end{array}
$$

The repetition of the remainder 3 at this point makes it clear that we have the repeating decimal expression

$$
\frac{14}{11} = 1.272727\cdots = 1.\overline{27}.
$$

A terminating decimal expression may be regarded as a repeating pattern where zeros repeat endlessly. For example,

$$
\frac{9}{8} = 1.125000\cdots = 1.125\overline{0}.
$$

With this point of view, the decimal expression for any rational number a/b will always have a repeating pattern. This can be seen from the long division algorithm: each remainder satisfies $0 \le r < b$, so there are only b distinct possibilities for the remainders, and the expression starts repeating whenever a remainder occurs for the second time.

Rational numbers that have a terminating decimal expression can be represented in another way by changing the range on the remainders in the long division from $0 \le r < b$ to $0 < r \le b$. If we perform the long division for 9/8 in this way, it appears as follows:

$$
\begin{array}{r}
1.1249 \\
8\overline{\smash{\big)}9.0000} \\
\underline{8} \\
10 \\
\underline{8} \\
20 \\
\underline{16} \\
40 \\
\underline{32} \\
80 \\
\underline{72} \\
8
\end{array}
$$

At this point, the remainder 8 has occurred twice, and the repeating pattern is seen to be

$$\frac{9}{8} = 1.124999\cdots = 1.124\overline{9}.$$

It can be proved that every repeating decimal expression represents a rational number, but we shall not go into that here. The next example provides some insight into why this assertion is true.

EXAMPLE 3   We shall express $2.1\overline{34}$ as a quotient of integers. If

$$x = 2.1343434\cdots,$$

then

$$10x = 21.343434\cdots$$

and

$$1000x = 2134.3434\cdots.$$

Thus

$$1000x - 10x = 2134.\overline{34} - 21.\overline{34}$$
$$990x = 2113$$
$$x = \frac{2113}{990}.$$ □

This discussion of decimal representations is not intended to be a rigorous presentation. Its purpose is to make the following remarks appear plausible:

1. Each real number can be represented by a decimal expression.
2. Decimal expressions that repeat or terminate represent rational numbers.
3. Decimal expressions that do not repeat and do not terminate represent irrational numbers.

EXERCISES 7.1

Find the decimal representation for each of the numbers in Problems 1–6.

1 $\dfrac{5}{9}$ 2 $\dfrac{7}{33}$ 3 $\dfrac{80}{81}$

4 $\dfrac{16}{7}$ 5 $\dfrac{22}{7}$ 6 $\dfrac{19}{11}$

Express each of the numbers in Problems 7–12 as a quotient of integers, reduced to lowest terms.

7 $3.\overline{4}$ **8** $1.\overline{6}$ **9** $0.\overline{12}$

10 $0.6\overline{3}$ **11** $2.5\overline{1}$ **12** $3.21\overline{321}$

13   Prove that $\sqrt{3}$ is irrational. (That is, prove there is no rational number x such that $x^2 = 3$.)

14   Prove that $\sqrt[3]{2}$ is irrational.

15   Prove that if p is a prime integer, then $\sqrt{p}$ is irrational.

16   Prove that if a is rational and b is irrational, then $a + b$ is irrational.

17   Prove that if a is a nonzero rational number and b is irrational, then ab is irrational.

18   Prove that if a is a nonzero irrational number then a^{-1} is an irrational number.

19   Prove that if a is a nonzero rational number and ab is irrational, then b is irrational.

20   Give counterexamples for the following statements.
(a) If a and b are irrational, then $a + b$ is irrational.
(b) If a and b are irrational, then ab is irrational.

21   Let S be a nonempty subset of an ordered field F.
(a) Write definitions for **lower bound** of S and **greatest lower bound** of S.
(b) Prove that if F is a complete ordered field and the nonempty subset S has a lower bound in F, then S has a greatest lower bound in F.

22   Prove that if F is an ordered field with F^+ as its set of positive elements, then $F^+ \supseteq \{ne \mid n \in \mathbf{Z}^+\}$, where e denotes the multiplicative identity in F. [Hint: See Theorem 5.31 and its proof.]

23   If F is an ordered field, prove that F contains a subring that is isomorphic to $\mathbf{Z}$. [Hint: See Theorem 5.31 and its proof.]

24   Prove that any ordered field must contain a subfield that is isomorphic to the field $\mathbf{Q}$ of rational numbers.

25   If a and b are positive real numbers, prove that there exists a positive integer n such that $na > b$. This property is called the **Archimedean Property** of the real numbers. [Hint: If $ma \le b$ for all $m \in \mathbf{Z}^+$, then b is an upper bound for the set $S = \{ma \mid m \in \mathbf{Z}^+\}$. Use the completeness property of $\mathscr{R}$ to arrive at a contradiction.]

26   Prove that if a and b are real numbers such that $a > b$, then there exists a rational number m/n such that $a > m/n > b$. [Hint: Use Problem 25 to obtain $n \in \mathbf{Z}^+$ such that $a - b > 1/n$. Then choose m to be the least integer such that $m > nb$. With these choices of m and n, show that $(m - 1)/n \le b$ and then that $a > m/n > b$.]

7.2   THE COMPLEX NUMBERS

The fact that negative real numbers do not have square roots in $\mathscr{R}$ is a serious deficiency of the field of real numbers, but it is one that can be overcome by the introduction of complex numbers.

Although we do not present a characterization of the field of complex numbers until Section 8.4, it is possible to construct the complex numbers from the real numbers. Such a construction is the main purpose of this section.

In our construction, complex numbers appear first as ordered pairs (a, b) and later in the more familiar form $a + bi$. The operations given in the following definition will seem more natural if they are compared with the usual operations on complex numbers in the form $a + bi$.

DEFINITION 7.5   Let $\mathscr{C}$ be the set of all ordered pairs (a, b) of real numbers a and b. Equality, addition, and multiplication are defined in $\mathscr{C}$ by

$(a, b) = (c, d)$ if and only if $a = c$ and $b = d$

$(a, b) + (c, d) = (a + c, b + d)$

$(a, b)(c, d) = (ac - bd, ad + bc)$.

The elements of $\mathscr{C}$ are called **complex numbers**.

It is easy to see that the stated rules for addition and multiplication do in fact define binary operations on $\mathscr{C}$.

THEOREM 7.6   With addition and multiplication as given in Definition 7.5, $\mathscr{C}$ is a field. The set of all elements of the form $(a, 0)$ in $\mathscr{C}$ forms a subfield of $\mathscr{C}$ that is isomorphic to the field $\mathscr{R}$ of real numbers.

Proof   Closure of $\mathscr{C}$ under addition follows at once from the fact that $\mathscr{R}$ is closed under addition. It is left as exercises to prove that addition is associative and commutative, that $(0, 0)$ is the additive identity in $\mathscr{C}$, and that the additive inverse of $(a, b) \in \mathscr{C}$ is $(-a, -b) \in \mathscr{C}$.

Since $\mathscr{R}$ is closed under multiplication and addition, each of $ac - bd$ and $ad + bc$ is in $\mathscr{R}$ whenever (a, b) and (c, d) are in $\mathscr{C}$. Thus $\mathscr{C}$ is closed under multiplication.

For the remainder of the proof, let (a, b), (c, d), and (e, f) represent arbitrary elements of $\mathscr{C}$. The associative property of multiplication is verified by the following computations:

$$
\begin{aligned}
(a, b)[(c, d)(e, f)] &= (a, b)(ce - df, cf + de) \\
&= [a(ce - df) - b(cf + de), a(cf + de) + b(ce - df)] \\
&= (ace - adf - bcf - bde, acf + ade + bce - bdf) \\
&= [(ac - bd)e - (ad + bc)f, (ac - bd)f + (ad + bc)e] \\
&= (ac - bd, ad + bc)(e, f) \\
&= [(a, b)(c, d)](e, f).
\end{aligned}
$$

211

Before considering the distributive laws, we shall show that multiplication is commutative in $\mathscr{C}$. This follows from

$$(c,d)(a,b) = (ca - db, cb + da)$$
$$= (ca - db, da + cb)$$
$$= (ac - bd, ad + bc)$$
$$= (a,b)(c,d).$$

We shall verify the left distributive property and leave the proof of the right distributive property as an exercise:

$$(a,b)[(c,d) + (e,f)] = (a,b)(c + e, d + f)$$
$$= [a(c + e) - b(d + f), a(d + f) + b(c + e)]$$
$$= (ac + ae - bd - bf, ad + af + bc + be)$$
$$= (ac - bd, ad + bc) + (ae - bf, af + be)$$
$$= (a,b)(c,d) + (a,b)(e,f).$$

To this point, we have established that $\mathscr{C}$ is a commutative ring. The computation

$$(1,0)(a,b) = (1 \cdot a - 0 \cdot b, 1 \cdot b + 0 \cdot a)$$
$$= (a,b)$$

shows that $(1,0)$ is a left identity for multiplication in $\mathscr{C}$. Since multiplication in $\mathscr{C}$ is commutative, it follows that $(1,0)$ is a nonzero unity in $\mathscr{C}$.

If $(a,b) \neq (0,0)$ in $\mathscr{C}$, then at least one of the real numbers a or b is nonzero, and it follows that $a^2 + b^2$ is a positive real number. Hence

$$\left(\frac{a}{a^2 + b^2}, \frac{-b}{a^2 + b^2} \right)$$

is an element of $\mathscr{C}$. The multiplication

$$(a,b)\left(\frac{a}{a^2 + b^2}, \frac{-b}{a^2 + b^2} \right) = \left(\frac{a^2 + b^2}{a^2 + b^2}, \frac{-ab + ba}{a^2 + b^2} \right) = (1,0)$$

shows that

$$(a,b)^{-1} = \left(\frac{a}{a^2 + b^2}, \frac{-b}{a^2 + b^2} \right)$$

since multiplication is commutative in $\mathscr{C}$. This completes the proof that $\mathscr{C}$ is a field.

Consider now the set R' that consists of all elements of $\mathscr{C}$ that have the form $(a,0)$:

$$R' = \{(a,0) \,|\, a \in \mathscr{R}\}.$$

The proof that R' is a subfield of $\mathscr{C}$ is left as an exercise. The mapping $\theta : \mathscr{R} \to R'$ defined by

$$\theta(a) = (a, 0)$$

is a bijection since $(a, 0) = (b, 0)$ if and only if $a = b$. For arbitrary a and b in $\mathscr{R}$,

$$\theta(a + b) = (a + b, 0)$$
$$= (a, 0) + (b, 0)$$
$$= \theta(a) + \theta(b)$$

and

$$\theta(ab) = (ab, 0)$$
$$= (a, 0)(b, 0)$$
$$= \theta(a)\theta(b).$$

Thus θ preserves both operations and is an isomorphism from $\mathscr{R}$ to R'. ∎

We shall use the isomorphism θ in the preceding proof to identify $a \in \mathscr{R}$ with $(a, 0)$ in R'. We write a instead of $(a, 0)$, and consider $\mathscr{R}$ to be a subset of $\mathscr{C}$. The calculation

$$(0, 1)(0, 1) = (0 \cdot 0 - 1 \cdot 1, 0 \cdot 1 + 1 \cdot 0)$$
$$= (-1, 0)$$
$$= -1$$

shows that the equation $x^2 = -1$ has a solution $x = (0, 1)$ in $\mathscr{C}$.

To obtain the customary notation for complex numbers, we define the number i by

$$i = (0, 1).$$

This makes i a number such that $i^2 = -1$. We now note that any $(a, b) \in \mathscr{C}$ can be written in the form

$$(a, b) = (a, 0) + (0, b)$$
$$= (a, 0) + b(0, 1)$$
$$= a + bi,$$

and this gives us the familiar form for complex numbers.

Using the field properties freely, we may rewrite the rules for addition and multiplication in $\mathscr{C}$ as follows:

$$(a + bi) + (c + di) = a + c + bi + di$$
$$= (a + c) + (b + d)i$$

and

$$(a + bi)(c + di) = (a + bi)c + (a + bi)di$$
$$= ac + bci + adi + bdi^2$$
$$= (ac - bd) + (ad + bc)i,$$

where the last step was obtained by replacing i^2 by -1.

The fact that $i^2 = -1$ was used in Section 5.4 to prove that it is impossible to impose an order relation on $\mathscr{C}$. Hence $\mathscr{C}$ is not an ordered field.

It is easy to show that all negative real numbers have square roots in $\mathscr{C}$. For any positive real number a, the negative real number $-a$ has both $\sqrt{a}i$ and $-\sqrt{a}i$ as square roots since

$$(\sqrt{a}i)^2 = (\sqrt{a})^2i^2 = a(-1) = -a$$

and

$$(-\sqrt{a}i)^2 = (-\sqrt{a})^2i^2 = a(-1) = -a.$$

We shall see later in this chapter that every nonzero complex number has two distinct square roots in $\mathscr{C}$.

EXAMPLE 1   The following results illustrate some calculations with complex numbers.

(a) $(1 + 2i)(3 - 5i) = 3 + 6i - 5i - 10i^2 = 13 + i$
(b) $(2 + 3i)(2 - 3i) = 4 - 9i^2 = 13$
(c) $(-3 + 4i)(3 + 4i) = -9 + 16i^2 = -25$
(d) $(1 - i)^2 = 1 - 2i + i^2 = -2i$
(e) $i^4 = (i^2)^2 = (-1)^2 = 1$ $\square$

In connection with part (b) of Example 1, we note that

$$(a + bi)(a - bi) = a^2 - b^2i^2$$
$$= a^2 + b^2$$

for any complex number $a + bi$. The number $a^2 + b^2$ is always real, and it is positive if $a + bi$ is nonzero.

DEFINITION 7.7   For any a, b in $\mathscr{R}$, the **conjugate** of the complex number $a + bi$ is the number $a - bi$.

Division of complex numbers may be accomplished by multiplying the numerator and denominator of a quotient by the conjugate of the denominator.

EXAMPLE 2   We have the following illustrations of division.

(a) $\dfrac{3+7i}{2-3i} = \dfrac{3+7i}{2-3i}\cdot\dfrac{2+3i}{2+3i} = \dfrac{6+23i-21}{4+9} = -\dfrac{15}{13}+\dfrac{23}{13}i$

(b) $\dfrac{1}{2+i} = \dfrac{1}{2+i}\cdot\dfrac{2-i}{2-i} = \dfrac{2-i}{5} = \dfrac{2}{5}-\dfrac{1}{5}i$ $\square$

By using the techniques illustrated in Examples 1 and 2, we can write the result of any calculation involving the field operations with complex numbers in the form $a + bi$, with a and b real numbers. This form is called the **standard form** of the complex number. If $b \neq 0$, the number is called **imaginary**. If $a = 0$, the number is called **pure imaginary**.

EXERCISES 7.2

Perform the computations in Problems 1–12 and express the results in standard form $a + bi$.

1   $(2 - 3i)(-1 + 4i)$ 2 $(5 - 3i)(2 - 4i)$

3   i^{15} 4 i^{87}

5   $(2 - i)^3$ 6 $i(2 + i)^2$

7   $\dfrac{1}{2-i}$ 8 $\dfrac{1}{3+i}$

9   $\dfrac{2-i}{8-6i}$ 10 $\dfrac{1-i}{1+3i}$

11   $\dfrac{5+2i}{5-2i}$ 12 $\dfrac{4-3i}{4+3i}$

13   Find two square roots of each given number.
   (a) -9 (b) -16 (c) -25
   (d) -36 (e) -13 (f) -8

14   With addition as given in Definition 7.5, prove the following statements.
   (a) Addition is associative in $\mathscr{C}$.
   (b) Addition is commutative in $\mathscr{C}$.
   (c) $(0,0)$ is the additive identity in $\mathscr{C}$.
   (d) The additive inverse of $(a, b) \in \mathscr{C}$ is $(-a, b) \in \mathscr{C}$.

15   With addition and multiplication as in Definition 7.5, prove that the right-distributive property holds in $\mathscr{C}$.

16   With $\mathscr{C}$ as given in Definition 7.5, prove that $R' = \{(a, 0) \mid a \in \mathscr{R}\}$ is a subfield of $\mathscr{C}$.

17   Let θ be the mapping $\theta: \mathscr{C} \to \mathscr{C}$ defined for $z = a + bi$ in standard form by

$\theta(z) = a - bi.$

Prove that θ is a ring isomorphism.

18   Let $\bar{z} = a - bi$ denote the conjugate of a complex number $z = a + bi$ (in standard form). It follows from Problem 17 that $\overline{z_1 + z_2} = \bar{z}_1 + \bar{z}_2$ and $\overline{z_1 z_2} = \bar{z}_1 \bar{z}_2$ for all z_1, z_2 in $\mathscr{C}$. Prove the following statements concerning conjugates of complex numbers.

(a) $\overline{(\bar{z})} = z$
(b) If $z \neq 0, (\bar{z})^{-1} = \overline{(z^{-1})}$
(c) $z + \bar{z} \in \mathscr{R}$
(d) $z = \bar{z}$ if and only if $z \in \mathscr{R}$.

19   Assume that $\theta : \mathscr{C} \to \mathscr{C}$ is an isomorphism and $\theta(a) = a$ for all $a \in \mathscr{R}$. Prove that if θ is not the identity mapping, then $\theta(z) = \bar{z}$ for all $z \in \mathscr{C}$.

20   (See Example 8 of Section 5.1.) Show that the mapping θ defined by

$$\theta(a + bi) = \begin{bmatrix} a & -b \\ b & a \end{bmatrix} \quad \text{for } a, b \in \mathscr{R}$$

is an isomorphism from $\mathscr{C}$ to a subring of the ring of all 2×2 matrices over $\mathscr{R}$.

7.3   DE MOIVRE'S THEOREM AND ROOTS OF COMPLEX NUMBERS

We have seen that real numbers may be represented geometrically by the points in a straight line. In much the same way it is possible to represent complex numbers by the points in a plane. We begin with a conventional rectangular coordinate system in the plane (see Figure 7.3). With each complex number $x + yi$ in standard form, we associate the point that has coordinates (x, y). This association establishes a one-to-one correspondence from the set $\mathscr{C}$ of complex numbers to the set of all points in the plane.

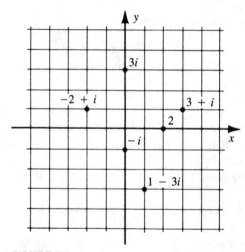

FIGURE 7.3

The point in the plane that corresponds to a complex number is called the **graph** of the number and the complex number that corresponds to a point in the plane is called the **coordinate** of the point. Points on the horizontal axis have coordinates $a + 0i$ that are real numbers, and consequently the horizontal axis is referred to as the **real axis**. Points on the vertical axis have coordinates $0 + bi$ that are pure imaginary numbers, so the vertical axis is called the **imaginary axis**. Several points are labeled with their coordinates in Figure 7.3.

Complex numbers are sometimes represented geometrically by directed line segments called **vectors**. In this approach, the complex number $a + bi$ is represented by the directed line segment from the origin of the coordinate system to the point with rectangular coordinates (a, b), or by any directed line segment with the same length and direction as this one. This is shown in Figure 7.4.

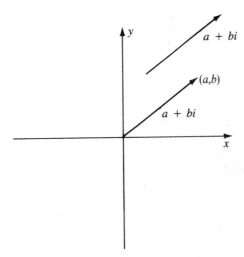

FIGURE 7.4

In this book we have little use for the vector representation of complex numbers. We simply note that in this interpretation, addition of complex numbers corresponds to the usual "parallelogram rule" for adding vectors. This is illustrated in Figure 7.5.

Returning now to the representation of complex numbers by points in the plane, we observe that any point P in the plane can be located by designating its *distance r* from the origin O and an *angle* θ in standard position that has OP as its terminal side. Figure 7.6 shows r and θ for a complex number $x + yi$ in standard form.

From Figure 7.6, we see that r and θ are related to x and y by the equations

$$x = r\cos\theta, \qquad y = r\sin\theta, \qquad r = \sqrt{x^2 + y^2}.$$

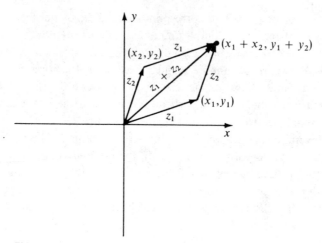

FIGURE 7.5

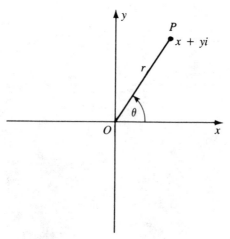

FIGURE 7.6

The complex number $x + yi$ can thus be written in the form

$$x + yi = r(\cos \theta + i \sin \theta).$$

DEFINITION 7.8   When a complex number in standard form $x + yi$ is written as

$$x + yi = r(\cos \theta + i \sin \theta),$$

the expression[†] $r(\cos \theta + i \sin \theta)$ is called the **trigonometric form** (or

---

[†] The expression $\cos \theta + i \sin \theta$ is sometimes abbreviated as cis θ.

polar form) of $x + yi$. The number $r = \sqrt{x^2 + y^2}$ is called the **absolute value** (or **modulus**) of $x + yi$, and the angle θ is called the **argument** (or **amplitude**) of $x + yi$.

The usual notation is used for the absolute value of a complex number:

$$|x + yi| = r = \sqrt{x^2 + y^2}.$$

The absolute value, r, is unique, but the angle θ is not unique since there are many angles in standard position with P on their terminal side. This is illustrated in the next example.

EXAMPLE 1 Expressing the complex number $-1 - i$ in trigonometric form[†], we have

$$-1 - i = \sqrt{2}\left(-\frac{1}{\sqrt{2}} - \frac{1}{\sqrt{2}}i\right)$$

$$= \sqrt{2}\left(\cos\frac{5\pi}{4} + i\sin\frac{5\pi}{4}\right)$$

$$= \sqrt{2}\left[\cos\left(-\frac{3\pi}{4}\right) + i\sin\left(-\frac{3\pi}{4}\right)\right]$$

$$= \sqrt{2}\left(\cos\frac{13\pi}{4} + i\sin\frac{13\pi}{4}\right).$$

Many other such expressions are possible. $\square$

Although the argument θ is not unique, an equation of the form

$$r_1(\cos\theta_1 + i\sin\theta_1) = r_2(\cos\theta_2 + i\sin\theta_2)$$

does require that $r_1 = r_2$ and that θ_1 and θ_2 be coterminal. Hence

$$\theta_2 = \theta_1 + k(2\pi)$$

for some integer k.

The next theorem gives a hint as to the usefulness of the trigonometric form of complex numbers. In proving the theorem, we shall use the following identities from trigonometry:

$$\cos(A + B) = \cos A \cos B - \sin A \sin B$$

$$\sin(A + B) = \sin A \cos B + \cos A \sin B.$$

[†] We choose to use radian measure for angles. Degree measure could just as well be chosen.

THEOREM 7.9   If

$$z_1 = r_1(\cos\theta_1 + i\sin\theta_1)$$

and

$$z_2 = r_2(\cos\theta_2 + i\sin\theta_2)$$

are arbitrary complex numbers in trigonometric form, then

$$z_1 z_2 = r_1 r_2[\cos(\theta_1 + \theta_2) + i\sin(\theta_1 + \theta_2)].$$

In words, the absolute value of the product of two complex numbers is the product of their absolute values, and an argument of the product is the sum of their arguments.

Proof   The statement of the theorem follows from

$$
\begin{aligned}
z_1 z_2 &= [r_1(\cos\theta_1 + i\sin\theta_1)][r_2(\cos\theta_2 + i\sin\theta_2)] \\
&= r_1 r_2[(\cos\theta_1 \cos\theta_2 - \sin\theta_1 \sin\theta_2) \\
&\quad + i(\cos\theta_1 \sin\theta_2 + \sin\theta_1 \cos\theta_2)] \\
&= r_1 r_2[\cos(\theta_1 + \theta_2) + i\sin(\theta_1 + \theta_2)].
\end{aligned}
$$
∎

The preceding result generalizes to the next theorem, which begins to reveal the true usefulness of the trigonometric form.

THEOREM 7.10   (**De Moivre's Theorem**) If n is a positive integer and

$$z = r(\cos\theta + i\sin\theta)$$

is a complex number in trigonometric form, then

$$z^n = r^n(\cos n\theta + i\sin n\theta).$$

Proof   For $n = 1$, the statement is trivial. Assume that it is true for $n = k$, that is, that

$$z^k = r^k(\cos k\theta + i\sin k\theta).$$

Using Theorem 7.9, we have

$$
\begin{aligned}
z^{k+1} &= z^k \cdot z \\
&= [r^k(\cos k\theta + i\sin k\theta)][r(\cos\theta + i\sin\theta)] \\
&= r^{k+1}[\cos(k\theta + \theta) + i\sin(k\theta + \theta)] \\
&= r^{k+1}[\cos(k + 1)\theta + i\sin(k + 1)\theta].
\end{aligned}
$$

Thus the theorem is true for $n = k + 1$, and it follows by induction that the theorem is true for all positive integers. ∎

EXAMPLE 2   Some applications of De Moivre's Theorem are shown in the following computations.

(a) $(-2 + 2i)^4 = \left[2\sqrt{2} \left(-\frac{1}{\sqrt{2}} + \frac{1}{\sqrt{2}} i \right) \right]^4$

$= \left[2\sqrt{2} \left(\cos \frac{3\pi}{4} + i \sin \frac{3\pi}{4} \right) \right]^4$

$= 64(\cos 3\pi + i \sin 3\pi)$

$= 64(-1 + 0i)$

$= -64.$

(b) $\left(\frac{\sqrt{3}}{2} + \frac{1}{2} i \right)^{40} = \left[1 \left(\cos \frac{\pi}{6} + i \sin \frac{\pi}{6} \right) \right]^{40}$

$= 1^{40} \left(\cos \frac{20\pi}{3} + i \sin \frac{20\pi}{3} \right)$

$= \cos \left(\frac{2\pi}{3} + 6\pi \right) + i \sin \left(\frac{2\pi}{3} + 6\pi \right)$

$= \cos \frac{2\pi}{3} + i \sin \frac{2\pi}{3}$

$= -\frac{1}{2} + \frac{\sqrt{3}}{2} i.$ $\square$

If n is a positive integer greater than 1 and $u^n = z$ for complex numbers u and z, then u is called an **nth root** of z. We shall prove that every nonzero complex number has exactly n nth roots in $\mathscr{C}$.

THEOREM 7.11   For each integer $n \geq 1$, any nonzero complex number

$z = r(\cos \theta + i \sin \theta)$

has exactly n distinct nth roots in $\mathscr{C}$, and these are given by

$r^{1/n} \left(\cos \frac{\theta + 2k\pi}{n} + i \sin \frac{\theta + 2k\pi}{n} \right), \qquad k = 0, 1, 2, \ldots, n - 1,$

where $r^{1/n} = \sqrt[n]{r}$ denotes the positive real nth root of r.

Proof   For an arbitrary integer k, let

$v = r^{1/n} \left(\cos \frac{\theta + 2k\pi}{n} + i \sin \frac{\theta + 2k\pi}{n} \right).$

Then

$$v^n = (r^{1/n})^n \left[\cos \frac{n(\theta + 2k\pi)}{n} + i \sin \frac{n(\theta + 2k\pi)}{n} \right]$$
$$= r[\cos(\theta + 2k\pi) + i \sin(\theta + 2k\pi)]$$
$$= r(\cos \theta + i \sin \theta) = z,$$

and v is an nth root of z. The n angles

$$\frac{\theta}{n}, \quad \frac{\theta + 2\pi}{n}, \quad \frac{\theta + 2(2\pi)}{n}, \quad \dots, \quad \frac{\theta + (n-1)(2\pi)}{n}$$

are equally spaced $\dfrac{2\pi}{n}$ radians apart, so no two of them have the same terminal side. Thus the n values of v obtained by letting $k = 0, 1, 2, \dots, n-1$ are distinct, and we have shown that z has at least n distinct nth roots in $\mathscr{C}$.

To show that there are no other nth roots of z in $\mathscr{C}$, suppose $v = t(\cos \phi + i \sin \phi)$ is the trigonometric form of a complex number v such that $v^n = z$. Then

$$t^n(\cos n\phi + i \sin n\phi) = r(\cos \theta + i \sin \theta)$$

by De Moivre's Theorem. It follows from this that

$$t^n = r, \qquad \cos n\phi = \cos \theta, \quad \text{and} \quad \sin n\phi = \sin \theta.$$

Since r and t are positive, it must be that $t = r^{1/n}$. The other equations require that $n\phi$ and θ be coterminal, and hence they differ by a multiple of 2π:

$$n\phi = \theta + m(2\pi)$$

for some integer m. By the Division Algorithm,

$$m = qn + k,$$

where $k \in \{0, 1, 2, \dots, n-1\}$. Thus

$$n\phi = \theta + (qn + k)(2\pi)$$

and

$$\phi = \frac{\theta + 2k\pi}{n} + q(2\pi).$$

This equation shows that ϕ is coterminal with the angle $\dfrac{(\theta + 2k\pi)}{n}$, and hence v is one of the nth roots listed in the statement of the theorem. ∎

EXAMPLE 3   We shall find the three cube roots of $8i$ and express each in standard form $a + bi$. Expressing $8i$ in trigonometric form, we have

$$8i = 8\left(\cos \frac{\pi}{2} + i \sin \frac{\pi}{2} \right).$$

By the formula in Theorem 7.11, the cube roots of $z = 8i$ are given by

$$8^{1/3} \left(\cos \frac{\frac{\pi}{2} + 2k\pi}{3} + i \sin \frac{\frac{\pi}{2} + 2k\pi}{3} \right), \qquad k = 0, 1, 2.$$

Each of these roots has absolute value $8^{1/3} = 2$, and they are equally spaced $\frac{2\pi}{3}$ radians apart with the first one at $\frac{\pi}{6}$. Thus the three cube roots of $8i$ are

$$2\left(\cos \frac{\pi}{6} + i \sin \frac{\pi}{6} \right) = 2\left(\frac{\sqrt{3}}{2} + \frac{1}{2}i \right) = \sqrt{3} + i$$

$$2\left(\cos \frac{5\pi}{6} + i \sin \frac{5\pi}{6} \right) = 2\left(-\frac{\sqrt{3}}{2} + \frac{1}{2}i \right) = -\sqrt{3} + i$$

$$2\left(\cos \frac{3\pi}{2} + i \sin \frac{3\pi}{2} \right) = 2(0 - i) = -2i.$$

These results may be checked by direct multiplication.   □

EXERCISES 7.3

1   Graph each of the following complex numbers and express each in trigonometric form.
   (a) $-2 + 2\sqrt{3}i$ (b) $2 + 2i$
   (c) $3 - 3i$ (d) $\sqrt{3} + i$
   (e) $1 + \sqrt{3}i$ (f) $-1 - i$
   (g) -4 (h) $-5i$

2   Find each of the following products. Write each result in both trigonometric and standard form.

   (a) $\left[4\left(\cos \frac{\pi}{8} + i \sin \frac{\pi}{8} \right) \right]\left[\cos \frac{5\pi}{8} + i \sin \frac{5\pi}{8} \right]$

   (b) $\left[3\left(\cos \frac{7\pi}{6} + i \sin \frac{7\pi}{6} \right) \right]\left[\cos \frac{2\pi}{3} + i \sin \frac{2\pi}{3} \right]$

   (c) $\left[2\left(\cos \frac{5\pi}{6} + i \sin \frac{5\pi}{6} \right) \right]\left\{ 3\left[\cos\left(-\frac{\pi}{6} \right) + i \sin\left(-\frac{\pi}{6} \right) \right] \right\}$

   (d) $\left[6\left(\cos \frac{5\pi}{3} + i \sin \frac{5\pi}{3} \right) \right]\left\{ 5\left[\cos\left(-\frac{\pi}{3} \right) + i \sin\left(-\frac{\pi}{3} \right) \right] \right\}$

3   Use De Moivre's Theorem to find the value of each of the following. Leave your answers in standard form $a + bi$.

(a) $(\sqrt{3} + i)^7$

(b) $\left(\dfrac{\sqrt{3}}{2} + \dfrac{1}{2}i\right)^{21}$

(c) $(-\sqrt{3} + i)^{10}$

(d) $\left(\dfrac{\sqrt{3}}{2} - \dfrac{1}{2}i\right)^{18}$

(e) $\left(-\dfrac{1}{2} + \dfrac{\sqrt{3}}{2}i\right)^{18}$

(f) $(\sqrt{2} + \sqrt{2}i)^9$

(g) $(1 - i\sqrt{3})^8$

(h) $\left(-\dfrac{1}{2} + \dfrac{\sqrt{3}}{2}i\right)^{12}$

4   Show that the n distinct nth roots of 1 are equally spaced around a circle with center at the origin and radius 1.

5   If $w = \cos\dfrac{2\pi}{n} + i\sin\dfrac{2\pi}{n}$, show that the nth roots of 1 are given by $w, w^2, \ldots, w^{n-1}, w^n = 1$.

6   Find the indicated roots of 1 in standard form $a + bi$ and graph them on a unit circle with center at the origin.
(a) Cube roots of 1
(b) Fourth roots of 1
(c) Eighth roots of 1
(d) Sixth roots of 1

7   Find all the indicated roots of the given number. Leave your results in trigonometric form.

(a) Cube roots of $\dfrac{\sqrt{3}}{2} + \dfrac{1}{2}i$

(b) Cube roots of $-1 + i$

(c) Fourth roots of $-\dfrac{\sqrt{3}}{2} + \dfrac{1}{2}i$

(d) Fourth roots of $\dfrac{1}{2} - \dfrac{\sqrt{3}}{2}i$

(e) Fifth roots of $-16\sqrt{2} - 16i\sqrt{2}$

(f) Sixth roots of $32\sqrt{3} - 32i$

8   Find all complex numbers that are solutions of the given equation. Leave your answers in standard form $a + bi$.

(a) $z^3 + 27 = 0$

(b) $z^8 - 16 = 0$

(c) $z^3 - i = 0$

(d) $z^3 + 8i = 0$

(e) $z^4 + \dfrac{1}{2} - \dfrac{\sqrt{3}}{2}i = 0$

(f) $z^4 + 1 - i\sqrt{3} = 0$

(g) $z^4 + \dfrac{1}{2} + \dfrac{\sqrt{3}}{2}i = 0$

(h) $z^4 + 8 + 8i\sqrt{3} = 0$

9   If $w = \cos\dfrac{2\pi}{n} + i\sin\dfrac{2\pi}{n}$ and u is any nth root of $z \in \mathscr{C}$, show that the nth roots of z are given by

$$wu, w^2u, \ldots, w^{n-1}u, w^nu = u.$$

10 Prove that, for a fixed value of n, the set of all nth roots of 1 forms a group with respect to multiplication.

11 Prove that the group in Problem 10 is cyclic, with $w = \cos\dfrac{2\pi}{n} + i\sin\dfrac{2\pi}{n}$ as a generator.

12 Any generator of the group in Problem 10 is called a **primitive** nth root of 1. Prove that

$$\cos\frac{2k\pi}{n} + i\sin\frac{2k\pi}{n}$$

is a primitive nth root of 1 if and only if k and n are relatively prime.

13 (a) Find all primitive sixth roots of 1.
(b) Find all primitive eighth roots of 1.

14 Prove that the set of *all* roots of 1 forms a group with respect to multiplication.

15 Prove the following statements concerning absolute values of complex numbers. (As in Problem 18 of Exercises 7.2, $\bar{z}$ denotes the conjugate of z.)
(a) $|\bar{z}| = |z|$
(b) $z\bar{z} = |z|^2$
(c) *If $z \neq 0$, then* $z^{-1} = \dfrac{\bar{z}}{|z|^2}$.
(d) If $z_2 \neq 0$, then $\left|\dfrac{z_1}{z_2}\right| = \dfrac{|z_1|}{|z_2|}$.

(e) $|z_1 + z_2| \leq |z_1| + |z_2|$

16 Prove that the set of all complex numbers that have absolute value 1 forms a group with respect to multiplication.

17 Prove that if $z = r(\cos\theta + i\sin\theta)$ is a nonzero complex number in trigonometric form, then $z^{-1} = r^{-1}[\cos(-\theta) + i\sin(-\theta)]$.

18 Prove that if n is a positive integer and $z = r(\cos\theta + i\sin\theta)$ is a nonzero complex number in trigonometric form, then

$$z^{-n} = r^{-n}[\cos(-n\theta) + i\sin(-n\theta)].$$

19 Prove that if

$$z_1 = r_1(\cos\theta_1 + i\sin\theta_1)$$

and

$$z_2 = r_2(\cos\theta_2 + i\sin\theta_2)$$

are complex numbers in trigonometric form and z_2 is nonzero, then

$$\frac{z_1}{z_2} = \frac{r_1}{r_2}[\cos(\theta_1 - \theta_2) + i\sin(\theta_1 - \theta_2)].$$

20 In the ordered field $\mathscr{R}$, absolute value is defined according to Problem 7 of Exercises 5.4 by

$$|a| = \begin{cases} a & \text{if } a \geq 0 \\ -a & \text{if } a < 0. \end{cases}$$

For $a \in \mathscr{R}$, show that the absolute value of $a + 0i$ according to Definition 7.8 agrees with the definition from Chapter 5. (Keep in mind, however, that $\mathscr{C}$ is not an ordered field, as was shown in Section 5.4.)

225

Key Words and Phrases

Upper bound

Least upper bound

Complete ordered field

Real numbers

Rational numbers

Irrational numbers

Decimal representations

Complex numbers

Conjugate of a complex number

Standard form of a complex number

Imaginary number

Pure imaginary number

Trigonometric form

Polar form

Absolute value

Modulus

Argument

Amplitude

De Moivre's Theorem

nth roots

References

Ball, Richard W. *Principles of Abstract Algebra.* New York: Holt, Rinehart and Winston, 1963.

Birkhoff, Garrett and Saunders MacLane. *A Survey of Modern Algebra* (4th ed.). New York: Macmillan, 1977.

Durbin, John R. *Modern Algebra.* New York: Wiley, 1979.

McCoy, Neal H. *Introduction to Modern Algebra* (3rd ed.). Boston: Allyn and Bacon, 1975.

Mostow, George D., Joseph H. Sampson and Jean-Pierre Meyer. *Fundamental Structures of Algebra.* New York: McGraw-Hill, 1963.

8 POLYNOMIALS

8.1 POLYNOMIALS OVER A RING

Starting with beginning algebra courses, a great deal of time is devoted to developing skills in various manipulations with polynomials. Procedures are learned for the basic operations of addition, subtraction, multiplication, and division of polynomials. By the time a student begins an abstract algebra course, polynomials are a very familiar topic.

Much of this prior experience involved polynomials in a single letter, such as $5 + 4t + t^2$, where the letter usually represented a variable with domain a subset of the real numbers. In this section, our point of view is very different. We wish to start with a commutative ring R with unity[†] 1 and *construct* a ring that contains both R and a given element x. More precisely, we want to construct a *smallest* ring that contains R and x in this sense: any ring that contains both R and x would necessarily contain the constructed ring. We assume that x is *not* an element of R, but nothing more than this. For the time being, the letter x will be a formal symbol subject only to the definitions that are made as we proceed. The letter x is referred to as an **indeterminate** in order to emphasize its role here. Later, we shall consider other possible roles for x.

[†] Throughout this chapter, the unity is denoted by 1 rather than e. A similar construction can be made with fewer restrictions on R, but such generality results in complications that are avoided here.

DEFINITION 8.1   Let R be a commutative ring with unity 1 and let x be an indeterminate. A **polynomial in x with coefficients in R**, or a **polynomial in x over R**, is an expression of the form

$$a_0 x^0 + a_1 x^1 + a_2 x^2 + \cdots + a_n x^n,$$

where n is a nonnegative integer and each a_i is an element of R. The set of all polynomials in x over R is denoted by $R[x]$.

The construction that we shall carry out will be guided by our previous experience with polynomials. Consistent with this, we adopt the familiar language of elementary algebra and refer to the parts $a_i x^i$ of the above expression as **terms** of the polynomial and to a_i as the **coefficient** of x^i in the term $a_i x^i$. As a notational convenience, we shall use functional notations such as $f(x)$ for shorthand names of polynomials. That is, we shall write things such as

$$f(x) = a_0 x^0 + a_1 x^1 + \cdots + a_n x^n,$$

but this indicates only that $f(x)$ is a symbolic name for the polynomial. It does *not* indicate a function or a function value.

EXAMPLE 1   Some examples of polynomials in x over the ring $\mathbf{Z}$ of integers are listed below.

(a) $f(x) = 2x^0 + (-4)x^1 + 0x^2 + 5x^3$
(b) $g(x) = 1x^0 + 2x^1 + (-1)x^2$
(c) $h(x) = (-5)x^0 + 0x^1 + 0x^2$ □

We have not yet defined equality of polynomials. (The preceding use of $=$ only indicated that certain polynomials had been given shorthand names.) To be consistent with prior experience, it is desirable to define equality of polynomials so that terms with zero coefficients can be deleted with equality retained. With this goal in mind, we make the following somewhat cumbersome definition.

DEFINITION 8.2(A)   Suppose that R is a commutative ring with unity, x is an indeterminate, and that

$$f(x) = a_0 x^0 + a_1 x^1 + \cdots + a_n x^n$$

and

$$g(x) = b_0 x^0 + b_1 x^1 + \cdots + b_m x^m$$

are polynomials in x over R. Then $f(x)$ and $g(x)$ are **equal polynomials**,

$f(x) = g(x)$, if and only if the following conditions hold for all i that occur as a subscript on a coefficient in either $f(x)$ or $g(x)$:
(1)   If one of a_i, b_i is zero, then the other is either omitted or is also. zero.
(2)   If one of a_i, b_i is not zero, then the other is not omitted and $a_i = b_i$.

EXAMPLE 2   According to Definition 8.2(A), the following equalities are valid in the set $\mathbf{Z}[x]$ of all polynomials in x over $\mathbf{Z}$:

(a) $2x^0 + (-4)x^1 + 0x^2 + 5x^3 = 2x^0 + (-4)x^1 + 5x^3$
(b) $(-5)x^0 + 0x^1 + 0x^2 = (-5)x^0$ $\square$

The compact sigma notation is useful when we work with polynomials. In the sigma notation, a capital Greek letter $\sum$ is used to indicate a sum:

$$\sum_{i=1}^{n} a_i = a_1 + a_2 + \cdots + a_n.$$

The variable i is called the *index of summation*, and the notations below and above the sigma indicate the values of i where the sum starts and where it ends. For example,

$$\sum_{i=3}^{5} b_i = b_3 + b_4 + b_5.$$

The index of summation is sometimes called a "dummy variable" because the value of the sum is unaffected if the index is changed to a different letter:

$$\sum_{i=0}^{3} a_i = \sum_{j=0}^{3} a_j = \sum_{k=0}^{3} a_k = a_0 + a_1 + a_2 + a_3.$$

The polynomial

$$f(x) = a_0 x^0 + a_1 x^1 + \cdots + a_n x^n$$

may be written compactly by use of the sigma notation as

$$f(x) = \sum_{i=0}^{n} a_i x^i.$$

After the convention concerning zero coefficients has been clarified and agreed upon as stated in conditions (1) and (2) of Definition 8.2(A), the definition of *equality of polynomials* may be shortened as follows.

DEFINITION 8.2(B)   If R is a commutative ring with unity and $f(x) = \sum_{i=0}^{n} a_i x^i$ and $g(x) = \sum_{i=0}^{m} b_i x^i$ are polynomials in x over R, then $f(x) = g(x)$ if and only if $a_i = b_i$ for all i.

229

It is understood, of course, that any polynomial over R has only a finite number of nonzero terms. The notational agreements that have been made allow us to make concise definitions of addition and multiplication in $R[x]$.

DEFINITION 8.3   Let R be a commutative ring with unity. For any $f(x) = \sum_{i=0}^{n} a_i x^i$ and $g(x) = \sum_{i=0}^{m} b_i x^i$ in $R[x]$, we define **addition** in $R[x]$ by

$$f(x) + g(x) = \sum_{i=0}^{k} (a_i + b_i)x^i,$$

where k is the larger of the two integers n, m. We define **multiplication** in $R[x]$ by

$$f(x)g(x) = \sum_{i=0}^{n+m} c_i x^i,$$

where $c_i = \sum_{j=0}^{i} a_j b_{i-j}$.

The expanded expression for c_i appears as

$$c_i = a_0 b_i + a_1 b_{i-1} + a_2 b_{i-2} + \cdots + a_{i-2} b_2 + a_{i-1} b_1 + a_i b_0.$$

We shall see presently that this formula agrees with previous experience in the multiplication of polynomials.

To introduce some novelty in our next example, we consider the sum and product of two polynomials over the ring $\mathbf{Z}_6$.

EXAMPLE 3   We shall follow a convention that has been used on some earlier occasions, and write a for $[a]$ in $\mathbf{Z}_6$. Let

$$f(x) = \sum_{i=0}^{3} a_i x^i = 1x^0 + 5x^1 + 3x^3$$

and

$$g(x) = \sum_{i=0}^{1} b_i x^i = 4x^0 + 2x^1$$

in $\mathbf{Z}_6[x]$. According to our agreement regarding zero coefficients, these polynomials may be written as

$$f(x) = 1x^0 + 5x^1 + 0x^2 + 3x^3$$
$$g(x) = 4x^0 + 2x^1 + 0x^2 + 0x^3$$

and the definition of addition yields

$$f(x) + g(x) = \sum_{i=0}^{3} (a_i + b_i)x^i$$
$$= (1 + 4)x^0 + (5 + 2)x^1 + (0 + 0)x^2 + (3 + 0)x^3$$
$$= 5x^0 + 1x^1 + 0x^2 + 3x^3$$
$$= 5x^0 + 1x^1 + 3x^3$$

since $5 + 2 = 1$ in $\mathbf{Z}_6$. The definition of multiplication gives

$$f(x)g(x) = \sum_{i=1}^{4} c_i x^i$$

where

$$c_0 = a_0 b_0 = 1 \cdot 4 = 4$$
$$c_1 = a_0 b_1 + a_1 b_0 = 1 \cdot 2 + 5 \cdot 4 = 2 + 2 = 4$$
$$c_2 = a_0 b_2 + a_1 b_1 + a_2 b_0 = 1 \cdot 0 + 5 \cdot 2 + 0 \cdot 4 = 4$$
$$c_3 = a_0 b_3 + a_1 b_2 + a_2 b_1 + a_3 b_0 = 1 \cdot 0 + 5 \cdot 0 + 0 \cdot 2 + 3 \cdot 4 = 0$$
$$c_4 = a_0 b_4 + a_1 b_3 + a_2 b_2 + a_3 b_1 + a_4 b_0$$
$$= 1 \cdot 0 + 5 \cdot 0 + 0 \cdot 0 + 3 \cdot 2 + 0 \cdot 4 = 0.$$

Thus

$$f(x)g(x) = (1x^0 + 5x^1 + 3x^3)(4x^0 + 2x^1)$$
$$= 4x^0 + 4x^1 + 4x^2 + 0x^3 + 0x^4$$
$$= 4x^0 + 4x^1 + 4x^2$$

in $\mathbf{Z}_6[x]$. This product, obtained by using Definition 8.3, agrees with the result obtained by the usual multiplication procedure based on the distributive laws:

$$f(x)g(x) = (1x^0 + 5x^1 + 3x^3)(4x^0) + (1x^0 + 5x^1 + 3x^3)(2x^1)$$
$$= (4x^0 + 2x^1 + 0x^3) + (2x^1 + 4x^2 + 0x^4)$$
$$= 4x^0 + 4x^1 + 4x^2. \qquad \square$$

The expanded forms of the c_i in Example 3 illustrate how the coefficient of x^i in the product is the sum of all products of the form $a_p b_q$ with $p + q = i$. In general it is true that

$$c_i = \sum_{j=0}^{i} a_j b_{i-j}$$
$$= a_0 b_i + a_1 b_{i-1} + a_2 b_{i-2} + \cdots + a_{i-1} b_1 + a_i b_0$$
$$= \sum_{p+q=i} a_p b_q.$$

This observation is useful in the proof of our next theorem.

THEOREM 8.4 Let R be a commutative ring with unity. With addition and multiplication as given in Definition 8.3, $R[x]$ forms a commutative ring with unity.

Proof Let

$$f(x) = \sum_{i=0}^{n} a_i x^i, \qquad g(x) = \sum_{i=0}^{m} b_i x^i, \qquad h(x) = \sum_{i=0}^{k} c_i x^i$$

represent arbitrary elements of $R[x]$, and let s be the greatest of the integers n, m, and k.

It follows immediately from Definition 8.3 that the sum $f(x) + g(x)$ is a well-defined element of $R[x]$, and $R[x]$ is closed under addition. Addition in $R[x]$ is associative since

$$f(x) + [g(x) + h(x)] = \sum_{i=0}^{n} a_i x^i + \sum_{i=0}^{s} (b_i + c_i) x^i$$

$$= \sum_{i=0}^{s} [a_i + (b_i + c_i)] x^i$$

$$= \sum_{i=0}^{s} [(a_i + b_i) + c_i] x^i \qquad \text{since addition is}$$
$$\text{associative in } R$$

$$= \sum_{i=0}^{s} (a_i + b_i) x^i + \sum_{i=0}^{k} c_i x^i$$

$$= [f(x) + g(x)] + h(x).$$

The polynomial $0x^0$ is an additive identity in $R[x]$ since $f(x) + 0x^0 = 0x^0 + f(x) = f(x)$ for all $f(x)$ in $R[x]$. The additive inverse of $f(x)$ is $\sum_{i=0}^{n} (-a_i) x^i$ since

$$f(x) + \sum_{i=0}^{n} (-a_i) x^i = \sum_{i=0}^{n} [a_i + (-a_i)] x^i = 0x^0$$

and $\sum_{i=0}^{n} (-a_i) x^i + f(x) = 0x^0$ in similar fashion. Addition in $R[x]$ is commutative since

$$f(x) + g(x) = \sum_{i=0}^{s} (a_i + b_i) x^i = \sum_{i=0}^{s} (b_i + a_i) x^i = g(x) + f(x).$$

Thus $R[x]$ is an abelian group with respect to addition.

It is clear from Definition 8.3 that $R[x]$ is closed under the binary operation of multiplication. To see that multiplication is associative in $R[x]$, we first note that the coefficient of x^i in $f(x)[g(x)h(x)]$ is given by

$$\sum_{p+q+r=i} a_p(b_q c_r),$$

the sum of all products $a_p(b_q c_r)$ of coefficients a_p, b_q, c_r such that the subscripts sum to i. Similarly in $[f(x)g(x)]h(x)$, the coefficient of x^i is

$$\sum_{p+q+r=i} (a_p b_q)c_r.$$

Now $a_p(b_q c_r) = (a_p b_q)c_r$ since multiplication is associative in R, and therefore $f(x)[g(x)h(x)] = [f(x)g(x)]h(x)$.

Before considering the distributive laws, we shall establish that multiplication in $R[x]$ is commutative. This follows from the equalities

$$f(x)g(x) = \sum_{i=0}^{n+m} \left(\sum_{p+q=i} a_p b_q \right) x^i$$

$$= \sum_{i=0}^{m+n} \left(\sum_{q+p=i} b_q a_p \right) x^i \quad \text{since multiplication is commutative}$$
$$\text{in } R$$

$$= g(x)f(x).$$

Let t be the greater of the integers m and k, and consider the left distributive law. We have

$$f(x)[g(x) + h(x)] = \sum_{i=0}^{n} a_i x^i \left[\sum_{i=0}^{t} (b_i + c_i)x^i \right]$$

$$= \sum_{i=0}^{n+t} \left[\sum_{p+q=i} a_p(b_q + c_q) \right] x^i$$

$$= \sum_{i=0}^{n+t} \left[\sum_{p+q=i} (a_p b_q + a_p c_q) \right] x^i$$

$$= \sum_{i=0}^{n+t} \left(\sum_{p+q=i} a_p b_q \right) x^i + \sum_{i=0}^{n+t} \left(\sum_{p+q=i} a_p c_q \right) x^i$$

$$= \sum_{i=0}^{n+m} \left(\sum_{p+q=i} a_p b_q \right) x^i + \sum_{i=0}^{n+k} \left(\sum_{p+q=i} a_p c_q \right) x^i$$

$$= f(x)g(x) + f(x)h(x),$$

and the left distributive property is established. The right distributive property is now easy to prove:

$$[f(x) + g(x)]h(x) = h(x)[f(x) + g(x)] \quad \text{since multiplication is}$$
$$\text{commutative in } R[x]$$

$$= h(x)f(x) + h(x)g(x) \quad \text{by the left distributive}$$
$$\text{law}$$

$$= f(x)h(x) + g(x)h(x) \quad \text{since multiplication is}$$
$$\text{commutative in } R[x]$$

The element $1x^0$ is a unity in $R[x]$ since

$$1x^0 \cdot f(x) = f(x) \cdot 1x^0 = \sum_{i=0}^{n} (a_i \cdot 1)x^i = \sum_{i=0}^{n} a_i x^i = f(x).$$

This completes the proof that $R[x]$ is a commutative ring with unity.

∎

Theorem 8.4 justifies referring to $R[x]$ as the **ring of polynomials over R**, or as the **ring of polynomials with coefficients in R**.

THEOREM 8.5   For any commutative ring R with unity, the ring $R[x]$ of polynomials over R contains a subring R' that is isomorphic to R.

Proof   Let R' be the subset of $R[x]$ that consists of all elements of the form ax^0. We shall show that R' is a subring by utilizing Theorem 5.4.

The subset R' contains elements such as the additive identity $0x^0$ and the unity $1x^0$ of $R[x]$. For arbitrary ax^0 and bx^0 in R',

$$ax^0 - bx^0 = (a - b)x^0$$

and

$$(ax^0)(bx^0) = (ab)x^0$$

are in R', and therefore R' is a subring of $R[x]$ by Theorem 5.4.

Guided by our previous experience with polynomials, we define $\theta : R \to R'$ by

$$\theta(a) = ax^0$$

for all $a \in R$. This rule defines a bijection since

$$\theta(a) = \theta(b) \Leftrightarrow ax^0 = bx^0 \Leftrightarrow a = b.$$

Moreover, θ is an isomorphism since

$$\theta(a + b) = (a + b)x^0 = ax^0 + bx^0 = \theta(a) + \theta(b)$$

and

$$\theta(ab) = (ab)x^0 = (ax^0)(bx^0) = \theta(a)\theta(b).$$ ∎

Thus R is embedded in $R[x]$. We can use the isomorphism θ to identify $a \in R$ with ax^0 in $R[x]$, and from now on we shall write a in place of ax^0. In particular, 0 may denote the zero polynomial $0x^0$ and 1 may denote the unity $1x^0$

in $R[x]$. We write an arbitrary polynomial

$$f(x) = a_0 x^0 + a_1 x^1 + a_2 x^2 + \cdots + a_n x^n$$

as

$$f(x) = a_0 + a_1 x^1 + a_2 x^2 + \cdots + a_n x^n.$$

Actually, we want to carry this notational simplification a bit farther, writing x for x^1, x^i for $1x^i$, and $-a_i x^i$ for $(-a_i)x^i$. This allows us to use all the conventional polynomial notations for the elements of $R[x]$. Also, we can now regard each term $a_i x^i$ with $i \geq 1$ as a product:

$$a_i x^i = a_i \cdot x \cdot x \cdot \cdots \cdot x,$$

with i factors of x in the product.

Having made the agreements described in the last paragraph, we may observe that our major goal for this section has been achieved. We have constructed a "smallest" ring $R[x]$ that contains R and x. It is "smallest" since any ring that contained both R and x would have to contain all polynomials

$$f(x) = a_0 + a_1 x + a_2 x^2 + \cdots + a_n x^n$$

as a consequence of the closure properties.

It is appropriate now to pick up some more of the language that is customarily used in work with polynomials.

DEFINITION 8.6 Let R be a commutative ring with unity, and let

$$f(x) = a_0 + a_1 x + \cdots + a_n x^n$$

be a *nonzero* element of $R[x]$. Then the **degree of $f(x)$** is the largest integer k such that the coefficient of x^k is not zero, and this coefficient a_k is called the **leading coefficient** of $f(x)$. The term a_0 of $f(x)$ is called the **constant term** of $f(x)$, and elements of R are referred to as **constant polynomials**.

The degree of $f(x)$ will be abbreviated $\deg f(x)$. Note that degree is *not* defined for the zero polynomial. (The reason for this will be clear later.) Note also that the *polynomials of degree zero* are the same as the *nonzero elements of R*.

EXAMPLE 4 The polynomials $f(x)$ and $g(x)$ in Example 3 can now be written as

$$f(x) = 1 + 5x + 3x^3 = 3x^3 + 5x + 1$$
$$g(x) = 4 + 2x = 2x + 4.$$

(a) The constant term of $f(x)$ is 1 and the leading coefficient of $f(x)$ is 3.

(b)   The polynomial $g(x)$ has constant term 4 and leading coefficient 2.

(c) $\deg f(x) = 3$ and $\deg g(x) = 1$.

(d)   In Example 3, we found that

$$f(x)g(x) = 4 + 4x + 4x^2,$$

so that $\deg(f(x)g(x)) = 2$. In connection with the next theorem, we note that

$$\deg(f(x)g(x)) \neq \deg f(x) + \deg g(x)$$

in this instance.   □

THEOREM 8.7   If R is an integral domain and $f(x)$ and $g(x)$ are nonzero elements of $R[x]$, then

$$\deg(f(x)g(x)) = \deg f(x) + \deg g(x).$$

Proof   Let R be an integral domain, and suppose that

$$f(x) = \sum_{i=0}^{n} a_i x^i \quad \text{has degree } n$$

and

$$g(x) = \sum_{i=0}^{m} b_i x^i \quad \text{has degree } m$$

in $R[x]$. Then $a_n \neq 0$ and $b_m \neq 0$, and this implies that $a_n b_m \neq 0$ since R is an integral domain. But $a_n b_m$ is the leading coefficient in $f(x)g(x)$ since

$$f(x)g(x) = \sum_{i=0}^{n+m} \left(\sum_{j=0}^{i} a_j b_{i-j} \right) x^i$$

by Definition 8.3. Therefore

$$\deg(f(x)g(x)) = n + m = \deg f(x) + \deg g(x).$$ ■

COROLLARY 8.8 $R[x]$ is an integral domain if and only if R is an integral domain.

Proof   Assume that R is an integral domain. If $f(x)$ and $g(x)$ are arbitrary nonzero elements of $R[x]$, then both $f(x)$ and $g(x)$ have degrees. According to Theorem 8.7, $f(x)g(x)$ has a degree that is the sum of $\deg f(x)$ and $\deg g(x)$. Therefore $f(x)g(x)$ is not the zero polynomial, and this shows that $R[x]$ is an integral domain.

If $R[x]$ is an integral domain, however, then R must also be an integral domain since R is a commutative ring with unity and $R \subseteq R[x]$. ∎

We make some final observations concerning Theorem 8.7. Since the product of the zero polynomial and any polynomial always yields the zero polynomial, the equation in Theorem 8.7 cannot hold when one of the factors is a zero polynomial. This is justification for not defining degree for the zero polynomial. We also note that the reason the conclusion of Theorem 8.7 fails to hold in Example 4 is that Z_6 is *not* an integral domain.

EXERCISES 8.1

1   Write the following polynomials in expanded form.

(a) $\sum_{i=0}^{3} c_i x^i$ (b) $\sum_{j=0}^{4} d_j x^j$

(c) $\sum_{k=1}^{3} a_k x^k$ (d) $\sum_{k=2}^{4} x^k$

2   Express the following polynomials by using sigma notation.
(a) $c_0 x^0 + c_1 x^1 + c_2 x^2$ (b) $d_2 x^2 + d_3 x^3 + d_4 x^4$
(c) $x + x^2 + x^3 + x^4$ (d) $x^3 + x^4 + x^5$

3   Consider the following polynomials over Z_8, where a is written for $[a]$ in Z_8:

$$f(x) = 2x^3 + 7x + 4, \qquad g(x) = 4x^2 + 4x + 6, \qquad h(x) = 6x^2 + 3.$$

Find each of the following polynomials, with all coefficients in Z_8.
(a) $f(x) + g(x)$ (b) $g(x) + h(x)$
(c) $f(x)g(x)$ (d) $g(x)h(x)$
(e) $f(x)g(x) + h(x)$ (f) $f(x) + g(x)h(x)$
(g) $f(x)g(x) + f(x)h(x)$ (h) $f(x)h(x) + g(x)h(x)$

4   Consider the following polynomials over Z_7, where a is written for $[a]$ in Z_7:

$$f(x) = 2x^3 + 7x + 4, \qquad g(x) = 4x^2 + 4x + 6, \qquad h(x) = 6x^2 + 3.$$

Find each of the following polynomials, with all coefficients in Z_7.
(a) $f(x) + g(x)$ (b) $g(x) + h(x)$
(c) $f(x)g(x)$ (d) $g(x)h(x)$
(e) $f(x)g(x) + h(x)$ (f) $f(x) + g(x)h(x)$
(g) $f(x)g(x) + f(x)h(x)$ (h) $f(x)h(x) + g(x)h(x)$

5   Decide if each of the following subsets is a subring of $R[x]$, and justify your decision in each case.
(a) The set of all polynomials with zero constant term
(b) The set of all polynomials that have zero coefficients for all *even* powers of x
(c) The set of all polynomials that have zero coefficients for all *odd* powers of x
(d) The set consisting of the zero polynomial together with all polynomials that have degree 2 or less

6   Determine which of the subsets in Problem 5 are ideals of $R[x]$, and justify your choices.

7   Let R be a commutative ring with unity. Prove that

$$\deg(f(x)g(x)) \leq \deg f(x) + \deg g(x)$$

for all $f(x), g(x)$ in $R[x]$, even if R is not an integral domain.

8   List all the polynomials in $\mathbf{Z}_3[x]$ that have degree 2.

9   (a) How many polynomials of degree 2 are there in $\mathbf{Z}_n[x]$?
(b) If m is a positive integer, how many polynomials of degree m are there in $\mathbf{Z}_n[x]$?

10 (a) Suppose that R is a commutative ring with unity, and define $\theta: R[x] \to R$ by

$$\theta(a_0 + a_1 x + \cdots + a_n x^n) = a_0$$

for all $a_0 + a_1 x + \cdots + a_n x^n$ in $R[x]$. Prove that θ is an epimorphism from $R[x]$ to R.
(b) Describe the kernel of the epimorphism in part (a).

11 Let R be a commutative ring with unity and let I be the principal ideal $I = (x)$ in $R[x]$. Prove that $R[x]/I$ is isomorphic to R.

12 In the integral domain $\mathbf{Z}[x]$, let $(\mathbf{Z}[x])^+$ denote the set of all $f(x)$ in $\mathbf{Z}[x]$ that have a positive integer as a leading coefficient. Prove that $\mathbf{Z}[x]$ is an ordered integral domain by proving that $(\mathbf{Z}[x])^+$ is a set of positive elements for $\mathbf{Z}[x]$.

13 Consider the mapping $\phi: \mathbf{Z}[x] \to \mathbf{Z}_k[x]$ defined by

$$\phi(a_0 + a_1 x + \cdots + a_n x^n) = [a_0] + [a_1]x + \cdots + [a_n]x^n,$$

where $[a_i]$ denotes the congruence class of $\mathbf{Z}_k$ that contains a_i. Prove that ϕ is an epimorphism from $\mathbf{Z}[x]$ to $\mathbf{Z}_k[x]$.

14 Describe the kernel of the epimorphism ϕ in Problem 13.

15 Assume that each of R and S is a commutative ring with unity, and that $\theta: R \to S$ is an epimorphism from R to S. Let $\phi: R[x] \to S[x]$ be defined by

$$\phi(a_0 + a_1 x + \cdots + a_n x^n) = \theta(a_0) + \theta(a_1)x + \cdots + \theta(a_n)x^n.$$

Prove that ϕ is an epimorphism from $R[x]$ to $S[x]$.

16 Describe the kernel of the epimorphism ϕ in Problem 15.

17 For each $f(x) = \sum_{i=0}^{n} a_i x^i$ in $R[x]$, the **derivative of $f(x)$** is the polynomial

$$f'(x) = \sum_{i=1}^{n} i a_i x^{i-1}.$$

(For $n = 0$, $f'(x) = 0$ by definition.)
(a) Prove that $[f(x) + g(x)]' = f'(x) + g'(x)$.
(b) Prove that $[f(x)g(x)]' = f(x)g'(x) + f'(x)g(x)$.

8.2   DIVISIBILITY AND GREATEST COMMON DIVISOR

If a ring R is not an integral domain, the division of polynomials over R is not a very satisfactory subject for study because of the possible presence of zero divisors. In order to obtain the results that we need on division of polynomials, the ring of coefficients actually needs to be a field. For this reason, we confine our attention for the rest of this chapter to rings of polynomials $F[x]$ where F is a *field*. This assures us that $F[x]$ is an integral domain (Corollary 8.8), and that every nonzero element of F has a multiplicative inverse.

The definitions, theorems, and even the proofs in this section are very similar to corresponding statements in Chapter 2 concerning division in the integral domain **Z**.

> **DEFINITION 8.9** If $f(x)$ and $g(x)$ are in $F[x]$, then $f(x)$ **divides** $g(x)$ if there exists $h(x)$ in $F[x]$ such that $g(x) = f(x)h(x)$.
>
> If $f(x)$ divides $g(x)$, we write $f(x)\,|\,g(x)$, and we say that $g(x)$ is a **multiple** of $f(x)$, that $f(x)$ is a **factor** of $g(x)$, or that $f(x)$ is a **divisor** of $g(x)$. We write $f(x) \nmid g(x)$ to indicate that $f(x)$ does not divide $g(x)$.

Polynomials of degree zero (the nonzero elements of F) have two special properties that are worth noting. First, any nonzero element a of F is a factor of every $f(x) \in F[x]$ since $a^{-1}f(x)$ is in $F[x]$ and

$$f(x) = a[a^{-1}f(x)].$$

Second, if $f(x)\,|\,g(x)$, then $af(x)\,|\,g(x)$ for all nonzero $a \in F$ since the equation

$$g(x) = f(x)h(x)$$

implies that

$$g(x) = [af(x)][a^{-1}h(x)].$$

The Division Algorithm for integers has the following analogue in $F[x]$.

> **THEOREM 8.10** (**The Division Algorithm**) Let $f(x)$ and $g(x)$ be elements of $F[x]$ with $f(x)$ a nonzero polynomial. There exist unique elements $q(x)$ and $r(x)$ in $F[x]$ such that
>
> $$g(x) = f(x)q(x) + r(x),$$
>
> with either $r(x) = 0$ or $\deg r(x) < \deg f(x)$.

Proof We postpone the proof of uniqueness until existence of the required $q(x)$ and $r(x)$ in $F[x]$ has been proven. There are two trivial cases that we shall dispose of first.

1. If $g(x) = 0$ or if $\deg g(x) < \deg f(x)$, then we see from the equality

$$g(x) = f(x) \cdot 0 + g(x)$$

that $q(x) = 0$ and $r(x) = g(x)$ satisfy the required conditions.

2. If $\deg f(x) = 0$, then $f(x) = c$ for some nonzero constant c. The equality

$$g(x) = c[c^{-1}g(x)] + 0$$

shows that $q(x) = c^{-1}g(x)$ and $r(x) = 0$ satisfy the required conditions.

Suppose now that $g(x) \neq 0$ and $1 \le \deg f(x) \le \deg g(x)$. The proof is by induction on $n = \deg g(x)$, using the second principle of finite induction. For each positive integer n, let S_n be the statement that if $g(x) \in F[x]$ has degree n and $1 \le \deg f(x) \le \deg g(x)$, then there exist $q(x)$ and $r(x) \in F[x]$ such that $g(x) = f(x)q(x) + r(x)$, with either $r(x) = 0$ or $\deg r(x) < \deg f(x)$.

If $n = 1$, then the condition $1 \le \deg f(x) \le \deg g(x) = n$ requires that both $f(x)$ and $g(x)$ have degree 1, say

$$f(x) = ax + b, \qquad g(x) = cx + d,$$

where $a \neq 0$ and $c \neq 0$. The equality

$$cx + d = (ax + b)(ca^{-1}) + (d - bca^{-1})$$

shows that $q(x) = ca^{-1}$ and $r(x) = d - bca^{-1}$ satisfy the required conditions, and S_1 is true.

Now assume that k is a positive integer such that S_m is true for all positive integers $m < k$. To prove that S_k is true, let $g(x) \in F[x]$ with $\deg g(x) = k$ and $f(x) \in F[x]$ with $1 \le \deg f(x) \le \deg g(x)$. Then

$$f(x) = ax^j + \cdots, \qquad g(x) = cx^k + \cdots,$$

with $a \neq 0$, $c \neq 0$, and $j \le k$. The first step in the usual long division of $g(x)$ by $f(x)$ is shown in Figure 8.1.

$$
\begin{array}{r}
ca^{-1}x^{k-j} \\
\hline
ax^j + \cdots \, \big)\; cx^k + \cdots \\
ca^{-1}x^{k-j}f(x) \\
\hline
g(x) - ca^{-1}x^{k-j}f(x)
\end{array}
$$

FIGURE 8.1

This first step in long division yields

$$g(x) = ca^{-1}x^{k-j}f(x) + [g(x) - ca^{-1}x^{k-j}f(x)].$$

Let $h(x) = g(x) - ca^{-1}x^{k-j}f(x)$. Then the coefficient of x^k in $h(x)$ is zero, and $\deg h(x) < k$. By the induction hypothesis, there exist polynomials $q_0(x)$ and $r(x)$ such that

$$h(x) = f(x)q_0(x) + r(x)$$

with either $r(x) = 0$ or $\deg r(x) < \deg f(x)$. This gives the equality

$$g(x) = ca^{-1}x^{k-j}f(x) + h(x)$$
$$= ca^{-1}x^{k-j}f(x) + f(x)q_0(x) + r(x)$$
$$= f(x)[ca^{-1}x^{k-j} + q_0(x)] + r(x),$$

which shows that $q(x) = ca^{-1}x^{k-j} + q_0(x)$ and $r(x)$ are polynomials that satisfy the required conditions. Therefore S_k is true, and the existence part of the theorem follows from the second principle of finite induction.

To prove uniqueness, suppose that $g(x) = f(x)q_1(x) + r_1(x)$ and $g(x) = f(x)q_2(x) + r_2(x)$, where either $r_i(x) = 0$ or $\deg r_i(x) < \deg f(x)$ for $i = 1, 2$. Then

$$r_1(x) - r_2(x) = [g(x) - f(x)q_1(x)] - [g(x) - f(x)q_2(x)]$$
$$= f(x)[q_2(x) - q_1(x)].$$

The right member of this equation, $f(x)[q_2(x) - q_1(x)]$, is either zero or has degree greater than or equal to $\deg f(x)$, by Theorem 8.7. However, the left member, $r_1(x) - r_2(x)$, is either zero or has degree less than $\deg f(x)$ since $\deg r_1(x) < \deg f(x)$ and $\deg r_2(x) < \deg f(x)$. Therefore both members must be zero, and this requires that $r_1(x) = r_2(x)$ and $q_1(x) = q_2(x)$ since $f(x)$ is nonzero. Therefore $q(x)$ and $r(x)$ are unique, and the proof is complete. ∎

In the Division Algorithm, the polynomial $q(x)$ is called the **quotient** and $r(x)$ is called the **remainder** in the division of $g(x)$ by $f(x)$. For any field F, the quotient and remainder in $F[x]$ can be found by the familiar long division procedure. An illustration is given in the next example.

EXAMPLE 1   Let $f(x) = 3x^2 + 2$ and $g(x) = 4x^4 + 2x^3 + 6x^2 + 4x +$ 5 in $Z_7[x]$. We shall find $q(x)$ and $r(x)$ by the long division procedure. Referring to Figure 8.1, we have $a = 3$ in $f(x)$, $c = 4$ in $g(x)$, and $ca^{-1} = 3(4^{-1}) = 3(2) = 6$ in the first step.

$$
\begin{array}{r}
6x^2 + 3x + 5 \\
3x^2 + 2 \overline{\smash{\big)}\, 4x^4 + 2x^3 + 6x^2 + 4x + 5} \\
\underline{4x^4 + 5x^2} \\
2x^3 + x^2 \\
\underline{2x^3 + 6x} \\
x^2 + 5x \\
\underline{x^2 + 3} \\
5x + 2
\end{array}
$$

Thus the quotient is $q(x) = 6x^2 + 3x + 5$ and the remainder is $r(x) = 5x + 2$ in the division of $g(x)$ by $f(x)$. □

Our next objective in this section is to prove that any two nonzero polynomials over F have a greatest common divisor in $F[x]$. We saw earlier that if $f(x)$ is a divisor of $g(x)$, then $af(x)$ is also a divisor of $g(x)$, for every nonzero $a \in F$. By choosing a to be the multiplicative inverse of the leading coefficient of $f(x)$, the leading coefficient in $af(x)$ can be made equal to 1. This means that in the consideration of common divisors of two polynomials, there is no loss of generality if attention is restricted to polynomials with 1 as their leading coefficient.

DEFINITION 8.11   A polynomial with 1 as its leading coefficient is called a **monic** polynomial.

One of the conditions that we place on a greatest common divisor of two polynomials is that it be monic. Without this condition, the greatest common divisor of two polynomials would not be unique.

DEFINITION 8.12   Let $f(x)$ and $g(x)$ be nonzero polynomials in $F[x]$. A polynomial $d(x)$ in $F[x]$ is a **greatest common divisor** of $f(x)$ and $g(x)$ if these conditions are satisfied:
(1) $d(x)$ is a monic polynomial.
(2) $d(x) \mid f(x)$ and $d(x) \mid g(x)$.
(3) $h(x) \mid f(x)$ and $h(x) \mid g(x)$ imply that $h(x) \mid d(x)$.

The next theorem shows that any two nonzero elements $f(x), g(x)$ of $F[x]$ have a unique greatest common divisor $d(x)$. The proof, which is obtained by making minor adjustments in the proof of Theorem 2.12, shows that $d(x)$ is a **linear combination** of $f(x)$ and $g(x)$; that is, $d(x)$ can be written in the form

$$d(x) = f(x)s(x) + g(x)t(x)$$

for some $s(x), t(x) \in F[x]$.

THEOREM 8.13   Let $f(x)$ and $g(x)$ be nonzero polynomials over F. Then there exists a unique greatest common divisor $d(x)$ of $f(x)$ and $g(x)$ in $F[x]$. Moreover, $d(x)$ can be expressed as

$$d(x) = f(x)s(x) + g(x)t(x)$$

for $s(x)$ and $t(x)$ in $F[x]$, and $d(x)$ is the monic polynomial of least degree that can be written in this form.

Proof Consider the set S of all polynomials in $F[x]$ that can be written in the form

$$f(x)u(x) + g(x)v(x)$$

with $u(x)$ and $v(x)$ in $F[x]$. Since $f(x) = f(x) \cdot 1 + g(x) \cdot 0 \neq 0$, the set of nonzero polynomials in S is nonempty. Let

$$d_1(x) = f(x)u_1(x) + g(x)v_1(x)$$

be a polynomial of least degree among the nonzero elements of S. If c is the leading coefficient of $d_1(x)$, then

$$d(x) = c^{-1}d_1(x) = f(x)[c^{-1}u_1(x)] + g(x)[c^{-1}v_1(x)]$$

is a monic polynomial of least degree in S. Letting $s(x) = c^{-1}u_1(x)$ and $t(x) = c^{-1}v_1(x)$, we have a polynomial

$$d(x) = f(x)s(x) + g(x)t(x),$$

which is expressed in the required form and satisfies the first condition in Definition 8.12.

We shall show that $d(x) \mid f(x)$. By the Division Algorithm, there are elements $q(x)$ and $r(x)$ of $F[x]$ such that

$$f(x) = d(x)q(x) + r(x),$$

with either $r(x) = 0$ or $\deg r(x) < \deg d(x)$. Since

$$
\begin{aligned}
r(x) &= f(x) - d(x)q(x) \\
&= f(x) - [f(x)s(x) + g(x)t(x)]q(x) \\
&= f(x)[1 - s(x)q(x)] + g(x)[-t(x)q(x)],
\end{aligned}
$$

$r(x)$ is an element of S. By choice of $d(x)$ as having smallest possible degree among the nonzero elements of S, it cannot be true that $\deg r(x) < \deg d(x)$. Therefore $r(x) = 0$ and $d(x) \mid f(x)$. A similar argument shows that $d(x) \mid g(x)$, and hence $d(x)$ satisfies condition (2) in Definition 8.12.

If $h(x) \mid f(x)$ and $h(x) \mid g(x)$, then $f(x) = h(x)p_1(x)$ and $g(x) = h(x)p_2(x)$ for $p_i(x) \in F[x]$. Therefore

$$
\begin{aligned}
d(x) &= f(x)s(x) + g(x)t(x) \\
&= h(x)p_1(x)s(x) + h(x)p_2(x)t(x) \\
&= h(x)[p_1(x)s(x) + p_2(x)t(x)],
\end{aligned}
$$

and this shows that $h(x) \mid d(x)$. By Definition 8.12, $d(x)$ is a greatest common divisor of $f(x)$ and $g(x)$.

To show uniqueness, suppose that $d_1(x)$ and $d_2(x)$ are both greatest common divisors of $f(x)$ and $g(x)$. Then $d_1(x) \mid d_2(x)$ and also

$d_2(x) | d_1(x)$. Since both $d_1(x)$ and $d_2(x)$ are monic polynomials, this means that $d_1(x) = d_2(x)$. (See Problem 16 of the exercises.)   ■

If $f(x)$ and $g(x)$ are nonzero polynomials such that $f(x)|g(x)$, the greatest common divisor of $f(x)$ and $g(x)$ is simply the product of $f(x)$ and the multiplicative inverse of its leading coefficient. If $f(x) \nmid g(x)$, the Euclidean Algorithm extends readily to polynomials, furnishing a systematic method for finding the greatest common divisor of $f(x)$ and $g(x)$ and for finding $s(x)$ and $t(x)$ in the equation

$$d(x) = f(x)s(x) + g(x)t(x).$$

The Euclidean Algorithm consists of repeated application of the Division Algorithm to yield the following sequence, where $r_n(x)$ is the last nonzero remainder:

$$g(x) = f(x)q_0(x) + r_1(x), \qquad \deg r_1(x) < \deg f(x)$$
$$f(x) = r_1(x)q_1(x) + r_2(x), \qquad \deg r_2(x) < \deg r_1(x)$$
$$r_1(x) = r_2(x)q_2(x) + r_3(x), \qquad \deg r_3(x) < \deg r_2(x)$$
$$\vdots \qquad\qquad\qquad \vdots$$
$$r_{n-2}(x) = r_{n-1}(x)q_{n-1}(x) + r_n(x), \quad \deg r_n(x) < \deg r_{n-1}(x)$$
$$r_{n-1}(x) = r_n(x)q_n(x)$$

Suppose that a is the leading coefficient of the last nonzero remainder, $r_n(x)$. It is left as an exercise to prove that $a^{-1}r_n(x)$ is the greatest common divisor of $f(x)$ and $g(x)$.

EXAMPLE 2   We shall find the greatest common divisor of $f(x) = 3x^3 + 5x^2 + 6x$ and $g(x) = 4x^4 + 2x^3 + 6x^2 + 4x + 5$ in $\mathbf{Z}_7[x]$. Long division of $g(x)$ by $f(x)$ yields a quotient of $q_0(x) = 6x$ and a remainder of $r_1(x) = 5x^2 + 4x + 5$, so we have

$$g(x) = f(x) \cdot (6x) + (5x^2 + 4x + 5).$$

Dividing $f(x)$ by $r_1(x)$, we obtain

$$f(x) = r_1(x) \cdot (2x + 5) + (4x + 3),$$

so $q_1(x) = 2x + 5$ and $r_2(x) = 4x + 3$ in the Euclidean Algorithm. Division of $r_1(x)$ by $r_2(x)$ then yields

$$r_1(x) = r_2(x) \cdot (3x + 4).$$

Thus $r_2(x) = 4x + 3$ is the last nonzero remainder, and the greatest common divisor of $f(x)$ and $g(x)$ in $\mathbf{Z}_7[x]$ is

$$d(x) = 4^{-1}(4x + 3)$$
$$= 2(4x + 3)$$
$$= x + 6.$$

☐

As mentioned earlier, the Euclidean Algorithm can also be used to find polynomials $s(x)$ and $t(x)$ such that

$$d(x) = f(x)s(x) + g(x)t(x).$$

This is illustrated in Example 3.

EXAMPLE 3   As in Example 2, let $f(x) = 3x^3 + 5x^2 + 6x$ and $g(x) = 4x^4 + 2x^3 + 6x^2 + 4x + 5$. From Example 2, the greatest common divisor of $f(x)$ and $g(x)$ is $d(x) = x + 6$. To find polynomials $s(x)$ and $t(x)$ such that

$$d(x) = f(x)s(x) + g(x)t(x),$$

we first solve for the remainders in the Euclidean Algorithm (see Example 2) as follows:

$$r_2(x) = f(x) - r_1(x)(2x + 5)$$
$$r_1(x) = g(x) - f(x)(6x).$$

Substituting for $r_1(x)$ in the first equation, we have

$$r_2(x) = f(x) - [g(x) - f(x)(6x)](2x + 5)$$
$$= f(x) + f(x)(6x)(2x + 5) - g(x)(2x + 5)$$
$$= f(x)[1 + (6x)(2x + 5)] + g(x)(-2x - 5)$$
$$= f(x)(5x^2 + 2x + 1) + g(x)(5x + 2).$$

To express $d(x) = 4^{-1}r_2(x) = 2r_2(x)$ as a linear combination of $f(x)$ and $g(x)$, we multiply both members of the last equation by $4^{-1} = 2$:

$$d(x) = 2r_2(x) = f(x)(2)(5x^2 + 2x + 1) + g(x)(2)(5x + 2)$$

$$d(x) = f(x)(3x^2 + 4x + 2) + g(x)(3x + 4).$$

The desired polynomials are given by $s(x) = 3x^2 + 4x + 2$ and $t(x) = 3x + 4$.

☐

EXERCISES 8.2

For $f(x)$, $g(x)$, and $\mathbf{Z}_n[x]$ given in Problems 1–6, find $q(x)$ and $r(x)$ in $\mathbf{Z}_n[x]$ that satisfy the conditions in the Division Algorithm.

1    $f(x) = 3x + 1$, $g(x) = 2x^3 + 3x^2 + 4x + 1$, in $\mathbf{Z}_5[x]$
2    $f(x) = 2x + 2$, $g(x) = x^3 + 2x^2 + 2$, in $\mathbf{Z}_3[x]$
3    $f(x) = x^3 + x^2 + 2x + 2$, $g(x) = x^4 + 2x^2 + x + 1$, in $\mathbf{Z}_3[x]$

4 $f(x) = x^3 + 2x^2 + 2, g(x) = 2x^5 + 2x^4 + x^2 + 2$, in $\mathbf{Z}_3[x]$

5 $f(x) = 3x^2 + 2, g(x) = x^4 + 5x^2 + 2x + 2$, in $\mathbf{Z}_7[x]$

6 $f(x) = 3x^2 + 2, g(x) = 4x^4 + 2x^3 + 6x^2 + 4x + 5$, in $\mathbf{Z}_7[x]$

For $f(x)$, $g(x)$, and $\mathbf{Z}_n[x]$ given in Problems 7–10, find the greatest common divisor $d(x)$ of $f(x)$ and $g(x)$ in $\mathbf{Z}_n[x]$.

7 $f(x) = x^3 + x^2 + 2x + 2, g(x) = x^4 + 2x^2 + x + 1$, in $\mathbf{Z}_3[x]$

8 $f(x) = x^3 + 2x^2 + 2, g(x) = 2x^5 + 2x^4 + x^2 + 2$, in $\mathbf{Z}_3[x]$

9 $f(x) = 3x^2 + 2, g(x) = x^4 + 5x^2 + 2x + 2$, in $\mathbf{Z}_7[x]$

10 $f(x) = 3x^2 + 2, g(x) = 4x^4 + 2x^3 + 6x^2 + 4x + 5$, in $\mathbf{Z}_7[x]$

For $f(x)$, $g(x)$, and $\mathbf{Z}_n[x]$ given in Problems 11–14, find $s(x)$ and $t(x)$ in $\mathbf{Z}_n[x]$ such that $d(x) = f(x)s(x) + g(x)t(x)$ is the greatest common divisor of $f(x)$ and $g(x)$.

11 $f(x) = 2x^3 + 2x^2 + x + 1, g(x) = x^4 + 2x^2 + x + 1$, in $\mathbf{Z}_3[x]$

12 $f(x) = 2x^3 + x^2 + 1, g(x) = x^5 + x^4 + 2x^2 + 1$ in $\mathbf{Z}_3[x]$

13 $f(x) = 3x^2 + 2, g(x) = x^4 + 5x^2 + 2x + 2$, in $\mathbf{Z}_7[x]$

14 $f(x) = 3x^2 + 2, g(x) = 4x^4 + 2x^3 + 6x^2 + 4x + 5$, in $\mathbf{Z}_7[x]$

15 Prove that if $f(x)$ and $g(x)$ are nonzero elements of $F[x]$ such that $f(x)\,|\,g(x)$ and $g(x)\,|\,f(x)$, then $f(x) = ag(x)$ for some nonzero $a \in F$.

16 Prove that if $d_1(x)$ and $d_2(x)$ are monic polynomials over the field F such that $d_1(x)\,|\,d_2(x)$ and $d_2(x)\,|\,d_1(x)$, then $d_1(x) = d_2(x)$.

17 Prove that if $h(x)\,|\,f(x)$ and $h(x)\,|\,g(x)$ in $F[x]$, then $h(x)\,|\,[f(x)u(x) + g(x)v(x)]$ for all $u(x)$ and $v(x)$ in $F[x]$.

18 In the statement of the Division Algorithm (Theorem 8.10), prove that the greatest common divisor of $g(x)$ and $f(x)$ is equal to the greatest common divisor of $f(x)$ and $r(x)$.

19 With the notation used in the description of the Euclidean Algorithm, prove that $a^{-1}r_n(x)$ is the greatest common divisor of $f(x)$ and $g(x)$.

20 Prove that every nonzero remainder $r_j(x)$ in the Euclidean Algorithm is a linear combination of $f(x)$ and $g(x)$: $r_j(x) = f(x)s_j(x) + g(x)t_j(x)$ for some $s_j(x)$ and $t_j(x)$ in $F[x]$.

21 Prove that the only elements of $F[x]$ that have multiplicative inverses are the nonzero elements of the field F.

8.3 FACTORIZATION IN F[x]

Let $f(x) = a_0 + a_1 x + a_2 x^2 + \cdots + a_n x^n$ denote an arbitrary polynomial over the field F. For any $c \in F$, $f(c)$ is defined by the equation

$$f(c) = a_0 + a_1 c + a_2 c^2 + \cdots + a_n c^n.$$

That is, $f(c)$ is obtained by replacing the indeterminate x in $f(x)$ by the element c. For each $c \in F$, this replacement rule yields a unique value $f(c) \in F$, and hence the pairing $(c, f(c))$ defines a mapping from F to F. A mapping obtained in this manner is called a **polynomial mapping**, or a **polynomial function**, from F to F.

DEFINITION 8.14   Let $f(x)$ be a polynomial over the field F. If c is an element of F such that $f(c) = 0$, then c is called a **zero** of $f(x)$, and we say that c is a **root**, or a **solution**, of the equation $f(x) = 0$.

EXAMPLE 1   Consider $f(x) = x^2 + 1$ in $\mathbf{Z}_5[x]$. Since

$$f(2) = 2^2 + 1 = 0$$

in $\mathbf{Z}_5$, 2 is a *zero* of $x^2 + 1$. Also, 2 is a *root*, or a *solution*, of $x^2 + 1 = 0$ over $\mathbf{Z}_5$. $\square$

For arbitrary polynomials $f(x)$ and $g(x)$ over a field F, let $h(x) = f(x) + g(x)$ and $p(x) = f(x)g(x)$. Two consequences of the definitions of addition and multiplication in $F[x]$ are that

$$h(c) = f(c) + g(c) \quad \text{and} \quad p(c) = f(c)g(c)$$

for all c in F. We shall use these results quite freely, with their justifications left as exercises.

The following example is of some interest in connection with the discussion in the preceding paragraph.

EXAMPLE 2   Consider the polynomials $f(x) = 3x^5 - 4x^2$ and $g(x) = x^2 + 3x$ in $\mathbf{Z}_5[x]$. By direct computation, we find that

$$f(0) = 0 = g(0) \quad f(1) = 4 = g(1) \quad f(2) = 0 = g(2)$$
$$f(3) = 3 = g(3) \quad f(4) = 3 = g(4).$$

Thus $f(c) = g(c)$ for all c in $\mathbf{Z}_5$, but $f(x) \neq g(x)$ in $\mathbf{Z}_5[x]$. $\square$

The next two theorems are two of the simplest and most useful results on factorization in $F[x]$.

THEOREM 8.15   **(The Remainder Theorem)** If $f(x)$ is a polynomial over the field F and $c \in F$, the remainder in the division of $f(x)$ by $x - c$ is $f(c)$.

Proof   Since $x - c$ has degree one, the remainder r in

$$f(x) = (x - c)q(x) + r$$

is a constant. Replacing x by c, we obtain

$$f(c) = (c - c)q(c) + r$$
$$= 0 \cdot q(c) + r$$
$$= r.$$

Thus $r = f(c)$.   ■

THEOREM 8.16   (The Factor Theorem) A polynomial $f(x)$ over the field F has a factor $x - c$ in $F[x]$ if and only if $c \in F$ is a zero of $f(x)$.

Proof   From the Remainder Theorem, we have

$$f(x) = (x - c)q(x) + f(c).$$

Thus $x - c$ is a factor of $f(x)$ if and only if $f(c) = 0$.   ■

The Factor Theorem can be extended as follows.

THEOREM 8.17   Let $f(x)$ be a polynomial over the field F that has positive degree n and leading coefficient a. If $c_1, c_2, \ldots, c_n$ are n distinct zeros of $f(x)$ in F, then

$$f(x) = a(x - c_1)(x - c_2) \cdots (x - c_n).$$

Proof   The proof is by induction on $n = \deg f(x)$. For each positive integer n, let S_n be the statement of the theorem.

For $n = 1$, suppose that $f(x)$ has degree 1 and leading coefficient a, and let c_1 be a zero of $f(x)$ in F. Then $f(x) = ax + b$, where $a \neq 0$ and $f(c_1) = 0$. This implies that $ac_1 + b = 0$ and $b = -ac_1$. Therefore $f(x) = ax - ac_1 = a(x - c_1)$, and S_1 is true.

Assume now that S_k is true, and let $f(x)$ be a polynomial with leading coefficient a and degree $k + 1$ that has $k + 1$ distinct zeros $c_1, c_2, \ldots, c_k, c_{k+1}$ in F. Since c_{k+1} is zero of $f(x)$,

$$f(x) = (x - c_{k+1})q(x)$$

by the Factor Theorem. By Theorem 8.7, $q(x)$ must have degree k. Since the factor $x - c_{k+1}$ is monic, $q(x)$ and $f(x)$ have the same leading coefficient. For $i = 1, 2, \ldots, k$, we have

$$(c_i - c_{k+1})q(c_i) = f(c_i) = 0,$$

where $c_i - c_{k+1} \neq 0$ since the zeros $c_1, c_2, \ldots, c_k, c_{k+1}$ are distinct. Therefore $q(c_i) = 0$ for $i = 1, 2, \ldots, k$. That is, $c_1, c_2, \ldots, c_k$ are k distinct zeros of $q(x)$ in F. By the induction hypothesis,

$$q(x) = a(x - c_1)(x - c_2) \cdots (x - c_k).$$

Substitution of this factored expression for $q(x)$ in $f(x) = (x - c_{k+1})q(x)$ yields

$$f(x) = a(x - c_1)(x - c_2)\cdots(x - c_k)(x - c_{k+1}).$$

Therefore S_{k+1} is true whenever S_k is true, and it follows by induction that S_n is true for all positive integers n. ∎

The proof of the following corollary is left as an exercise.

COROLLARY 8.18 A polynomial of positive degree n over the field F has at most n distinct zeros in F.

In the factorization of polynomials over a field F, the concept of an irreducible polynomial is analogous to the concept of a prime integer in the factorization of integers.

DEFINITION 8.19 A polynomial $f(x)$ in $F[x]$ is **irreducible** (or **prime**) over F if $f(x)$ has positive degree and $f(x)$ *cannot* be expressed as a product $f(x) = g(x)h(x)$ with both $g(x)$ and $h(x)$ of positive degree in $F[x]$.

EXAMPLE 3 Note that whether or not a given polynomial is irreducible over F depends on the field F. For instance, $x^2 + 1$ is irreducible over the field of real numbers, but it is not irreducible over the field $\mathscr{C}$ of complex numbers since $x^2 + 1$ can be factored as

$$x^2 + 1 = (x - i)(x + i)$$

in $\mathscr{C}[x]$. □

If $g(x)$ and $h(x)$ are polynomials of positive degree, their product $g(x)h(x)$ has degree at least two. Therefore all polynomials of degree one are irreducible. Constant polynomials, however, are never irreducible since they do not have positive degree.

It is usually not easy to decide whether or not a given polynomial is irreducible over a certain field. However, the following theorem is sometimes quite helpful for polynomials with degree less than four.

THEOREM 8.20 If $f(x)$ is a polynomial of degree two or three over the field F, then $f(x)$ is irreducible over F if and only if $f(x)$ has no zeros in F.

Proof Let $f(x)$ be a polynomial of degree two or three over the field F. We shall prove the theorem in this form: $f(x)$ is not irreducible over F if and only if $f(x)$ has at least one zero in F.

Suppose first that $f(x)$ has a zero c in F. By the Factor Theorem,

$$f(x) = (x - c)q(x),$$

where $q(x)$ has degree one less than that of $f(x)$, by Theorem 8.7. This factorization shows that $f(x)$ is not irreducible over F.

Assume conversely that $f(x)$ is not irreducible over F. That is, there are polynomials $g(x)$ and $h(x)$ in $F[x]$ such that $f(x) = g(x)h(x)$, with both $g(x)$ and $h(x)$ of positive degree. By Theorem 8.7,

$$\deg f(x) = \deg g(x) + \deg h(x).$$

Since $\deg f(x)$ is either two or three, one of the factors $g(x)$ and $h(x)$ must have degree one. Without loss of generality, we may assume that this factor is $g(x)$, and we have

$$f(x) = (ax + b)h(x),$$

where $a \neq 0$. It follows at once from this equation that $-a^{-1}b$ is a zero of $f(x)$ in F, and the proof is complete. ∎

EXAMPLE 4 Let us determine whether or not each of the following polynomials is irreducible over Z_5:

(a) $f(x) = x^5 + 2x^4 - 3x^2 + 2$.
(b) $g(x) = x^2 + 3x + 4$.

Routine computations show that

$$f(0) = 2, \quad f(1) = 2, \quad f(2) = 4, \quad f(3) = 0, \quad f(4) = 0.$$

Thus 3 and 4 are zeros of $f(x)$ in Z_5, and $f(x)$ is not irreducible over Z_5. However, $g(x)$ is irreducible over Z_5 since $g(x)$ has no zeros in Z_5:

$$g(0) = 4, \quad g(1) = 3, \quad g(2) = 4, \quad g(3) = 2, \quad g(4) = 2. \quad \square$$

Irreducible polynomials play a role in the factorization of polynomials corresponding to the role prime integers play in the factorization of integers. This is illustrated by the next theorem.

THEOREM 8.21 If $p(x)$ is an irreducible polynomial over the field F and $p(x)$ divides $f(x)g(x)$ in $F[x]$, then either $p(x)\,|\,f(x)$ or $p(x)\,|\,g(x)$ in $F[x]$.

Proof   Assume that $p(x)$ is irreducible over F and that $p(x)$ divides $f(x)g(x)$, say

$$f(x)g(x) = p(x)q(x)$$

for some $q(x)$ in $F[x]$. If $p(x)|f(x)$, the conclusion is satisfied. Suppose, then, that $p(x)$ does not divide $f(x)$. This means that 1 is the greatest common divisor of $f(x)$ and $p(x)$ since the only divisors of $p(x)$ with positive degree are constant multiples of $p(x)$. By Theorem 8.13, there exist $s(x)$ and $t(x)$ in $F[x]$ such that

$$1 = f(x)s(x) + p(x)t(x),$$

and this implies that

$$\begin{aligned}
g(x) &= g(x)[f(x)s(x) + p(x)t(x)] \\
&= f(x)g(x)s(x) + p(x)g(x)t(x) \\
&= p(x)q(x)s(x) + p(x)g(x)t(x)
\end{aligned}$$

since $f(x)g(x) = p(x)q(x)$. Factoring $p(x)$ from the two terms in the right member, we see that $p(x)|g(x)$:

$$g(x) = p(x)[q(x)s(x) + g(x)t(x)].$$

Thus $p(x)$ divides $g(x)$ if it does not divide $f(x)$.   ∎

A comparison of Theorem 8.21 with Theorem 2.14 provides an indication of how closely the theory of divisibility in $F[x]$ resembles the theory of divisibility in the integers. This analogy carries over to the proofs as well. For this reason, the proofs of the remaining results in this section are left as exercises.

THEOREM 8.22   If $p(x)$ is an irreducible polynomial over the field F and $p(x)$ divides a product $f_1(x)f_2(x)\cdots f_n(x)$ in $F[x]$, then $p(x)$ divides some $f_j(x)$.

Just as with integers, two nonzero polynomials $f(x)$ and $g(x)$ over the field F are called **relatively prime** over F if their greatest common divisor in $F[x]$ is 1.

THEOREM 8.23   If $f(x)$ and $g(x)$ are relatively prime polynomials over the field F and if $f(x)|g(x)h(x)$ in $F[x]$, then $f(x)|h(x)$ in $F[x]$.

THEOREM 8.24   (**Unique Factorization Theorem**) Every polynomial of positive degree over the field F can be expressed as a product of its leading coefficient and a finite number of monic irreducible polynomials over F. This factorization is unique except for the order of the factors.

The monic irreducible polynomials involved in the factorization of $f(x)$ over F may not all be distinct, of course. If $p_1(x), p_2(x), \ldots, p_r(x)$ are the *distinct* monic irreducible factors of $f(x)$, then all repeated factors may be collected together and expressed by use of exponents to yield

$$f(x) = a[p_1(x)]^{m_1}[p_2(x)]^{m_2}\cdots[p_r(x)]^{m_r},$$

where each m_i is a positive integer.

In the last expression for $f(x)$, m_i is called the **multiplicity** of the factor $p_i(x)$. More generally, if $g(x)$ is an arbitrary polynomial of positive degree such that $[g(x)]^m$ divides $f(x)$ and no higher power of $g(x)$ divides $f(x)$ in $F[x]$, then $g(x)$ is said to be a factor of $f(x)$ over $F[x]$ with **multiplicity** m. Also, if c is an element of the field F such that $(x - c)^m$ divides $f(x)$ for some positive integer m but no higher power of $x - c$ divides $f(x)$, then c is called a **zero of multiplicity** m.

EXAMPLE 5   We shall find the factorization that is described in the Unique Factorization Theorem for the polynomial

$$f(x) = 2x^4 + x^3 + 3x^2 + 2x + 4$$

over the field $\mathbf{Z}_5$.

We first determine the zeros of $f(x)$ in $\mathbf{Z}_5$:

$$f(0) = 4, \qquad f(1) = 2, \qquad f(2) = 0, \qquad f(3) = 1, \qquad f(4) = 1.$$

Thus 2 is the only zero of $f(x)$ in $\mathbf{Z}_5$, and the Factor Theorem assures us that $x - 2$ is a factor of $f(x)$. Dividing by $x - 2$, we get

$$f(x) = (x - 2)(2x^3 + 3x + 3).$$

By Problem 10 of the exercises for this section, the zeros of $f(x)$ are 2 and the zeros of $g(x) = 2x^3 + 3x + 3$. We therefore need to determine the zeros of $g(x)$, and the only possibility is 2 since this is the only zero of $f(x)$ in $\mathbf{Z}_5$. We find that $g(2) = 0$, and this indicates that $x - 2$ is a factor of $g(x)$. Performing the required division, we obtain

$$2x^3 + 3x + 3 = (x - 2)(2x^2 + 4x + 1)$$

and

$$f(x) = (x - 2)(x - 2)(2x^2 + 4x + 1)$$
$$= (x - 2)^2(2x^2 + 4x + 1).$$

We now find that $2x^2 + 4x + 1$ is irreducible over $\mathbf{Z}_5$ since it has no zeros in $\mathbf{Z}_5$. To arrive at the desired factorization, we only need to factor the leading coefficient of $f(x)$ from the factor $2x^2 + 4x + 1$:

$$f(x) = (x - 2)^2(2x^2 + 4x + 1)$$
$$= (x - 2)^2[2x^2 + 4x + (2)(3)]$$
$$= 2(x - 2)^2(x^2 + 2x + 3).$$ $\square$

EXERCISES 8.3

1   Let $\mathbf{Q}$ denote the field of rational numbers, $\mathscr{R}$ the field of real numbers, and $\mathscr{C}$ the field of complex numbers. Determine whether or not each of the following polynomials is irreducible over each of the indicated fields.
(a) $x^2 - 2$ over $\mathbf{Q}$, $\mathscr{R}$, and $\mathscr{C}$
(b) $x^2 + 1$ over $\mathbf{Q}$, $\mathscr{R}$, and $\mathscr{C}$
(c) $x^2 + x - 2$ over $\mathbf{Q}$, $\mathscr{R}$, and $\mathscr{C}$
(d) $x^2 + 2x + 2$ over $\mathbf{Q}$, $\mathscr{R}$, and $\mathscr{C}$
(e) $x^2 + x - 2$ over $\mathbf{Z}_3$, $\mathbf{Z}_5$, and $\mathbf{Z}_7$
(f) $x^2 + 2x + 2$ over $\mathbf{Z}_3$, $\mathbf{Z}_5$, and $\mathbf{Z}_7$
(g) $x^3 + x^2 + 2x + 2$ over $\mathbf{Z}_3$, $\mathbf{Z}_5$, and $\mathbf{Z}_7$
(h) $x^4 + 2x^2 + 1$ over $\mathbf{Z}_3$, $\mathbf{Z}_5$, and $\mathbf{Z}_7$

2   Find all monic irreducible polynomials of degree two over $\mathbf{Z}_3$.

3   Write each of the following polynomials as a product of its leading coefficient and a finite number of monic irreducible polynomials over $\mathbf{Z}_5$.
(a) $2x^3 + 1$ (b) $3x^3 + 2x^2 + x + 2$
(c) $3x^3 + x^2 + 2x + 4$ (d) $2x^3 + 4x^2 + 3x + 1$
(e) $2x^4 + x^3 + 3x + 2$ (f) $3x^4 + 3x^3 + x + 3$
(g) $x^4 + x^3 + x^2 + 2x + 3$ (h) $x^4 + x^3 + 2x^2 + 3x + 2$

4   Prove Corollary 8.18: A polynomial of positive degree n over the field F has at most n distinct zeros in F.

5   Let $f(x)$ and $g(x)$ be two polynomials over the field F, both of degree n or less. Prove that if $m > n$ and if there exist m distinct elements $c_1, c_2, \ldots, c_m$ of F such that $f(c_i) = g(c_i)$ for $i = 1, 2, \ldots, m$, then $f(x) = g(x)$.

6   Let p be a prime integer and consider the polynomials $f(x) = x^p$ and $g(x) = x$ over the field $\mathbf{Z}_p$. Prove that $f(c) = g(c)$ for all c in $\mathbf{Z}_p$. (This result is known as **Fermat's Theorem**: $c^p \equiv c \pmod{p}$. To prove it, consider the multiplicative group of nonzero elements of $\mathbf{Z}_p$.)

7   Give an example of a polynomial of degree four over the field $\mathscr{R}$ of real numbers that is *not* irreducible over $\mathscr{R}$ and yet has no zeros in the real numbers.

8   If $f(x)$ and $g(x)$ are polynomials over the field F and $h(x) = f(x) + g(x)$, prove that $h(c) = f(c) + g(c)$ for all c in F.

9   If $f(x)$ and $g(x)$ are polynomials over the field F and $p(x) = f(x)g(x)$, prove that $p(c) = f(c)g(c)$ for all c in F.

10   Let $f(x)$ be a polynomial of positive degree n over the field F, and assume that $f(x) = (x - c)q(x)$ for some $c \in F$ and $q(x)$ in $F[x]$. Prove that (a) c and the zeros of $q(x)$ in F are zeros of $f(x)$, and (b) $f(x)$ has no other zeros in F.

11   Suppose that $f(x)$, $g(x)$, and $h(x)$ are polynomials over the field F, each of which has positive degree, and that $f(x) = g(x)h(x)$. Prove that the zeros of $f(x)$ in F consist of the zeros of $g(x)$ in F together with the zeros of $h(x)$ in F.

12   Prove that a polynomial $f(x)$ of positive degree n over the field F has at most n (not necessarily distinct) zeros in F.

13   Prove Theorem 8.22: If $p(x)$ is an irreducible polynomial over the field F and $p(x)$ divides a product $f_1(x)f_2(x)\cdots f_n(x)$ in $F[x]$, then $p(x)$ divides some $f_j(x)$.

14   Prove Theorem 8.23: If $f(x)$ and $g(x)$ are relatively prime polynomials over the field F and if $f(x)\,|\,g(x)h(x)$ in $F[x]$, then $f(x)\,|\,h(x)$ in $F[x]$.

15   Prove the Unique Factorization Theorem in $F[x]$ (Theorem 8.24).

8.4   ZEROS OF A POLYNOMIAL

We now focus our interest on polynomials that have their coefficients in either the field $\mathscr{C}$ of complex numbers, the field $\mathscr{R}$ of real numbers, or the field $\mathbf{Q}$ of rational numbers. Our results are concerned with the zeros of these polynomials and the related property of irreducibility over these fields.

The statement in Theorem 8.25 is so important that it is known as the Fundamental Theorem of Algebra. It was first proved in 1799 by the great German mathematician C. F. Gauss (1777–1855). Unfortunately, all known proofs of this theorem require elaborate theories that we do not have at our disposal. For this reason, we are forced to accept the theorem without proof.

THEOREM 8.25   (The Fundamental Theorem of Algebra) If $f(x)$ is a polynomial of positive degree over the field of complex numbers, then $f(x)$ has a zero in the complex numbers.

The Fundamental Theorem opens the door to a complete decomposition of any polynomial over $\mathscr{C}$, as described in the following theorem.

THEOREM 8.26   If $f(x)$ is a polynomial of positive degree n over the field $\mathscr{C}$ of complex numbers, then $f(x)$ can be factored as

$$f(x) = a(x - c_1)(x - c_2)\cdots(x - c_n),$$

where a is the leading coefficient of $f(x)$ and $c_1, c_2, \ldots, c_n$ are n (not necessarily distinct) complex numbers that are zeros of $f(x)$.

Proof   For each positive integer n, let S_n be the statement of the theorem.

If $n = 1$, then $f(x) = ax + b$, where $a \neq 0$. The complex number $c_1 = -a^{-1}b$ is a zero of $f(x)$, and

$$f(x) = ax + b = ax - ac_1 = a(x - c_1).$$

Thus S_1 is true.

Assume that S_k is true, and let $f(x)$ be a polynomial of degree $k + 1$ over $\mathscr{C}$. By the Fundamental Theorem of Algebra, $f(x)$ has a zero c_1 in the complex numbers, and the Factor Theorem asserts that

$$f(x) = (x - c_1)q(x)$$

for some polynomial $q(x)$ over $\mathscr{C}$. Since $x - c_1$ is monic, $q(x)$ has the same leading coefficient as $f(x)$, and Theorem 8.7 implies that $q(x)$ has degree k. By the induction hypothesis, $q(x)$ can be factored as the product of its leading coefficient and k factors of the form $x - c_i$:

$$q(x) = a(x - c_2)(x - c_3) \cdots (x - c_{k+1}).$$

Therefore

$$\begin{aligned} f(x) &= (x - c_1)q(x) \\ &= a(x - c_1)(x - c_2) \cdots (x - c_{k+1}), \end{aligned}$$

and S_{k+1} is true. It follows that the theorem is true for all positive integers n. ∎

As noted in the statement of Theorem 8.26, the zeros c_i are not necessarily distinct in the factorization of $f(x)$ that is described there. If the repeated factors are collected together, we have

$$f(x) = a(x - c_1)^{m_1}(x - c_2)^{m_2} \cdots (x - c_r)^{m_r}.$$

as a standard form for the unique factorization of a polynomial over the complex numbers. In particular, we observe that *the only irreducible polynomials over $\mathscr{C}$ are the first-degree polynomials.*

With such a simple description of the irreducible polynomials over $\mathscr{C}$, it is natural to ask which polynomials are irreducible over the real numbers. For polynomials of degree two (quadratic polynomials), an answer to this question is readily available from the *quadratic formula*. According to the **quadratic formula**, the zeros of a polynomial

$$f(x) = ax^2 + bx + c$$

with real coefficients[†] and $a \neq 0$ are given by

$$r_1 = \frac{-b + \sqrt{b^2 - 4ac}}{2a} \quad \text{and} \quad r_2 = \frac{-b - \sqrt{b^2 - 4ac}}{2a}.$$

These zeros are not real numbers if and only if the *discriminant*, $b^2 - 4ac$, is negative. Thus a quadratic polynomial is irreducible over the real numbers if and only if it has a negative discriminant.

[†] The quadratic formula is also valid if the coefficients are complex numbers, but at the moment we are interested only in the real case.

If we introduce some appropriate terminology, a meaningful characterization of the field of complex numbers can now be formulated. If F and E are fields such that $F \subseteq E$, then E is called an **extension** of F. An element $a \in E$ is called **algebraic** over F if a is the zero of a polynomial $f(x)$ with coefficients in F, and E is an **algebraic extension** of F if every element of E is algebraic over F. E is **algebraically closed** if every polynomial over E has a zero in E.

The field $\mathscr{C}$ of complex numbers can be characterized as a field with the following properties:

1. $\mathscr{C}$ is an algebraic extension of the field $\mathscr{R}$ of real numbers.
2. $\mathscr{C}$ is algebraically closed.

If $z = a + bi$ with $a, b \in \mathscr{R}$, then z is a zero of the polynomial

$$f(x) = [x - (a + bi)][x - (a - bi)]$$
$$= x^2 - 2ax + (a^2 + b^2)$$

over $\mathscr{R}$. Thus z is algebraic over $\mathscr{R}$ and property 1 is established. The Fundamental Theorem of Algebra (Theorem 8.25) asserts that $\mathscr{C}$ is algebraically closed. It can be proved that any field that is an algebraic extension of $\mathscr{R}$ and is algebraically closed must be isomorphic to $\mathscr{C}$. The proof of this assertion is beyond the scope of this text.

If a and b are real numbers, the *conjugate* of the complex number $z = a + bi$ is the complex number $\bar{z} = a - bi$. Note that the zeros r_1 and r_2 given by the quadratic formula are conjugates of each other when the coefficients are real and $b^2 - 4ac < 0$.

In the exercises at the end of this section, proofs are requested for the following facts concerning conjugates:

$$\overline{z_1 + z_2 + \cdots + z_n} = \bar{z}_1 + \bar{z}_2 + \cdots + \bar{z}_n$$
$$\overline{z_1 \cdot z_2 \cdot \cdots \cdot z_n} = \bar{z}_1 \cdot \bar{z}_2 \cdot \cdots \cdot z_n.$$

That is, the conjugate of a sum of terms is the sum of the conjugates of the individual terms, and the conjugate of a product of factors is the product of the conjugates of the individual factors. As a special case for products,

$$\overline{(z^n)} = (\bar{z})^n.$$

These properties of conjugates are used in the proof of the next theorem.

THEOREM 8.27   Suppose that $f(x)$ is a polynomial that has all its coefficients in the real numbers. If the complex number z is a zero of $f(x)$, then its conjugate $\bar{z}$ is also a zero of $f(x)$.

Proof   Let $f(x) = \sum_{i=0}^{n} a_i x^i$, where all a_i are real, and assume that z is a zero of $f(x)$. Then $f(z) = 0$, and therefore

$$
\begin{aligned}
0 = \overline{f(z)} \\
= \overline{a_0 + a_1 z + a_2 z^2 + \cdots + a_n z^n} \\
= \overline{a_0} + \overline{a_1 z} + \overline{a_2 z^2} + \cdots + \overline{a_n z^n} \\
= \bar{a}_0 + \bar{a}_1 \bar{z} + \bar{a}_2 (\bar{z})^2 + \cdots + \bar{a}_n (\bar{z})^n \\
= a_0 + a_1 \bar{z} + a_2 (\bar{z})^2 + \cdots + a_n (\bar{z})^n,
\end{aligned}
$$

where the last equality follows from the fact that each a_i is a real number. Thus we have $f(\bar{z}) = 0$, and the theorem is proved. ■

EXAMPLE 1   The monic polynomial of least degree over the complex numbers that has $1 - i$ and $2i$ as zeros is

$$
\begin{aligned}
f(x) &= [x - (1 - i)][x - 2i] \\
&= x^2 - (1 + i)x + 2 + 2i.
\end{aligned}
$$

However, a polynomial *with real coefficients* that has $1 - i$ and $2i$ as zeros must also have $1 + i$ and $-2i$ as zeros. Thus the monic polynomial of least degree with real coefficients that has $1 - i$ and $2i$ as zeros is

$$
\begin{aligned}
g(x) &= [x - (1 - i)][x - (1 + i)][x - 2i][x + 2i] \\
&= (x^2 - 2x + 2)(x^2 + 4) \\
&= x^4 - 2x^3 + 6x^2 - 8x + 8.
\end{aligned}
$$

□

EXAMPLE 2   Suppose that it is known that $1 - 2i$ is a zero of the polynomial $f(x) = x^4 - 3x^3 + x^2 + 7x - 30$, and that we wish to find all the zeros of $f(x)$. From Theorem 8.27, we know that $1 + 2i$ is also a zero of $f(x)$. The Factor Theorem then assures us that $x - (1 - 2i)$ and $x - (1 + 2i)$ are factors of $f(x)$:

$$
f(x) = [x - (1 - 2i)][x - (1 + 2i)]q(x).
$$

To find $q(x)$, we divide $f(x)$ by the polynomial

$$
[x - (1 - 2i)][x - (1 + 2i)] = x^2 - 2x + 5
$$

and obtain $q(x) = x^2 - x - 6$. Thus

$$
\begin{aligned}
f(x) &= [x - (1 - 2i)][x - (1 + 2i)](x^2 - x - 6) \\
&= [x - (1 - 2i)][x - (1 + 2i)](x - 3)(x + 2).
\end{aligned}
$$

It is now evident that the zeros of $f(x)$ are $1 - 2i$, $1 + 2i$, 3, and -2. □

The results obtained thus far prepare for the next theorem, which describes a standard form for the unique factorization of a polynomial over the real numbers. The proof of this theorem is left as an exercise.

THEOREM 8.28   Every polynomial of positive degree over the field $\mathscr{R}$ of real numbers can be factored as the product of its leading coefficient and a finite number of monic irreducible polynomials over $\mathscr{R}$, each of which is either quadratic or of first degree.

We restrict our attention now to the rational zeros of polynomials with rational coefficients, and to the irreducibility of such polynomials. Neither the zeros of a polynomial nor its irreducibility are changed when it is multiplied by a nonzero constant, so we lose no generality by restricting our attention to polynomials with coefficients that are all integers.

THEOREM 8.29   Let

$$f(x) = a_0 + a_1 x + \cdots + a_{n-1} x^{n-1} + a_n x^n$$

be a polynomial of positive degree n with coefficients that are all integers, and let p/q be a rational number that has been written in lowest terms. If p/q is a zero of $f(x)$, then p divides a_0 and q divides a_n.

Proof   Suppose that p/q is a rational number in lowest terms that is a zero of $f(x) = \sum_{i=0}^{n} a_i x^i$. Then

$$a_0 + a_1 \left(\frac{p}{q}\right) + \cdots + a_{n-1} \left(\frac{p}{q}\right)^{n-1} + a_n \left(\frac{p}{q}\right)^n = 0.$$

Multiplying both sides of this equality by q^n gives

$$a_0 q^n + a_1 p q^{n-1} + \cdots + a_{n-1} p^{n-1} q + a_n p^n = 0.$$

Subtracting $a_n p^n$ from both sides, we have

$$a_0 q^n + a_1 p q^{n-1} + \cdots + a_{n-1} p^{n-1} q = -a_n p^n$$

and hence

$$q(a_0 q^{n-1} + a_1 p q^{n-2} + \cdots + a_{n-1} p^{n-1}) = -a_n p^n.$$

This shows that q divides $a_n p^n$, and therefore $q \mid a_n$ since q and p are relatively prime. Similarly, the equation

$$a_1 p q^{n-1} + \cdots + a_{n-1} p^{n-1} q + a_n p^n = -a_0 q^n$$

can be used to show that $p \mid a_0$.   ■

It is important to note that Theorem 8.29 only restricts the possibilities of the rational zeros. It does not guarantee that any of these possibilities is actually a zero of $f(x)$.

It may happen that when some of the rational zeros of a polynomial have been found, the remaining zeros may be obtained by use of the quadratic formula. This is illustrated in the next example.

EXAMPLE 3   We shall obtain all zeros of the polynomial

$$f(x) = 2x^4 - 5x^3 + 3x^2 + 4x - 6$$

by first finding the rational zeros of $f(x)$. According to Theorem 8.29, any rational zero p/q of $f(x)$ that is in lowest terms must have a numerator p that divides the constant term and a denominator q that divides the leading coefficient. This means that

$$p \in \{\pm 1, \pm 2, \pm 3, \pm 6\}$$
$$q \in \{\pm 1, \pm 2\}$$
$$\frac{p}{q} \in \left\{ \pm \frac{1}{2}, \pm 1, \pm \frac{3}{2}, \pm 2, \pm 3, \pm 6 \right\}.$$

Testing the positive possibilities systematically, we get

$$f\left(\frac{1}{2}\right) = -\frac{15}{4}, \qquad f(1) = -2, \qquad f\left(\frac{3}{2}\right) = 0.$$

We could continue to test the remaining possibilities, but chances are that it is worthwhile to divide $f(x)$ by $x - (3/2)$ and then work with the quotient. Performing the division, we obtain

$$f(x) = (x - \tfrac{3}{2})(2x^3 - 2x^2 + 4)$$
$$= (2x - 3)(x^3 - x^2 + 2).$$

From this factorization, we see that the other zeros of $f(x)$ are the zeros of the factor $q(x) = x^3 - x^2 + 2$. Since this factor is monic, the only possible rational zeros are the divisors of 2. We already know that 1 is not a zero since $f(1) = -2$. Thus the remaining possibilities are $2, -1$, and -2. We find that

$$q(2) = 6, \qquad q(-1) = 0.$$

Therefore $x + 1$ is a factor of $x^3 - x^2 + 2$. Division by $x + 1$ yields

$$x^3 - x^2 + 2 = (x + 1)(x^2 - 2x + 2)$$

and

$$f(x) = (2x - 3)(x + 1)(x^2 - 2x + 2).$$

The remaining zeros of $f(x)$ can be found by using the quadratic formula on the factor $x^2 - 2x + 2$:

$$x = \frac{2 \pm \sqrt{4-8}}{2} = 1 \pm i.$$

Thus the zeros of $f(x)$ are $3/2, -1, 1 + i$, and $1 - i$.   ☐

The results concerning irreducibility over the field **Q** of rational numbers are not nearly as neat or complete as those we have obtained for the fields $\mathscr{C}$ and $\mathscr{R}$. The best known result for **Q** is a theorem that states what is known as Eisenstein's Irreducibility Criterion. To establish this result is the goal of the rest of this section. We need the following definition and two intermediate theorems to reach our objective.

DEFINITION 8.30   Let $f(x) = \sum_{i=0}^{n} a_i x^i$ be a polynomial in which all coefficients are integers. Then $f(x)$ is a **primitive** polynomial if the greatest common divisor of $a_0, a_1, \ldots, a_n$ is 1.

That is, a polynomial is primitive if and only if there is no prime integer that divides all of its coefficients.

Our first intermediate result simply asserts that the product of two primitive polynomials is primitive.

THEOREM 8.31   If $g(x)$ and $h(x)$ are primitive polynomials, then $g(x)h(x)$ is a primitive polynomial.

Proof   We shall assume that the theorem is false and arrive at a contradiction. Suppose that $g(x)$ and $h(x)$ are primitive polynomials, but the product $f(x) = g(x)h(x)$ is not primitive. Then there is a prime integer p that divides every coefficient of $f(x) = \sum_{i=0}^{n} a_i x^i$. The mapping $\phi : \mathbf{Z}[x] \to \mathbf{Z}_p[x]$ defined by

$$\phi(a_0 + a_1 x + \cdots + a_n x^n) = [a_0] + [a_1]x + \cdots + [a_n]x^n$$

is an epimorphism from $\mathbf{Z}[x]$ to $\mathbf{Z}_p[x]$, by Problem 13 of Exercises 8.1. Since every coefficient of $f(x)$ is a multiple of p, $\phi(f(x)) = [0]$ in $\mathbf{Z}_p[x]$. Therefore

$$\begin{aligned}
\phi(g(x)) \cdot \phi(h(x)) &= \phi(g(x)h(x)) \\
&= \phi(f(x)) \\
&= [0]
\end{aligned}$$

in $\mathbf{Z}_p[x]$. Since p is a prime, $\mathbf{Z}_p[x]$ is an integral domain, and either $\phi(g(x)) = [0]$ or $\phi(h(x)) = [0]$. Consequently, either p divides every coefficient of $g(x)$, or p divides every coefficient of $h(x)$. In either case, we have a contradiction to the supposition that $g(x)$ and $h(x)$ are primitive polynomials. This contradiction establishes the theorem. ∎

The following theorem is credited to the same mathematician who first proved the Fundamental Theorem of Algebra.

THEOREM 8.32 (**Gauss' Lemma**) Let $f(x)$ be a primitive polynomial. If $f(x)$ can be factored as $f(x) = g(x)h(x)$, where $g(x)$ and $h(x)$ have rational coefficients and positive degree, then $f(x)$ can be factored as $f(x) = G(x)H(x)$, where $G(x)$ and $H(x)$ have integral coefficients and positive degree.

Proof   Suppose that $f(x) = g(x)h(x)$ as described in the hypothesis. Let b be the least common denominator of the coefficients of $g(x)$, so that $g(x)$ can be expressed as $g(x) = \dfrac{1}{b} g_1(x)$, where $g_1(x)$ has integral coefficients. Now let a be the greatest common divisor of the coefficients of $g_1(x)$, so that $g_1(x) = aG(x)$, where $G(x)$ is a primitive polynomial. Then we have $g(x) = \dfrac{a}{b} G(x)$, where a and b are integers and $G(x)$ is primitive and of the same degree as $g(x)$. Similarly, we may write $h(x) = \dfrac{c}{d} H(x)$, where c and d are integers and $H(x)$ is primitive and of the same degree as $h(x)$. Substituting these expressions for $g(x)$ and $h(x)$, we obtain

$$f(x) = \frac{a}{b} G(x) \cdot \frac{c}{d} H(x)$$

and therefore

$$bdf(x) = acG(x)H(x).$$

Since $f(x)$ is primitive, the greatest common divisor of the coefficients of the left member of this equation is bd. By Theorem 8.31, $G(x)H(x)$ is primitive, and therefore the greatest common divisor of the coefficients of the right member is ac. Hence $bd = ac$, and this implies that $f(x) = G(x)H(x)$, where $G(x)$ and $H(x)$ have integral coefficients and positive degrees.

We are now in a position to prove Eisenstein's result. ∎

THEOREM 8.33 (**Eisenstein's Irreducibility Criterion**) Let $f(x) = a_0 + a_1x + \cdots + a_nx^n$ be a polynomial of positive degree with integral coefficients. If there exists a prime integer p such that $p \mid a_i$ for $i = 0, 1, \ldots, n - 1$ but $p \nmid a_n$ and $p^2 \nmid a_0$, then $f(x)$ is irreducible over the field of rational numbers.

Proof Dividing out the greatest common divisor of the coefficients of a polynomial would have no effect on whether or not the criterion was satisfied by a prime p because of the requirement that $p \nmid a_n$. Therefore we may restrict our attention to the case where $f(x)$ is a primitive polynomial.

Let $f(x) = \sum_{i=0}^{n} a_ix^i$ be a primitive polynomial, and assume there exists a prime integer p that satisfies the hypothesis. At the same time, assume that the conclusion is false, so that $f(x)$ factors over the rational numbers as a product of two polynomials of positive degree. Then $f(x)$ can be factored as the product of two polynomials of positive degree that have integral coefficients, by Theorem 8.32. Suppose that

$$f(x) = (b_0 + b_1x + \cdots + b_rx^r)(c_0 + c_1x + \cdots + c_sx^s)$$

where all the coefficients are integers and $r > 0$, $s > 0$. Then $a_0 = b_0c_0$, and hence $p \mid b_0c_0$ but $p^2 \nmid b_0c_0$ by the hypothesis. This implies that either $p \mid b_0$ or $p \mid c_0$, but p does not divide both b_0 and c_0. Without loss of generality, we may assume that $p \mid b_0$ and $p \nmid c_0$. If all of the b_i were divisible by p, then p would divide all the coefficients in the product, $f(x)$. Since $p \nmid a_n$, some of the b_i are not divisible by p. Let k be the smallest subscript such that $p \nmid b_k$, and consider

$$a_k = b_0c_k + b_1c_{k-1} + \cdots + b_{k-1}c_1 + b_kc_0.$$

By the choice of k, p divides each of $b_0, b_1, \ldots, b_{k-1}$ and therefore

$$p \mid (b_0c_k + b_1c_{k-1} + \cdots + b_{k-1}c_1).$$

Also, $p \mid a_k$ since $k < n$. Hence p divides the difference:

$$p \mid [a_k - (b_0c_k + b_1c_{k-1} + \cdots + b_{k-1}c_1)].$$

That is, $p \mid b_kc_0$. This is impossible, however, since $p \nmid b_k$ and $p \nmid c_0$. We have arrived at a contradiction, and therefore $f(x)$ is irreducible over the rational numbers. ∎

EXAMPLE 4 Consider the polynomial

$$f(x) = 10 - 15x + 25x^2 - 7x^4.$$

The prime integer $p = 5$ divides all of the coefficients in $f(x)$ except the leading coefficient $a_n = -7$, and 5^2 does not divide the constant term $a_0 = 10$. Therefore, $f(x)$ is irreducible over the rational numbers, by Eisenstein's Criterion.   □

EXERCISES 8.4

1   Find a monic polynomial $f(x)$ of least degree over $\mathscr{C}$ that has the given numbers as zeros, and a monic polynomial $g(x)$ of least degree with real coefficients that has the given numbers as zeros.

(a) $2i, 3$ (b) $-3i, 4$
(c) $2, 1 - i$ (d) $3, 2 - i$
(e) $3i, 1 + 2i$ (f) $i, 2 - i$
(g) $2 + i, -i,$ and 1 (h) $3 - i, i,$ and 2

2   One of the zeros is given for each of the following polynomials. Find the other zeros in the field of complex numbers.

(a) $x^3 - 4x^2 + 6x - 4$; $1 - i$ is a zero.
(b) $x^3 + x^2 - 4x + 6$; $1 - i$ is a zero.
(c) $x^4 + x^3 + 2x^2 + x + 1$; $-i$ is a zero.
(d) $x^4 + 3x^3 + 6x^2 + 12x + 8$; $2i$ is a zero.

Find all rational zeros of each of the polynomials in Problems 3–6.

3   $2x^3 - x^2 - 8x - 5$ 4 $3x^3 + 19x^2 + 30x + 8$
5   $2x^4 - x^3 - x^2 - x - 3$ 6 $2x^4 + x^3 - 8x^2 + x - 10$

In Problems 7–12, find all zeros of the given polynomial.

7   $x^3 + x^2 - x + 2$ 8 $3x^3 - 7x^2 + 8x - 2$
9   $3x^3 + 2x^2 - 7x + 2$ 10 $3x^3 - 2x^2 - 7x - 2$
11   $6x^3 + 11x^2 + x - 4$ 12 $9x^3 + 27x^2 + 8x - 20$

Factor each of the polynomials in Problems 13–16 as a product of its leading coefficient and a finite number of monic irreducible polynomials over the field of rational numbers.

13   $x^4 - x^3 - 2x^2 + 6x - 4$ 14 $2x^4 - x^3 - 13x^2 + 5x + 15$
15   $2x^4 + 5x^3 - 7x^2 - 10x + 6$ 16 $6x^4 + x^3 + 3x^2 - 14x - 8$

17   Show that each of the following polynomials is irreducible over the field of rational numbers.

(a) $3 + 9x + x^3$ (b) $7 - 14x + 28x^2 + x^3$
(c) $3 - 27x^2 + 2x^5$ (d) $6 + 12x^2 - 27x^3 + 10x^5$

18   Prove that $\overline{z_1 + z_2 + \cdots + z_n} = \bar{z}_1 + \bar{z}_2 + \cdots + \bar{z}_n$ for complex numbers $z_1, z_2, \ldots, z_n$.

19   Prove that $\overline{z_1 \cdot z_2 \cdot \cdots \cdot z_n} = \bar{z}_1 \cdot \bar{z}_2 \cdot \cdots \cdot \bar{z}_n$ for complex numbers $z_1, z_2, \ldots, z_n$.

20   Prove that for every positive integer n there exist polynomials of degree n that are irreducible over the rational numbers. [Hint: Consider $x^n - 2$.)

21   Let $f(x) = a_0 + a_1 x + \cdots + a_{n-1} x^{n-1} + x^n$ be a monic polynomial of positive degree n with coefficients that are all integers. Prove that any rational zero of $f(x)$ is an integer that divides the constant term a_0.

22   Derive the quadratic formula for the zeros of $ax^2 + bx + c$, where a, b, and c are complex numbers and $a \neq 0$.

23   Prove Theorem 8.28. [Hint: In the factorization described in Theorem 8.26, pair those factors of the form $x - (a + bi)$ and $x - (a - bi)$.]

24   Prove that any polynomial of odd degree that has real coefficients must have a zero in the field of real numbers.

8.5   ALGEBRAIC EXTENSIONS OF A FIELD

Some of the results in Chapter 6 concerning ideals and quotient rings are put to good use in this section. Starting with an irreducible polynomial $p(x)$ over a field F, these results are used in the construction of a field that is an extension of F that contains a zero of $p(x)$.

As a special case of Definition 6.2, if $p(x)$ is a fixed polynomial over the field F, the *principal ideal* generated by $p(x)$ in $F[x]$ is the set

$$P = (p(x)) = \{f(x)p(x) \,|\, f(x) \in F[x]\},$$

that consists of all multiples of $p(x)$ by elements $f(x)$ of $F[x]$. Most of our work in this section is related to quotient rings of the form $F[x]/(p(x))$.

THEOREM 8.34   For any $p(x)$ of positive degree in $F[x]$, the quotient ring $F[x]/(p(x))$ is a commutative ring with unity that contains a subring that is isomorphic to F.

Proof   For a fixed polynomial $p(x)$ in $F[x]$, let $P = (p(x))$. According to Theorem 6.4, the set $F[x]/P$ forms a ring with respect to addition defined by

$$[f(x) + P] + [g(x) + P] = (f(x) + g(x)) + P$$

and multiplication defined by

$$[f(x) + P][g(x) + P] = f(x)g(x) + P.$$

The ring $F[x]/P$ is commutative since $f(x)g(x) = g(x)f(x)$ in $F[x]$, and $1 + P$ is the unity in $F[x]$.

Consider the nonempty subset F' of $F[x]/P$ that consists of all cosets of the form $a + P$, with $a \in F$:

$$F' = \{a + P \,|\, a \in F\}.$$

For arbitrary elements $a + P$ and $b + P$ of F', the elements

$$(a + P) - (b + P) = (a - b) + P$$

and

$$(a + P)(b + P) = ab + P$$

are in F' since $a - b$ and ab are in F. Thus F' is a subring of $F[x]/P$, by Theorem 5.4. The unity $1 + P$ is in F', and every nonzero element $a + P$ of F' has the multiplicative inverse $a^{-1} + P$ in F'. Hence F' is a field.

The mapping $\theta: F \to F'$ defined by

$$\theta(a) = a + P$$

is a homomorphism since

$$\theta(a + b) = (a + b) + P$$
$$= (a + P) + (b + P) = \theta(a) + \theta(b)$$

and

$$\theta(ab) = ab + P$$
$$= (a + P)(b + P) = \theta(a)\theta(b).$$

It follows from the definition of F' that θ is an epimorphism. Since $p(x)$ has positive degree, 0 is the only element of F that is contained in P, and therefore

$$\theta(a) = \theta(b) \Leftrightarrow a + P = b + P$$
$$\Leftrightarrow a - b \in P$$
$$\Leftrightarrow a = b.$$

Thus θ is an isomorphism from F to the subring F' of $F[x]/(p(x))$. ∎

As we have done in similar situations in the past, we can now use the isomorphism θ in the preceding proof to identify $a \in F$ with $a + P$ in $F[x]/(p(x))$. This identification allows us to regard F as a subset of $F[x]/(p(x))$. This point of view is especially advantageous when the quotient ring $F[x]/(p(x))$ is a field.

THEOREM 8.35   Let $p(x)$ be a polynomial of positive degree over the field F. Then the ring $F[x]/(p(x))$ is a field if and only if $p(x)$ is an irreducible polynomial over F.

Proof   As in the proof of Theorem 8.34, let $P = (p(x))$.

Assume first that $p(x)$ is an irreducible polynomial over F. In view of Theorem 8.34, we only need to show that any nonzero element $f(x) + P$ in $F[x]/P$ has a multiplicative inverse in $F[x]/P$. If $f(x) + P \neq P$, then $f(x)$ is not a multiple of $p(x)$, and this means that the greatest common divisor of $f(x)$ and $p(x)$ is 1 since $p(x)$ is irreducible. By Theorem 8.13, there exist $s(x)$ and $t(x)$ in $F[x]$ such that

$$f(x)s(x) + p(x)t(x) = 1.$$

Now $p(x)t(x) \in P$, so $p(x)t(x) + P = 0 + P$ and hence

$$\begin{aligned}
1 + P &= [f(x)s(x) + p(x)t(x)] + P \\
&= [f(x)s(x) + P] + [p(x)t(x) + P] \\
&= [f(x)s(x) + P] + [0 + P] \\
&= f(x)s(x) + P \\
&= [f(x) + P][s(x) + P].
\end{aligned}$$

Thus $s(x) + P = [f(x) + P]^{-1}$, and we have proved that $F[x]/P$ is a field.

Suppose now that $p(x)$ is not irreducible over F. Then there exist polynomials $g(x)$ and $h(x)$ of positive degree in $F[x]$ such that $p(x) = g(x)h(x)$. Since $\deg p(x) = \deg g(x) + \deg h(x)$ and all these degrees are positive, it must be that $\deg g(x) < \deg p(x)$ and $\deg h(x) < \deg p(x)$. Therefore neither $g(x)$ nor $h(x)$ is a multiple of $p(x)$. That is,

$$g(x) + P \neq P \quad \text{and} \quad h(x) + P \neq P,$$

but

$$\begin{aligned}
[g(x) + P][h(x) + P] &= g(x)h(x) + P \\
&= p(x) + P \\
&= P.
\end{aligned}$$

We have $g(x) + P$ and $h(x) + P$ as two nonzero elements of $F[x]/P$ whose product is zero. Hence $F[x]/P$ is not a field in this case, and the proof is complete.   ■

If F and E are fields such that $F \subseteq E$, then E is called an **extension field** of F. With the identification that we have made between F and F', the preceding theorem shows that $F[x]/(p(x))$ is an extension field of F if and only if $p(x)$ is an irreducible polynomial over F. The main significance of all this becomes clear in the proof of the next theorem, which is credited to the German mathematician Leopold Kronecker (1823–1891).

THEOREM 8.36   If $p(x)$ is an irreducible polynomial over the field F, there exists an extension field of F that contains a zero of $p(x)$.

Proof   For a given irreducible polynomial

$$p(x) = p_0 + p_1 x + p_2 x^2 + \cdots + p_n x^n$$

over the field F, let $P = (p(x))$ in $F[x]$ and let $\alpha = x + P$ in $F[x]/P$. From the definition of multiplication in $F[x]/P$, it follows that

$$\alpha^2 = (x + P)(x + P) = x^2 + P$$

and that

$$\alpha^i = x^i + P$$

for every positive integer i. By using the identification of $a \in F$ with $a + P$ in $F[x]/P$, the polynomial

$$p(x) = p_0 + p_1 x + p_2 x^2 + \cdots + p_n x^n$$

may be written in the form

$$p(x) = (p_0 + P) + (p_1 + P)x + (p_2 + P)x^2 + \cdots + (p_n + P)x^n.$$

Hence

$$
\begin{aligned}
p(\alpha) &= (p_0 + P) + (p_1 + P)\alpha + (p_2 + P)\alpha^2 + \cdots + (p_n + P)\alpha^n \\
&= (p_0 + P) + (p_1 + P)(x + P) + (p_2 + P)(x^2 + P) \\
&\quad + \cdots + (p_n + P)(x^n + P) \\
&= (p_0 + P) + (p_1 x + P) + (p_2 x^2 + P) + \cdots + (p_n x^n + P) \\
&= (p_0 + p_1 x + p_2 x^2 + \cdots + p_n x^n) + P \\
&= p(x) + P \\
&= 0 + P.
\end{aligned}
$$

Thus $p(\alpha)$ is the zero element of $F[x]/P$, and α is a zero of $p(x)$ in $F[x]/P$.   ■

For a particular polynomial $p(x)$, explicit standard forms for the elements of the ring $F[x]/(p(x))$ can be given. Before going into this, we note that the ring $F[x]/(p(x))$ is unchanged if $p(x)$ is replaced by a multiple of the form $cp(x)$, with $c \neq 0$ in F. This follows from the fact that the ideal $P = (p(x))$, which consists of the set of all multiples of $p(x)$ in $F[x]$, is the same as the set of all multiples of $cp(x)$ in $F[x]$. In particular, c can be chosen to be the multiplicative inverse of the leading coefficient of $p(x)$ and thereby obtain a monic polynomial that gives the same ring $F[x]/P$ as $p(x)$ does. Thus there is no loss of generality in assuming from now on that $p(x)$ is a *monic* polynomial over F.

Before considering the general situation, we examine some particular cases in the following examples.

EXAMPLE 1   Consider the monic irreducible polynomial

$$p(x) = x^2 + 2x + 2$$

over the field $\mathbf{Z}_3$. We shall determine all the elements of the field $\mathbf{Z}_3[x]/(p(x))$, and at the same time construct addition and multiplication tables for this field.

Let $P = (p(x))$ and $\alpha = x + P$ in $\mathbf{Z}_3[x]/P$. We start construction of the addition table for $\mathbf{Z}_3[x]/P$ with the elements $0 = 0 + P$, $1 = 1 + P$, $2 = 2 + P$, and α. Filling out the table until closure is obtained, we pick up the new elements

$\alpha + 1, \alpha + 2, 2\alpha, 2\alpha + 1$, and $2\alpha + 2$. The completed table in Figure 8.2 shows that the set

$$\{0, 1, 2, \alpha, \alpha + 1, \alpha + 2, 2\alpha, 2\alpha + 1, 2\alpha + 2\}$$

is closed under addition.

+	0	1	2	α	$\alpha + 1$	$\alpha + 2$	2α	$2\alpha + 1$	$2\alpha + 2$
0	0	1	2	α	$\alpha + 1$	$\alpha + 2$	2α	$2\alpha + 1$	$2\alpha + 2$
1	1	2	0	$\alpha + 1$	$\alpha + 2$	α	$2\alpha + 1$	$2\alpha + 2$	2α
2	2	0	1	$\alpha + 2$	α	$\alpha + 1$	$2\alpha + 2$	2α	$2\alpha + 1$
α	α	$\alpha + 1$	$\alpha + 2$	2α	$2\alpha + 1$	$2\alpha + 2$	0	1	2
$\alpha + 1$	$\alpha + 1$	$\alpha + 2$	α	$2\alpha + 1$	$2\alpha + 2$	2α	1	2	0
$\alpha + 2$	$\alpha + 2$	α	$\alpha + 1$	$2\alpha + 2$	2α	$2\alpha + 1$	2	0	1
2α	2α	$2\alpha + 1$	$2\alpha + 2$	0	1	2	α	$\alpha + 1$	$\alpha + 2$
$2\alpha + 1$	$2\alpha + 1$	$2\alpha + 2$	2α	1	2	0	$\alpha + 1$	$\alpha + 2$	α
$2\alpha + 2$	$2\alpha + 2$	2α	$2\alpha + 1$	2	0	1	$\alpha + 2$	α	$\alpha + 1$

FIGURE 8.2

Turning now to multiplication, we start with the same nine elements that occur in the addition table. In making this table, we use the fact that α is a zero of $p(x) = x^2 + 2x + 2$ in the following manner:

$$\alpha^2 + 2\alpha + 2 = 0 \Rightarrow \alpha^2 = -2\alpha - 2$$
$$= \alpha + 1.$$

That is, whenever α^2 occurs in a product, it is replaced by $\alpha + 1$. As an illustration we have

$$(2\alpha + 1)(\alpha + 2) = 2\alpha^2 + 2\alpha + 2$$
$$= 2(\alpha + 1) + 2\alpha + 2$$
$$= 2\alpha + 2 + 2\alpha + 2 = \alpha + 1.$$

The completed table is shown in Figure 8.3.

·	0	1	2	α	$\alpha + 1$	$\alpha + 2$	2α	$2\alpha + 1$	$2\alpha + 2$
0	0	0	0	0	0	0	0	0	0
1	0	1	2	α	$\alpha + 1$	$\alpha + 2$	2α	$2\alpha + 1$	$2\alpha + 2$
2	0	2	1	2α	$2\alpha + 2$	$2\alpha + 1$	α	$\alpha + 2$	$\alpha + 1$
α	0	α	2α	$\alpha + 1$	$2\alpha + 1$	1	$2\alpha + 2$	2	$\alpha + 2$
$\alpha + 1$	0	$\alpha + 1$	$2\alpha + 2$	$2\alpha + 1$	2	α	$\alpha + 2$	2α	1
$\alpha + 2$	0	$\alpha + 2$	$2\alpha + 1$	1	α	$2\alpha + 2$	2	$\alpha + 1$	2α
2α	0	2α	α	$2\alpha + 2$	$\alpha + 2$	2	$\alpha + 1$	1	$2\alpha + 1$
$2\alpha + 1$	0	$2\alpha + 1$	$\alpha + 2$	2	2α	$\alpha + 1$	1	$2\alpha + 2$	α
$2\alpha + 2$	0	$2\alpha + 2$	$\alpha + 1$	$\alpha + 2$	1	2α	$2\alpha + 1$	α	2

FIGURE 8.3

$\square$

EXAMPLE 2   The polynomial $p(x) = x^2 + 1$ is not irreducible over the field $\mathbf{Z}_2$ since $p(1) = 0$. We follow the same procedure as in Example 1, and construct addition and multiplication tables for the ring $\mathbf{Z}_2[x]/(p(x))$.

As before, let $P = (p(x))$ and $\alpha = x + P$ in $\mathbf{Z}_2[x]/P$. Extending an addition table until closure is obtained, we arrive at the table shown in Figure 8.4.

+	0	1	α	$\alpha + 1$
0	0	1	α	$\alpha + 1$
1	1	0	$\alpha + 1$	α
α	α	$\alpha + 1$	0	1
$\alpha + 1$	$\alpha + 1$	α	1	0

FIGURE 8.4

In making the multiplication table shown in Figure 8.5, we use the fact that $p(\alpha) = 0$ in this way:

$$\alpha^2 + 1 = 0 \Rightarrow \alpha^2 = -1$$
$$\Rightarrow \alpha^2 = 1.$$

$\cdot$	0	1	α	$\alpha + 1$
0	0	0	0	0
1	0	1	α	$\alpha + 1$
α	0	α	1	$\alpha + 1$
$\alpha + 1$	0	$\alpha + 1$	$\alpha + 1$	0

FIGURE 8.5

Theorem 8.35 assures us that $\mathbf{Z}_2[x]/P$ is not a field, and the multiplication table confirms this fact by showing that $\alpha + 1$ does not have a multiplicative inverse. $\square$

The next theorem and its corollary set forth the standard forms for the elements of the ring $F[x]/(p(x))$ that were referred to earlier.

THEOREM 8.37   Let $p(x)$ be a polynomial of positive degree n over the field F, and let $P = (p(x))$ in $F[x]$. Then each element of the ring $F[x]/P$ can be expressed uniquely in the form

$$(a_0 + a_1 x + a_2 x^2 + \cdots + a_{n-1}x^{n-1}) + P.$$

Proof   Assume the hypothesis, and let $f(x) + P$ be an arbitrary element in $F[x]/P$. By the Division Algorithm, there exist $q(x)$ and

$r(x)$ in $F[x]$ such that

$$f(x) = p(x)q(x) + r(x),$$

whether either $r(x) = 0$ or $\deg r(x) < n = \deg p(x)$. In either case, we may write

$$r(x) = a_0 + a_1 x + a_2 x^2 + \cdots + a_{n-1} x^{n-1}.$$

Since $p(x)q(x)$ is in P, $p(x)q(x) + P = 0 + P$, and therefore

$$
\begin{aligned}
f(x) + P &= [p(x)q(x) + P] + [r(x) + P] \\
&= [0 + P] + [r(x) + P] \\
&= r(x) + P \\
&= (a_0 + a_1 x + \cdots + a_{n-1} x^{n-1}) + P.
\end{aligned}
$$

To show uniqueness, suppose that $f(x) + P = r(x) + P$ as above, and also that $f(x) + P = g(x) + P$, where

$$g(x) = b_0 + b_1 x + b_2 x^2 + \cdots + b_{n-1} x^{n-1}.$$

Then $r(x) + P = g(x) + P$, and therefore $r(x) - g(x)$ is in P. Each of $r(x)$ and $g(x)$ is either zero or has degree less than n, and this implies that the difference $r(x) - g(x)$ is either zero or has degree less than n. Since $P = (p(x))$ contains no polynomials with degree less than n, it must be that $r(x) - g(x) = 0$, and $r(x) = g(x)$. ∎

COROLLARY 8.38 For a polynomial $p(x)$ of positive degree n over the field F, let $P = (p(x))$ in $F[x]$ and let $\alpha = x + P$ in $F[x]/P$. Then each element of the ring $F[x]/P$ can be expressed uniquely in the form

$$a_0 + a_1 \alpha + a_2 \alpha^2 + \cdots + a_{n-1} \alpha^{n-1}.$$

Proof From the theorem, each $f(x) + P$ in $F[x]/P$ can be expressed uniquely in the form

$$
\begin{aligned}
f(x) + P &= (a_0 + a_1 x + \cdots + a_{n-1} x^{n-1}) + P \\
&= (a_0 + P) + (a_1 + P)(x + P) + \cdots + (a_{n-1} + P)(x^{n-1} + P) \\
&= (a_0 + P) + (a_1 + P)\alpha + \cdots + (a_{n-1} + P)\alpha^{n-1} \\
&= a_0 + a_1 \alpha + \cdots + a_{n-1} \alpha^{n-1},
\end{aligned}
$$

where the last equality follows from the identification of a_i in F with $a_i + P$ in $F[x]/P$. ∎

In Example 1, the polynomials $f(x)$ in $\mathbf{Z}_3[x]$ and the cosets $f(x) + P$ in $\mathbf{Z}_3[x]/P$ receded into the background once the notation $\alpha = x + P$ was introduced, and we ended up with a field whose elements had the form $a_0 + a_1 \alpha$, with $a_i \in \mathbf{Z}_3$. This field $\mathbf{Z}_3(\alpha)$ of nine elements, given by

$$\mathbf{Z}_3(\alpha) = \{0, 1, 2, \alpha, \alpha + 1, \alpha + 2, 2\alpha, 2\alpha + 1, 2\alpha + 2\},$$

is called *the field obtained by adjoining a zero α of $x^2 + 2x + 2$ to $\mathbf{Z}_3$.*

In general, if $p(x)$ is an irreducible polynomial over the field F, the smallest field that contains both F and a zero α of $p(x)$ is denoted by $F(\alpha)$, and is referred to as **the field[†] obtained by adjoining α to the field F.** A field $F(\alpha)$ of this type is called a **simple algebraic extension** of F, and F is referred to as the **ground field.** Corollary 8.38 describes the standard form for the elements of $F(\alpha)$.

EXAMPLE 3 The polynomial $p(x) = x^3 + 2x^2 + 4x + 2$ is irreducible over $\mathbf{Z}_5$ since

$$p(0) = 2, \qquad p(1) = 4, \qquad p(2) = 1, \qquad p(3) = 4, \qquad p(4) = 4.$$

In the field $\mathbf{Z}_5(\alpha)$ obtained by adjoining a zero α of $p(x)$ to $\mathbf{Z}_5$, we shall obtain a formula for the product of two arbitrary elements $a_0 + a_1\alpha + a_2\alpha^2$ and $b_0 + b_1\alpha + b_2\alpha^2$.

In order to accomplish this objective, we first express α^3 and α^4 as polynomials in α with degrees less than 3. Since $p(\alpha) = 0$, we have

$$\alpha^3 + 2\alpha^2 + 4\alpha + 2 = 0 \Rightarrow \alpha^3 = -2\alpha^2 - 4\alpha - 2$$
$$= 3\alpha^2 + \alpha + 3.$$

Hence

$$\begin{aligned}
\alpha^4 &= \alpha(3\alpha^2 + \alpha + 3) \\
&= 3\alpha^3 + \alpha^2 + 3\alpha \\
&= 3(3\alpha^2 + \alpha + 3) + \alpha^2 + 3\alpha \\
&= 4\alpha^2 + 3\alpha + 4 + \alpha^2 + 3\alpha \\
&= \alpha + 4.
\end{aligned}$$

Using these results, we get

$$\begin{aligned}
(a_0 &+ a_1\alpha + a_2\alpha^2)(b_0 + b_1\alpha + b_2\alpha^2) \\
&= a_0 b_0 + (a_0 b_1 + a_1 b_0)\alpha + (a_0 b_2 + a_1 b_1 + a_2 b_0)\alpha^2 \\
&\quad + (a_1 b_2 + a_2 b_1)\alpha^3 + a_2 b_2 \alpha^4 \\
&= a_0 b_0 + (a_0 b_1 + a_1 b_0)\alpha + (a_0 b_2 + a_1 b_1 + a_2 b_0)\alpha^2 \\
&\quad + (a_1 b_2 + a_2 b_1)(3\alpha^2 + \alpha + 3) + a_2 b_2(\alpha + 4) \\
&= (a_0 b_0 + 3a_1 b_2 + 3a_2 b_1 + 4a_2 b_2) \\
&\quad + (a_0 b_1 + a_1 b_0 + a_1 b_2 + a_2 b_1 + a_2 b_2)\alpha \\
&\quad + (a_0 b_2 + a_1 b_1 + a_2 b_0 + 3a_1 b_2 + 3a_2 b_1)\alpha^2.
\end{aligned}$$

$\square$

[†] The existence of such a field $F(\alpha)$ is assured by Theorem 8.36.

EXAMPLE 4 With $\mathbf{Z}_5(\alpha)$ as in Example 3, suppose that we wish to find the multiplicative inverse of the element $\alpha^2 + 3\alpha + 1$ in the field $\mathbf{Z}_5(\alpha)$.

The polynomials $f(x) = x^2 + 3x + 1$ and $p(x) = x^3 + 2x^2 + 4x + 2$ are relatively prime over $\mathbf{Z}_5$, so there exist $s(x)$ and $t(x)$ in $\mathbf{Z}_5[x]$ such that

$$f(x)s(x) + p(x)t(x) = 1$$

by Theorem 8.13. Since $p(\alpha) = 0$, this means that

$$f(\alpha)s(\alpha) = 1$$

and that $(\alpha^2 + 3\alpha + 1)^{-1} = [f(\alpha)]^{-1} = s(\alpha)$. In order to find $s(x)$ and $t(x)$, we use the Euclidean Algorithm:

$$p(x) = f(x)(x + 4) + (x + 3)$$
$$f(x) = (x + 3)(x) + 1.$$

Thus

$$\begin{aligned}
1 &= f(x) - x(x + 3) \\
&= f(x) - x[p(x) - f(x)(x + 4)] \\
&= f(x)[1 + x(x + 4)] + p(x)(-x) \\
&= f(x)(x^2 + 4x + 1) + p(x)(-x),
\end{aligned}$$

so we have $s(x) = x^2 + 4x + 1$ and $t(x) = -x$. Therefore

$$(\alpha^2 + 3\alpha + 1)^{-1} = s(\alpha) = \alpha^2 + 4\alpha + 1.$$

This result may be checked by computing the product

$$(\alpha^2 + 3\alpha + 1)(\alpha^2 + 4\alpha + 1) \qquad \square$$

in $\mathbf{Z}_5(\alpha)$.

It is of some interest to consider an example similar to Example 4 but in a more familiar setting.

EXAMPLE 5 The polynomial $p(x) = x^2 - 2$ is irreducible over the field $\mathbf{Q}$ of rational numbers. In the field $\mathbf{Q}(\sqrt{2})$ obtained by adjoining a zero $\alpha = \sqrt{2}$ of $p(x)$ to $\mathbf{Q}$, let us find the multiplicative inverse of the element $4 + 3\sqrt{2}$ by the method employed in Example 4. The polynomials $f(x) = 3x + 4$ and $p(x) = x^2 - 2$ are relatively prime over $\mathbf{Q}$. To find $s(x)$ and $t(x)$ such that

$$f(x)s(x) + p(x)t(x) = 1,$$

we need only one step in the Euclidean Algorithm:

$$p(x) = f(x) \cdot \left(\frac{1}{3}x - \frac{4}{9}\right) + \left(-\frac{2}{9}\right).$$

Multiplying by $9/2$ and rewriting this equation, we obtain

$$f(x) \cdot \left(\frac{3}{2}x - 2\right) + p(x)\left(-\frac{9}{2}\right) = 1.$$

Since $p(\sqrt{2}) = 0$, this gives

$$f(\sqrt{2}) \cdot \left(\frac{3}{2}\sqrt{2} - 2\right) = 1$$

and

$$(4 + 3\sqrt{2})^{-1} = [f(\sqrt{2})]^{-1} = \frac{3}{2}\sqrt{2} - 2.$$

This agrees with the result obtained by the usual procedure of rationalizing the denominator:

$$\frac{1}{4 + 3\sqrt{2}} = \frac{(1)(4 - 3\sqrt{2})}{(4 + 3\sqrt{2})(4 - 3\sqrt{2})} = \frac{4 - 3\sqrt{2}}{-2} = \frac{3}{2}\sqrt{2} - 2. \qquad \square$$

The result in Theorem 8.36 generalizes to the following theorem.

THEOREM 8.39   If $p(x)$ is a polynomial of positive degree n over the field F, there exists an extension field E of F that contains n zeros of $p(x)$.

Proof   The proof is by induction on the degree n of $p(x)$. If $n = 1$, then $p(x)$ has the form $p(x) = ax + b$, with $a \neq 0$. Since $p(x)$ has the unique zero $-a^{-1}b$ in F, the theorem is true for $n = 1$.

Assume the theorem is true for all polynomials of degree less than k, and let $p(x)$ be a polynomial of degree k. We consider two cases, depending on whether $p(x)$ is irreducible or not.

If $p(x)$ is irreducible, then there exists an extension field E_1 of F that contains a zero α of $p(x)$, by Theorem 8.36. By the Factor Theorem,

$$p(x) = (x - \alpha)q(x),$$

where $q(x)$ must have degree $k - 1$, according to Theorem 8.7. Since $q(x)$ is a polynomial over E_1 that has degree less than k, the induction hypothesis applies to $q(x)$ over E_1, and there exists an extension field E of E_1 such that $q(x)$ has $k - 1$ zeros in E. By Problem 10 of Exercises 8.3, the zeros of $p(x)$ in E consist of α and the zeros of $q(x)$ in E. Thus $p(x)$ has k zeros in E.

If $p(x)$ is not irreducible, then $p(x)$ can be factored as a product $p(x) = g(x)h(x)$. where $n_1 = \deg g(x)$ and $n_2 = \deg h(x)$ are positive integers such that $n_1 + n_2 = k$. Since $n_1 < k$, the induction hypothesis applies to $g(x)$ over F, and there exists an extension field E_1 of F that contains n_1 zeros of $g(x)$. Now $h(x)$ is a polynomial of degree $n_2 < k$ over E_1, so the induction hypothesis applies again to $h(x)$ over E_1, and there exists an extension field E of E_1 such that $h(x)$ has n_2 zeros in E. By Problem 11 of Exercises 8.3, the zeros of $p(x)$ in E consist of the zeros of $g(x)$ in E together with the zeros of $h(x)$ in E. There are altogether $n_1 + n_2 = k$ of these zeros in E.

In either case, we have proved the existence of an extension field of F that contains k zeros of $p(x)$, and the theorem follows by induction. ∎

If E is a field that contains all the zeros of a polynomial $p(x)$, and if no proper subfield of E contains all of these zeros, then E is called the **splitting field** of $p(x)$ since it is the "smallest" field over which $p(x)$ "splits" into first-degree factors.

EXERCISES 8.5

1 Each of the following polynomials $p(x)$ is irreducible over Z_3. For each of these polynomials, find all the elements of $Z_3[x]/(p(x))$, and construct addition and multiplication tables for this field.
(a) $p(x) = x^2 + x + 2$
(b) $p(x) = x^2 + 1$

2 In each of the following parts, a polynomial $p(x)$ over a field F is given. Construct addition and multiplication tables for the ring $F[x]/(p(x))$ in each case, and decide whether or not this ring is a field.
(a) $p(x) = x^2 + x + 1$ over $F = Z_2$
(b) $p(x) = x^3 + 1$ over $F = Z_2$
(c) $p(x) = x^3 + x + 1$ over $F = Z_2$
(d) $p(x) = x^3 + x^2 + 1$ over $F = Z_2$
(e) $p(x) = x^2 + x + 1$ over $F = Z_3$
(f) $p(x) = x^2 + 2$ over $F = Z_3$

In Problems 3–6, a field F, a polynomial $p(x)$ over F, and an element of the field $F(\alpha)$ obtained by adjoining a zero α of $p(x)$ to F are given. In each case:
(a) Verify that $p(x)$ is irreducible over F.
(b) Find a formula for the product of two arbitrary elements $a_0 + a_1\alpha + a_2\alpha^2$ and $b_0 + b_1\alpha + b_2\alpha^2$ of $F(\alpha)$.
(c) Find the multiplicative inverse of the given element of $F(\alpha)$

3 $F = Z_3$, $p(x) = x^3 + 2x^2 + 1$, $\alpha^2 + \alpha + 2$

4 $F = Z_3$, $p(x) = x^3 + x^2 + 2x + 1$, $\alpha^2 + 2\alpha + 1$

5 $F = Z_5$, $p(x) = x^3 + x + 1$, $\alpha^2 + 4\alpha$

6 $F = Z_5$, $p(x) = x^3 + x^2 + 1$, $\alpha^2 + 2\alpha + 3$

7   For the given irreducible polynomial $p(x)$ over $\mathbf{Z}_3$, list all elements of the field $\mathbf{Z}_3(\alpha)$ that is obtained by adjoining a zero α of $p(x)$ to $\mathbf{Z}_3$.
   (a) $p(x) = x^3 + 2x^2 + 1$
   (b) $p(x) = x^3 + x^2 + 2x + 1$

8   If F is a finite field with k elements and $p(x)$ is a polynomial of positive degree n over F, find a formula for the number of elements in the ring $F[x]/(p(x))$.

9   Find the multiplicative inverse of $\sqrt[3]{4} - 2\sqrt[3]{2} - 2$ in $\mathbf{Q}(\sqrt[3]{2})$, where $\mathbf{Q}$ is the field of rational numbers.

10   An element u of a field F is a perfect square in F if there exists an element v in F such that $u = v^2$. The quadratic formula can be generalized in the following way: Suppose that $1 + 1 \neq 0$ in F, and let $p(x) = ax^2 + bx + c, a \neq 0$, be a quadratic polynomial over F.
   (a) Prove that $p(x)$ has a zero in F if and only if $b^2 - 4ac$ is a perfect square in F.
   (b) If $b^2 - 4ac$ is a perfect square in F, show that the zeros of $p(x)$ in F are given by

$$r_1 = \frac{-b + \sqrt{b^2 - 4ac}}{2a} \quad \text{and} \quad r_2 = \frac{-b - \sqrt{b^2 - 4ac}}{2a}$$

   and these zeros are distinct if $b^2 - 4ac \neq 0$.

11   Determine whether or not each of the following polynomials has a zero in the given field F. If a polynomial has zeros in the field, use the quadratic formula to find them.
   (a) $x^2 + 3x + 2,$ $F = \mathbf{Z}_5$
   (b) $x^2 + 3x + 3,$ $F = \mathbf{Z}_5$
   (c) $x^2 + 2x + 6,$ $F = \mathbf{Z}_7$
   (d) $x^2 + 3x + 1,$ $F = \mathbf{Z}_7$
   (e) $2x^2 + x + 1,$ $F = \mathbf{Z}_7$
   (f) $3x^2 + 2x - 1,$ $F = \mathbf{Z}_7$

12   (a) Find the value of c that will cause the polynomial $f(x) = x^2 + 3x + c$ to have 3 as a zero in the field $\mathbf{Z}_7$.
   (b) Find the other zero of $p(x)$ in $\mathbf{Z}_7$.

   Each of the polynomials $p(x)$ in Problems 13–16 is irreducible over the given field F. Find all zeros of $p(x)$ in the field $F(\alpha)$ obtained by adjoining a zero of $p(x)$ to F. (In Problems 15 and 16, $p(x)$ has three zeros in $F(\alpha)$.)

13   $p(x) = x^2 + 2x + 2,$ $F = \mathbf{Z}_3$
14   $p(x) = x^2 + x + 2,$ $F = \mathbf{Z}_3$
15   $p(x) = x^3 + x^2 + 1,$ $F = \mathbf{Z}_5$
16   $p(x) = x^3 + 2x^2 + 4x + 2,$ $F = \mathbf{Z}_5$

Key Words and Phrases

Indeterminate
Polynomial in x over R
Equality of polynomials
Addition of polynomials

Multiplication of polynomials
Ring of polynomials over R
Degree of a polynomial
Multiple of a polynomial

Factor (or divisor) of a polynomial
Division Algorithm
Quotient, remainder
Monic polynomial
Greatest common divisor
Linear combination
Polynomial mapping
Zero of a polynomial
Root (solution) of a polynomial
   equation
Remainder Theorem
Factor Theorem
Irreducible (prime) polynomial
Relatively prime polynomials
Multiplicity of a factor

Multiplicity of a zero
Fundamental Theorem of Algebra
Quadratic Formula
Algebraic element
Algebraic extension
Algebraically-closed field
Factorization of polynomials over $\mathscr{C}$
Factorization of polynomials over $\mathscr{R}$
Rational zeros
Primitive polynomial
Gauss' Lemma
Eisenstein's Irreducibility Criterion
Simple algebraic extension
Splitting field

References

Ball, Richard W. *Principles of Abstract Algebra*. New York: Holt, Rinehart and Winston, 1963.

Birkhoff, Garrett and Saunders MacLane. *A Survey of Modern Algebra* (4th ed.). New York: Macmillan, 1977.

Durbin, John R. *Modern Algebra*. New York: Wiley, 1979.

Fraleigh, John B. *A First Course in Abstract Algebra* (2nd ed.). Reading, Mass.: Addison-Wesley, 1976.

Hillman, Abraham P. and Gerald L. Alexanderson. *A First Undergraduate Course in Abstract Algebra* (2nd ed.). Belmont, Calif.: Wadsworth, 1978.

Keesee, John W. *Elementary Abstract Algebra*. Boston: Heath, 1965.

McCoy, Neal H. *Introduction to Modern Algebra* (3rd ed.). Boston: Allyn and Bacon, 1975.

Maxfield, John E. and Margaret W. Maxfield. *Abstract Algebra and Solution by Radicals*. New York: Dover, 1983.

Schilling, Otto F. G. and W. Stephen Piper. *Basic Abstract Algebra*. Boston: Allyn and Bacon, 1975.

APPENDIX

THE BASICS OF LOGIC

In any mathematical system, just as in any language, there must be some undefined terms. For example, the words "set" and "element" are undefined terms. We think of a set as a collection of objects, and the individual objects as elements of the set. We need to understand the word "set" to describe the word "element" and vice versa. Hence we must rely on our intuition to understand these undefined terms and feel comfortable using them to define new terms.

Postulates are statements (often expressed using undefined terms) that are assumed to be valid. Postulates and definitions are used to prove statements called **theorems**. Once a theorem is proved to be true it can be used to establish the truth of subsequent theorems. A **lemma** is itself a theorem whose major importance lies not in its own statement, but in its role as a stepping stone toward the statement or proof of a theorem. Finally, a **corollary** is also a theorem, but is not so named because it is usually either a direct consequence of or a special case of a preceding theorem. To avoid "stealing the thunder" of the more important theorem, it is labeled a corollary.

We now briefly discuss the basic concepts of logic that are essential to the mathematician for constructing proofs. First we describe a **proposition** as a statement that is either true or false but not both. We use the letters p, q, r, s, and so on to represent statements. Consider the following statements.

p: The sum of the angles in a triangle is $180°$.
q: $2^2 + 3^2 = (2 + 3)^2$
r: $x^2 + 1 = 0$
s: Beckie is pretty.

The statement p is a true proposition from plane geometry. The statement q is a false proposition, when we consider the usual multiplication and addition in the set of real numbers. The statement r is not a proposition since its truth or falsity cannot be determined unless the value of x is known. Also, s is not a proposition since its truth or falsity "lies in the eyes of the beholder" and depends upon which "Beckie" is under consideration.

The statement r given above can be clarified by placing restrictions on the variable x, such as "for every x," "for each x," " for all x," "for some x," or "there exists an x." The phrases "for every x," "for all x," and "for each x" mean the same thing and are often abbreviated by the symbol $\forall$, called the **universal quantifier**. Similarly, the phrases "for some x" and "there exists an x" mean the same thing and are abbreviated by the symbol $\exists$, called the **existential quantifier**. Another commonly used symbol is $\ni$, read "**such that**." Thus the statement

$$\forall x, \qquad x > 0$$

is read

"for every x, $x > 0$."

Similarly, the statement

$$\exists y \ni y^2 + 1 = 0$$

is read

"there exists a y such that $y^2 + 1 = 0$."

If p is a proposition, then the **negation of p** is denoted by $\sim p$ and is read "not p." If p is a true proposition the $\sim p$ must be false and vice versa. We illustrate the idea using a truth table (see Figure A.1) where T stands for true and F stands for false.

p	$\sim p$
T	F
F	T

Truth Table for $\sim p$
FIGURE A.1

The negation of statements involving the universal quantifier and the existential quantifier are given next. We use $p(x)$ to represent a statement involving the variable x. Then the statement

$$\sim(\forall x, p(x)) \quad \text{is} \quad \exists x \ni \sim p(x)$$

is read

"the negation of 'for every x, $p(x)$ is true'

is
'there exists an x such that $p(x)$ is false.' "

We also write

$$\sim(\exists x \ni p(x)) \quad \text{is} \quad \forall x, \sim p(x)$$

and read

"the negation of 'there exists an x such that $p(x)$ is true'
is
'for every x, $p(x)$ is false.' "

EXAMPLE 1   The negation of the statement

"all the students in the class are female,"
is
"there exists at least one student in the class who is not female."   □

EXAMPLE 2   The negation of the statement

"there is at least one student who passed the course"
is
"all the students failed the course."   □

Connectives are used to join propositions to make compound statements. Probably the most important connective is **implication**, denoted by $\Rightarrow$. Suppose p and q are propositions. Then

$$p \Rightarrow q$$

is read in several ways:

"p implies q,"
"if p then q,"
"p only if q,"
"p is sufficient for q," or
"q is necessary for p."

Let us consider the following situations. Algebra class meets only three days a week, on Monday, Wednesday, and Friday. Let p and q be the following propositions:

p: Today is Monday.
q: Algebra class meets today.

Consider the implication

$$p \Rightarrow q.$$

This implication is true if both p and q are true:

Today is Monday $\Rightarrow$ Algebra class meets today.

Suppose p is true and q is false. We consider the corresponding implication

Today is Monday $\Rightarrow$ Algebra class does not meet today

as false. Next suppose that p is false. The falsity of p does not affect the truth or falsity of the implication. That is,

Today is not Monday

does not give any information about whether algebra class meets today. Thus we conclude that

$$p \Rightarrow q$$

is false only when p is true and q is false. We record these results in the truth table in Figure A.2.

p	q	$p \Rightarrow q$
T	T	T
T	F	F
F	T	T
F	F	T

Truth Table for $p \Rightarrow q$
FIGURE A.2

Propositions p and q can be joined with the connective "and," commonly symbolized by $\wedge$ and called **conjunction**. We define $p \wedge q$ to be true only when both p is true and q is true. The corresponding truth table for $p \wedge q$ is given in Figure A.3.

p	q	$p \wedge q$
T	T	T
T	F	F
F	T	F
F	F	F

Truth Table for $p \wedge q$
FIGURE A.3

Similarly, propositions p and q can be joined with the connective "or," symbolized by $\vee$ and called **disjunction**. We define $p \vee q$ to be true when either p

is true or q is true, or both p and q are true. The truth table for $p \vee q$ is given in Figure A.4.

p	q	$p \vee q$
T	T	T
T	F	T
F	T	T
F	F	F

Truth Table for $p \vee q$
FIGURE A.4

Another prominent connective is the **biconditional** denoted by

$$p \Leftrightarrow q$$

and read in any one of three ways:

"p if and only if q,"

"p is necessary and sufficient for q,"

"p is equivalent to q."

The biconditional statement

$$p \Leftrightarrow q$$

can be expressed as the conjunction of two statements

$$(p \Rightarrow q) \wedge (q \Rightarrow p).$$

The truth table in Figure A.5 illustrates that the statement $p \Leftrightarrow q$ is true when p and q are both true or both false; otherwise $p \Leftrightarrow q$ is false.

p	q	$p \Rightarrow q$	$q \Rightarrow p$	$(p \Rightarrow q) \wedge (q \Rightarrow p)$ $p \Leftrightarrow q$
T	T	T	T	T
T	F	F	T	F
F	T	T	F	F
F	F	T	T	T

Truth Table for $p \Leftrightarrow q$
FIGURE A.5

If the truth tables for two propositions are identical, then the two propositions are said to be **logically equivalent** and we use the $\Leftrightarrow$ symbol to designate this.

EXAMPLE 3   Show that

$\sim (p \wedge q) \Leftrightarrow (\sim p) \vee (\sim q).$

Solution   We examine the two columns headed by $\sim (p \wedge q)$ and by $(\sim p) \vee (\sim q)$ in the truth table in Figure A.6 and note that they are identical.

p	q	$p \wedge q$	$\sim (p \wedge q)$	$\sim p$	$\sim q$	$(\sim p) \vee (\sim q)$
T	T	T	F	F	F	F
T	F	F	T	F	T	T
F	T	F	T	T	F	T
F	F	F	T	T	T	T

Truth Table for $\sim (p \wedge q) \Leftrightarrow (\sim p) \vee (\sim q)$
FIGURE A.6

The next example illustrates a truth table involving three propositions.

EXAMPLE 4   Show that

$r \wedge (p \vee q) \Leftrightarrow (r \wedge p) \vee (r \wedge q).$

Solution   We need eight rows in our truth table since there are 2^3 different ways to assign true and false to the three different statements (see Figure A.7).

r	p	q	$p \vee q$	$r \wedge (p \vee q)$	$r \wedge p$	$r \wedge q$	$(r \wedge p) \vee (r \wedge q)$
T	T	T	T	T	T	T	T
T	T	F	T	T	T	F	T
T	F	T	T	T	F	T	T
T	F	F	F	F	F	F	F
F	T	T	T	F	F	F	F
F	T	F	T	F	F	F	F
F	F	T	T	F	F	F	F
F	F	F	F	F	F	F	F

Truth Table for $r \wedge (p \vee q) \Leftrightarrow (r \wedge p) \vee (r \wedge q)$
FIGURE A.7

In this text we see some theorems whose statements involve an implication

$p \Rightarrow q.$

In some instances it is more convenient to prove a statement that is logically equivalent to the implication $p \Rightarrow q$. The truth table in Figure A.8 shows that the implication

$$p \Rightarrow q \quad \text{(implication)}$$

is logically equivalent to the statement

$$\sim q \Rightarrow \sim p \quad \text{(contrapositive)},$$

called the **contrapositive** of $p \Rightarrow q$.

p	q	$p \Rightarrow q$	$\sim q$	$\sim p$	$\sim q \Rightarrow \sim p$
T	T	T	F	F	T
T	F	F	T	F	F
F	T	T	F	T	T
F	F	T	T	T	T

Truth Table for $(p \Rightarrow q) \Leftrightarrow (\sim q \Rightarrow \sim p)$
FIGURE A.8

Two other variations of the implication $p \Rightarrow q$ are given special names. They are

$$q \Rightarrow p \quad \text{is the \textbf{converse} of } p \Rightarrow q$$

and

$$\sim p \Rightarrow \sim q \quad \text{is the \textbf{inverse} of } p \Rightarrow q.$$

We note that the converse and the inverse are logically equivalent, that is,

$$(q \Rightarrow p) \Leftrightarrow (\sim p \Rightarrow \sim q).$$

EXAMPLE 5   Let p and q be the following statements:

p: x is an even integer.
q: x is an integer.

In Figure A.9 we describe the implication $p \Rightarrow q$ and its variations.

Logically equivalent		Logically equivalent	
Implication	Contrapositive	Converse	Inverse
$p \Rightarrow q$	$\sim q \Rightarrow \sim p$	$q \Rightarrow p$	$\sim p \Rightarrow \sim q$
x is an even integer	x is not an integer	x is an integer	x is not an even integer
$\Rightarrow$	$\Rightarrow$	$\Rightarrow$	$\Rightarrow$
x is an integer	x is not an even integer	x is an even integer	x is not an integer
TRUE	TRUE	FALSE	FALSE

FIGURE A.9

EXAMPLE 6 Suppose p and q are the following statements:

p: The Broncos win this week.

q: The Broncos are in the playoffs next week.

Suppose the only way the Broncos go to the playoffs is if they win this week. Hence, if they do not win this week they will not go to the playoffs next week. In Figure A.10 we examine the implication $p \Rightarrow q$ and its variations.

Logically equivalent		Logically equivalent	
Implication $p \Rightarrow q$ Broncos win this week $\Rightarrow$ Broncos are in the playoffs next week TRUE	Contrapositive $\sim q \Rightarrow \sim p$ Broncos are not in the playoffs next week $\Rightarrow$ Broncos do not win this week TRUE	Converse $q \Rightarrow p$ Broncos are in the playoffs next week $\Rightarrow$ Broncos win this week TRUE	Inverse $\sim p \Rightarrow \sim q$ Broncos do not win this week $\Rightarrow$ Broncos are not in the playoffs next week TRUE

FIGURE A.10

Since the implication and its converse are true, we write

$$p \Leftrightarrow q.$$

☐

EXERCISES

Write the negation of each of the statements in Problems 1–8.

1   All the children received a gift from Santa Claus.

2   Every house has a fireplace.

3   Every senior graduated and received a job offer.

4   All the cheerleaders are tall and pretty.

5   There is a rotten apple in the basket.

6   There is a snake that is nonpoisonous.

7   There is a politician who is honest and trustworthy.

8   There is a cold medication that is safe and effective.

Construct truth tables for each of the statements in Problems 9–18.

9   $p \Rightarrow (p \vee q)$

10   $(p \wedge q) \Rightarrow p$

11   $p \Leftrightarrow \sim(\sim p)$

12   $p \vee (\sim p)$

13   $\sim(p \wedge (\sim p))$

14 $\sim(p \lor q) \Leftrightarrow (\sim p) \land (\sim q)$

15 $r \lor (p \land q) \Leftrightarrow (r \lor p) \land (r \lor q)$

16 $(p \land (p \Rightarrow q)) \Rightarrow q$

17 $(p \Rightarrow q) \land (q \Rightarrow r) \Rightarrow (p \Rightarrow r)$

18 $(p \Rightarrow q) \Leftrightarrow (p \land \sim q)$

In Problems 19–26, examine the implication $p \Rightarrow q$ and its variations (contrapositive, inverse, and converse) by writing each in English. Determine the truth or falsity of each.

19 p: My grade for this course is A.
    q: I can enroll in the next course.

20 p: My car ran out of gas.
    q: My car won't start.

21 p: The Saints win the Super Bowl.
    q: The Saints are the world champion football team.

22 p: I have completed all the requirements for a bachelor's degree.
    q: I can graduate with a bachelor's degree.

23 p: x is a positive real number.
    q: x is a nonnegative real number.

24 p: x is a positive real number.
    q: x^2 is a positive real number.

25 p: xy is even.
    q: x is even or y is even.

26 p: x is even and y is even.
    q: $x + y$ is even.

State the contrapositive, converse, and inverse of each of the implications in Problems 27–30.

27 $p \Rightarrow q \lor r$

28 $p \Rightarrow q \land r$

29 $p \Rightarrow \sim q$

30 $p \land \sim q \Rightarrow \sim p$

BIBLIOGRAPHY

Ames, Dennis B. *An Introduction to Abstract Algebra*. Scranton, Pa.: International Textbook, 1969.

Ball, Richard W. *Principles of Abstract Algebra*. New York: Holt, Rinehart and Winston, 1963.

Birkhoff, Garrett and Saunders MacLane. *A Survey of Modern Algebra* (4th ed.). New York: Macmillan, 1977.

Bundrick, Charles M. and John J. Leeson. *Essentials of Abstract Algebra*. Monterey, Calif.: Brooks-Cole, 1972.

Dubisch, Roy. *Introduction to Abstract Algebra*. New York: Wiley, 1965.

Durbin, John R. *Modern Algebra*. New York: Wiley, 1979.

Fraleigh, John B. *A First Course in Abstract Algebra* (2nd ed.). Reading, Mass.: Addison-Wesley, 1976.

Fuchs, Laszlo. *Infinite Abelian Groups*, Vols. I and II. New York: Academic Press, 1973.

Hall, Marshall, Jr. *The Theory of Groups* (2nd ed.). New York: Chelsea, 1961.

Herstein, I. N. *Topics in Algebra* (2nd ed.). New York: Wiley, 1975.

Hillman, Abraham P. and Gerald L. Alexanderson. *A First Undergraduate Course in Abstract Algebra* (2nd ed.). Belmont, Calif.: Wadsworth, 1978.

Jones, Burton W. *An Introduction to Modern Algebra*. New York: Macmillan, 1975.

Keesee, John W. *Elementary Abstract Algebra*. Boston: Heath, 1965.

Kurosh, A. *Theory of Groups*, Vols. I and II. Trans. K. A. Hirsch. New York: Chelsea, 1979.

Larsen, Max. D. *Introduction to Modern Algebraic Concepts*. Reading, Mass.: Addison-Wesley, 1969.

Lederman, Walter. *Introduction to the Theory of Finite Groups* (4th ed.). New York: Interscience, 1961.

McCoy, Neal H. *Fundamentals of Abstract Algebra*. Boston: Allyn and Bacon, 1972.

McCoy, Neal H. *Introduction to Modern Algebra* (3rd ed.). Boston: Allyn and Bacon, 1975.

McCoy, Neal H. *Rings and Ideals* (Carus Mathematical Monograph No. 8). Washington, D.C.: The Mathematical Association of America, 1968.

McCoy, Neal H. *The Theory of Rings*. New York: Chelsea, 1972.

Maxfield, John E. and Margaret W. Maxfield. *Abstract Algebra and Solution by Radicals*. New York: Dover, 1983.

Mitchell, A. Richard and Roger W. Mitchell. *An Introduction to Abstract Algebra*. Monterey, Calif: Brooks-Cole, 1970.

Mostow, George D., Joseph H. Sampson, and Jean-Pierre Meyer. *Fundamental Structures of Algebra*. New York: McGraw-Hill, 1963.

Niven, Ivan and Herbert S. Zuckerman. *An Introduction to the Theory of Numbers* (4th ed.). New York: Wiley, 1980.

Rotman, Joseph H. *The Theory of Groups: An Introduction* (2nd ed.). Boston: Allyn and Bacon, 1973.

Schilling, Otto F. G. and W. Stephen Piper. *Basic Abstract Algebra*. Boston: Allyn and Bacon, 1975.

Scott, W. R. *Group Theory*, Englewood Cliffs, N.J.: Prentice-Hall, 1964.

Shapiro, Louis. *Introduction to Abstract Algebra*. New York: McGraw-Hill, 1975.

Sierpinski, W. *Elementary Theory of Numbers*. Trans. A. Hulanicki. New York: Hafner, 1964.

Weiss, Marie J. and Roy Dubisch. *Higher Algebra for the Undergraduate* (2nd ed.). New York: Wiley, 1962.

INDEX